Crop genetic resources for today and tomorrow

THE INTERNATIONAL BIOLOGICAL PROGRAMME

The International Biological Programme was established by the International Council of Scientific Unions in 1964 as a counterpart of the International Geophysical Year. The subject of the IBP was defined as 'The Biological Basis of Productivity and Human Welfare', and the reason for its establishment was recognition that the rapidly increasing human population called for a better understanding of the environment as a basis for the rational management of natural resources. This could be achieved only on the basis of scientific knowledge, which in many fields of biology and in many parts of the world was felt to be inadequate. At the same time it was recognised that human activities were creating rapid and comprehensive changes in the environment. Thus, in terms of human welfare, the reason for the IBP lay in its promotion of basic knowledge relevant to the needs of man.

The IBP provided the first occasion on which biologists throughout the world were challenged to work together for a common cause. It involved an integrated and concerted examination of a wide range of problems. The Programme was co-ordinated through a series of seven sections representing the major subject areas of research. Four of these sections were concerned with the study of biological productivity on land, in freshwater, and in the seas, together with the processes of photosynthesis and nitrogen-fixation. Three sections were concerned with adaptability of human populations, conservation of ecosystems and the use of biological resources.

After a decade of work, the Programme terminated in June 1974 and this series of volumes brings together, in the form of syntheses, the results of national and international activities.

INTERNATIONAL BIOLOGICAL PROGRAMME 2

Crop genetic resources
for today and tomorrow

EDITED BY

O. H. Frankel
Senior Research Fellow
Division of Plant Industry, CSIRO, Canberra, Australia

AND

J. G. Hawkes
Professor of Botany, University of Birmingham, UK

CAMBRIDGE UNIVERSITY PRESS

CAMBRIDGE
LONDON · NEW YORK · MELBOURNE

Published by the Syndics of the Cambridge University Press
The Pitt Building, Trumpington Street, Cambridge CB2 1RP
Bentley House, 200 Euston Road, London NW1 2DB
32 East 57th Street, New York, NY 10022, USA
296 Beaconsfield Parade, Middle Park, Melbourne 3206, Australia

Library of Congress catalogue card number: 74–82586

ISBN: 0 521 20575 1

First published 1975

Printed in Great Britain
at the University Printing House, Cambridge
(Euan Phillips, University Printer)

Contents

Contents

vi

Table des matières

IIIième partie. Problèmes de l'évaluation

IVième partie. Conservation et stockage

Table des matières

Vième partie. Direction de la documentation et de l'information

VIIème partie. Centres de ressources génétiques

Содержание

Contenido

Contenido

Sección V. Manejo de documentación e información

Sección VI. Centros de recursos genéticos

Contributors

Alexander, D. E. Department of Agronomy, College of Agriculture, University of Illinois, Urbana, Illinois 61801, USA

Bradshaw, A. D. Department of Botany, University of Liverpool, P.O. Box 147, Liverpool L69 3BX, UK

Brezhnev, D. D. N. I. Vavilov All-Union Research Institute of Plant Industry, 44 Herzen Street, Leningrad 19000, USSR

Brown, A. H. D. CSIRO, P.O. Box 1600, Canberra City, A.C.T. 2601, Australia

Brown, W. L. Pioneer Hi-Bred International Inc., 1206 Mulberry Street, Des Moines, Iowa 50308, USA

Chang, T. T. International Rice Research Institute (IRRI), P.O. Box 583, Manila, Philippines

D'Amato, F. Institute of Genetics, University of Pisa, 56100 Pisa, Italy

Dinoor, A. Faculty of Agriculture, The Hebrew University of Jerusalem, P.O. Box 12, Rehovot, Israel

Dorofeev, V. F. Department of Wheats, N. I. Vavilov All-Union Research Institute of Plant Industry, 44 Herzen Street, Leningrad 19000, USSR

Frankel, O. H. CSIRO, P.O. Box 1600, Canberra City, A.C.T. 2601, Australia

Harlan, J. R. Department of Agronomy, S-516 Turner Hall, University of Illinois, Urbana, Illinois 61801, USA

Hawkes, J. G. Department of Botany, University of Birmingham, P.O. Box 363, Birmingham, B15 2TT, UK

Hegnauer, R. Laboratorium voor Experimentele Plantensystematiek, Rijksuniversiteit, Leiden, 5e Binnenvestgracht 8, Netherlands

Henshaw, G. G. Department of Botany, University of Birmingham, P.O. Box 363, Birmingham B15 2TT, UK

Contributors

Hersh, G. Taximetrics Laboratory, Department of EPO Biology, University of Colorado, Boulder, Colorado 80302, USA

Howard, B. East Malling Research Station, East Malling, Maidstone ME19 6BJ, Kent, UK

Hyland, H. Germplasm Resources Laboratory, USDA Agricultural Research Center, Beltsville, Maryland 20705, USA

Jain, S. K. Department of Agronomy, University of California, Davis, California 95616, USA

Kjellqvist, E. Agricultural Research and Introduction Centre, P.O. Box 9, Menemen, Izmir, Turkey

Kauffman, H. E. International Rice Research Institute, P.O. Box 583, Manila, Philippines

Ling, K. C. International Rice Research Institute, P.O. Box 583, Manila, Philippines

Loresto, G. International Rice Research Institute, P.O. Box 583, Manila, Philippines

Marshall, D. R. CSIRO, P.O. Box 1600, Canberra City, A.C.T. 2601, Australia

Martin, F. W. Federal Experiment Station, USDA, Mayaguez, Puerto Rico 00708

Mengesha, M. Haile Selassie I University, Addis Ababa, Ethiopia

Morel, G. Centre National de Récherches Agronomiques, Station de Physiologie Végétale, Etoile de Choisy, Route de St Cyr, 78, Versailles, France

Noshiro, M. Institute of Low Temperature Science, Hokkaido University, Sapporo, Japan

Ochoa, C. Centro Internacional de la Papa, Apartado 5969, Lima, Peru

Ou, S. H. International Rice Research Institute, P.O. Box 583, Manila, Philippines

Pathak, M. D. International Rice Research Institute, P.O. Box 583, Manila, Philippines

Perez, A. T. International Rice Research Institute, P.O. Box 583, Manila, Philippines

Qualset, C. O.   Department of Agronomy and Range Science, University of California, Davis, California 95616, USA

Röbbelen, G.   Institut für Pflanzenbau und Pflanzenzüchtung, 34 Göttingen, v. Siebold Strasse 8, German Federal Republic

Roberts, E. H.   Department of Agriculture and Horticulture, University of Reading, Earley Gate, Reading RG6 2AT, UK

Rogers, D. J.   Taximetrics Laboratory, Department of EPO Biology, University of Colorado, Boulder, Colorado 80302, USA

Sakai, A.   Institute of Low Temperature Science, Hokkaido University, Sapporo, Japan

Seidewitz, L.   Institut Pflanzenbau FAL, Braunschweig–Völkenrode, 33, Braunschweig, Bundesallee 50, German Federal Republic

Sencer, H. A.   Agricultural Research and Introduction Centre, P.O. Box 9, Menemen, Izmir, Turkey

Snoad, B.   Applied Genetics Department, John Innes Institute, Colney Lane, Norwich NOR 7OF, UK

Soria, J.   Tropical Crops and Soils Department, IICA, P.O. Box 74, Turrialba, Costa Rica

Sykes, J. T.*   Department of Horticultural Science, University of Guelph, Guelph, Ontario, Canada

Villareal, R. L.   International Rice Research Institute, P.O. Box 583, Manila, Philippines

Villiers, T. A.   Department of Biological Sciences, University of Natal, Durban, Natal, South Africa

* Present address: Crop Ecology and Genetic Resources Unit, FAO, Via delle Terme di Caracalla, 00100, Rome, Italy.

1. Genetic resources – the past ten years and the next

O. H. FRANKEL & J. G. HAWKES

When the International Biological Programme started in 1964, little was thought, talked or written about 'plant genetic resources'. Even the term had not been coined. There was a general awareness of their existence and potential value ever since Vavilov discovered the geographical centres of genetic diversity, but only a small number of specially interested scientists had become, concerned at the growing threat to the continued existence of the 'genetic treasuries'. The first draft programme of IBP recommended that 'a survey be made of available knowledge of genetic variation of major crop plants and their wild relatives to determine areas of the globe where intensive exploration should be made for additional germplasm material'. Emphasis was placed on information and availability, rather than on preservation in the face of impending loss or destruction. The recommendations of the first FAO Technical Conference on Plant Exploration and Introduction held in 1961 (Whyte & Julén, 1963), similarly failed to convey a sense of urgency, although ever since early warnings by Harlan & Martini (1936) individual scientists had noted the gradual disappearance of primitive crop varieties in some of the 'gene centres' (e.g. Julén in Whyte & Julén, *loc. cit.*).

A transformation came in the mid 1960s, from a variety of sources. First, warnings of a greatly accelerated rate of displacement of primitive crop varieties by locally selected or introduced cultivars came from scientists with experience in important gene centres, e.g. H. Kuckuck (Iran, Turkey, Ethiopia), J. R. Harlan and D. Zohary (Near East), E. Bennett (Turkey), J. G. Hawkes (Latin America). Second, there was a rapidly growing interest in the primitive cultivars and the immense range of variation they contain, and in the wild relatives of economic species, many as yet scarcely explored and exploited.

FAO–IBP partnership

Third, the then director-general of FAO, Dr B. R. Sen, initiated a review of FAO's activities and responsibilities in plant exploration and

1

introduction, and the chairman of the IBP gene pools committee was entrusted with carrying out this review. As a result the working partnership which had been inaugurated at an IBP meeting at FAO headquarters in 1965 led to a virtual integration of effort in the following years.

The 1967 Technical Conference was planned and prepared jointly by FAO and IBP, with the latter organizing the scientific programme, FAO the participation of member countries and the technical arrangements. A meeting convened by IBP at FAO Headquarters in October 1966, attended by members of the IBP gene pools committee and the FAO panel on plant exploration and introduction, greatly assisted in planning the scientific programme, which was intended to result in a much needed book on plant exploration and conservation. This book, the predecessor of the present volume, was published in the series of *IBP Handbooks* (Frankel & Bennett, 1970).

Scientific clarification

The 1967 conference, and the book which followed, were landmarks in developing the scientific background and methodology, and in defining goals and strategies. The multi-disciplinary nature of the problems was recognized by specialists in many of the relevant fields, from ecology, genetics and systematics, to plant breeding, agronomy and computer science. The methodologies of plant exploration and genetic conservation were examined and clarified. The primitive varieties of traditional agriculture were designated as primary targets for urgent exploration, with wild relatives, especially where threatened, as secondary ones. A survey of genetic resources in the field was to supply urgently required information on priorities for exploration. Long-term conservation of seeds, under appropriate conditions, was recommended, where applicable, as the safest and most economical method of conservation, and a co-operative network of seed storage laboratories was seen as the most efficient and economical form of organization. The documentation of genetic resources, in a manner which was generally approved and understood, was recognized as essential if they are to be put to practical use in research and in breeding practice, and the development of widely accepted documentation procedures became an important goal. Last but not least, co-ordination by FAO was regarded as essential for planning and executing an action programme with broad international participation.

Recent developments

In the years which followed there were several developments, and some of them will have great significance for the next decade. FAO set up a small but enthusiastic unit for genetic resources which became a reference point for genetic resources activities, and, with modest means, began to support exploration. Its *Genetic Resources Newsletter*, successor to the *Plant Introduction Newsletter*, became a valuable specialist journal and news sheet. The FAO Expert Panel on Plant Exploration and Introduction met in 1969, 1970 and 1973. Its reports (FAO, 1969; 1970; 1973) elaborated the major recommendations of the 1967 conference. The Panel named priority targets for exploration and urged that the proposed survey of threatened resources be executed. It drafted proposals for an international network of genetic resources centres and defined functions and responsibilities of participating institutions. It urged FAO to initiate international co-operation in seed conservation and drafted guidelines for participation (see Chapter 37). It urged that active steps be taken towards the development and adoption of appropriate documentation systems.

The survey of genetic resources in the field was organized by FAO and IBP and carried out by specialists in many countries in 1972 and published early in 1973 (see Chapter 6). It was disappointing that progress on the network of conservation centres (now termed 'base collections') was slow and inconclusive. This continues to be a most urgent task which, we are convinced, could be executed without undue difficulty. However, advances of a substantive kind were not lacking. Some examples must suffice. Two of the planned European gene banks, at Bari in Italy and at Braunschweig in the Federal Republic of Germany, were set up following the establishment of the Izmir centre in Turkey, some years earlier. IRRI, the International Rice Research Institute, greatly enlarged its collection by co-operative exploration in many countries with ancient rice culture. Taken together with its comprehensive evaluation, conservation and documentation programmes, IRRI has set a pattern for international activities in all phases of genetic resources work (see Chapters 13, 16 and 35). Furthermore, some of the existing collections were substantially enlarged by recent explorations, such as those of the Vavilov Institute in Leningrad or of the US Department of Agriculture in Beltsville and elsewhere.

Activity in plant exploration has grown, although not to the extent required to outpace the rate of loss which, in many of the gene centres,

3

is rapidly rising. Accounts of recent explorations are found in Chapters 10–15, and more, but by no means all, in *Plant Genetic Resources Newsletter*. It is to be hoped that the *Newsletter*, which goes to everyone concerned with plant exploration, will be more fully informed of all exploration activities than has been the case in the past.

Advances have also been made in the documentation area, and a good deal of agreement was reached among people concerned with documentation at a Conference on European and Regional Gene Banks held at Izmir in April 1972, and at a workshop meeting at the Department of Botany of the University of Birmingham in July 1972 (see Hawkes & Lange, 1973). Practical progress was made through a successful pilot project using TAXIR (see Chapter 32). The stage is now set for the introduction of widely applicable documentation systems which should bring much needed information to all users of genetic resources.

Education and training have made a big step forward with the introduction of a postgraduate course in Conservation and Utilization of Plant Genetic Resources at the Department of Botany of Birmingham University in 1969. 36 students from 21 countries have passed the course and numbers should be rising further with the job opportunities at new or enlarged genetic resources centres (see Chapter 37). There is a need for short practical training courses for technicians, especially in the regions of genetic diversity, and for high-level refresher courses such as the one held in Leningrad in 1971.

Awareness and concern

Perhaps the biggest transformation of the last decade has been the development of widespread awareness and concern. Ten years ago, as we said at the beginning of this chapter, the problem of 'genetic resources' was unknown. In the last few years it has sunk into the consciousness of scientists and administrators and of the many people who have become concerned about the resources of the earth. Information has spread from scientific papers and articles in scientific magazines (from Frankel, 1967 to Miller, 1973) into the news media. At the United Nations Environment Conference at Stockholm, 1972, with both FAO and IBP strongly involved, the problems of the world's genetic resources became an issue which attracted world attention. Strongly worded recommendations were carried urging governments and UN agencies to save and preserve irreplaceable genetic resources for the good of present and future generations; and many of the world's news media carried the message.

Towards the close of the decade three moves have occurred which usher in the coming decade as a new era of incisive action. First, the Governing Council of UNEP, the United Nations Environment Programme, decided to include support for the exploration and conservation of genetic resources in the environment programme, and to allocate a substantial sum in the first biennium. Second, the Consultative Group on International Agricultural Research, after two years of study and negotiation, adopted the proposal of its Technical Advisory Committee to support a global network of genetic resources centres. The plan is largely based on the report by a group of experts – several of them members of the FAO Panel of Experts – which met at Beltsville, USA, in March 1972. The network is to consist of all genetic resources centres willing to participate, wherever they are located. Special emphasis is given to support for centres in the regions of genetic diversity, and strategically placed institutions are to be developed as regional centres (see Chapter 37). Funds for exploration, conservation, documentation and training are to be allocated from a central fund on a project basis. An International Board for Plant Genetic Resources, with a secretariat associated with the FAO unit for genetic resources, will administer the fund, and the FAO Expert Panel is to act as its technical committee.

The third move is a considerable strengthening of FAO's own participation. At the Biennial Conference of FAO in November 1973, the Director-General, Dr A. H. Boerma, stated that he considered the salvage and use of the world's genetic resources as a subject of highest priority, and requested and obtained approval for a considerable strengthening of FAO's staff and activities in this area.

The present book

This book, like its predecessor, grew out of a technical conference planned jointly by FAO and IBP, with the contents of the book clearly in mind. Most of the chapters grew out of contributions to the Conference. The following were the conveners of conference sections: D. R. Marshall (Part I), J. R. Harlan (Part II*B*), J. Léon (Part II*C*), D. Bommer (Part III), J. G. Hawkes (Part IV), D. J. Rogers (Part V), O. H. Frankel (Part VI). Mrs Averil Zaniboni was conference secretary.

The book stands between the first decade and that about to begin.

Introduction

Like the previous volume it deals with the scientific and technical problems of exploration, use and conservation, and with practical problems and experiences. However, it differs from its predecessor in dealing with specific scientific, technical and organizational problems likely to be encountered in developing and implementing effective programmes to collect and conserve threatened gene pools, rather than with the field as a whole. In other words, it is more 'action orientated'. At the same time confinement in breadth made possible extension in depth.

This certainly is the case in Part I which deals with the distribution and organization of genetic variation in plant populations. The aims of this section are to summarize current knowledge of the kinds and amounts of genetic variation encountered in plant species and the distribution of this variation in nature, and to consider the implications of differences in the genetic structure of plant populations, both within and between species, in defining the optimum strategies for exploration and conservation. The emphasis in Jain's chapter is on the population structure in species with different breeding systems, and in Bradshaw's on the effects of selection on population structure. Both deal almost exclusively with wild (or weed) plants, even in a section by Jain entitled 'Population structure of plant species under domestication', largely because next to nothing is known of the population structure in primitive cultivated varieties of whatever breeding system. Such information, as is evident from Marshall and Brown's chapter on sampling strategies, would be of considerable practical value in designing the most appropriate and economical sampling patterns. It also would be of considerable scientific interest. Primitive varieties are neither as mobile as the 'multi-national' modern crop varieties, nor as relatively sessile as are wild plant populations. They shift from field to field within circumscribed environments, and some have done so over long periods of time. An insight into the structure of such populations could supply valuable information on the evolution and maintenance of heterogeneity under domestication, free from deliberate selection. Such information might even be suggestive for the structuring of multilines.

The paper by Marshall and Brown gave rise to a good deal of controversy at the conference, and their chapter may have similar effects. Yet its message of rationalizing sampling strategies on the basis of known or, for the most part, assumed population structures is wholly unexceptionable. Indeed, the lesson should pervade all genetic resource operations, from collecting in the field to maintenance and regeneration.

It is now generally accepted that conservation, like exploration, should be conducted on a population basis, and not, as has become clear from the careful examination of at least one important collection, on the basis of one or two representatives selected from a population sample (Konzak, personal communication). Yet in most plants it is not difficult to maintain a representative sample of the original population, provided that conservation methods are adequate (see Chapters 22–29) and environmental conditions and sampling techniques during regeneration are designed to maintain the population structure.

Brief references have already been made to chapters in Part II on exploration. Problems and difficulties in exploration are discussed in Chapters 7–9. They contain many take-home lessons from experienced plant collectors. One we particularly appreciate is Harlan's statement that 'the value of a collection cannot be assessed in the field' (Chapter 7), which points again to the significance of sampling strategies. Hawkes and Sykes deal with difficult areas, vegetatively propagated crops and tree crops respectively, each covering important practical topics. Chapters 10–15 provide a bird's-eye-view of a wide range of collecting activities – from the two largest organizations dealing with many crops, to intensive collections of a single crop.

Part III, on evaluation, also presents a wide spectrum. Chang and his colleagues report comprehensively on the evaluation for resistance to the major diseases and pests of the rice plant. Qualset – whose Chapter 5 fits into this part as well as into Part I – examines the distribution of resistance to the barley yellow dwarf virus in Ethiopia, the only known source of resistance, whereas Dinoor deals with principles and methods for a systematic search for disease resistance. The important question of screening for cold and drought resistance is discussed by Dorofeev. Alexander, Röbbelen and Hegnauer discuss problems presented by three different groups of compounds in plants.

The section on Conservation (Part IV) will perhaps be found particularly useful, in view of the current interest in conservation methods; it certainly gives the most comprehensive coverage of its subject. Chapter 22, by Roberts, is a review of current knowledge of the principles and methods for the storage of seed and pollen. It shows how far we have progressed in the six years since the previous book was written. Its lesson is that long-term seed storage – long-term, at least, in the sense of the human life span – is not only feasible, but a relatively simple and inexpensive operation in terms of technology, facilities, staff and operating expenses. On this basis seeds of most crop plants can be stored over long periods with a minimal risk of genetic damage, and regeneration should be a rare event. It remains to spell out the organizational and administrative consequences – the size of preserved stocks, access, and organization of distribution, and no doubt a great deal more. To make seed conservation available to all should be an early task for the coming decade.

Howard (Chapter 28) suggests that procedures for long-term storage of woody material could be developed from current methods of low-temperature storage, but a substantial extension of the storage time for roots and tubers cannot be expected (Chapter 29).

Chapters 23–27 present reviews and some original experimental evidence on novel approaches to conservation. Villiers' work (Chapter 23) on storage of fully imbibed seeds at normal temperatures is of considerable interest and may provide a solution for the 'recalcitrant' species which cannot be stored at low humidity. Application is dependent on a germination inhibitor, such as darkness in the case of lettuce which was used in the experiments reported here. It is known that many seeds have such mechanisms, hence there may be wide application for this approach. Sakai and Noshiro, also reporting original work, deal with storage at relatively high moisture content, but at the other end of the temperature scale, viz in liquid nitrogen. This approach may also have application for recalcitrant seeds.

Chapters 25–27 report on the current state of research on tissue culture which is of particular importance for the conservation of material which cannot be preserved in the form of seeds. Meristem culture, as the chapter by Morel shows, has advanced to practicability although not for more than a limited number of gene banks. D'Amato deals with problems of genetic stability, and Henshaw with the technical aspects of long-term tissue storage. Neither problem is as yet solved and much research is needed before standard techniques can be evolved.

In the final chapter of this section Dr Jain discusses the conditions for the survival of species in nature reserves. This is of special relevance for the preservation of relatives of economic species and of any with an economic potential.

The section on documentation also testifies to the advance and clarification which has taken place in recent years, and Chapter 32 clearly shows the way in which a regional, or even a global network can be instituted and operated.

The final section, which deals with genetic resources centres, has been referred to in earlier parts of this chapter. It begins with a national centre for Ethiopia, progresses through a regional link for the Near East to two crop-specific international centres, and finally to the global network which, we hope, will justify the preparatory labours of the past decade.

The contribution of IBP

This 'synthesis volume' signifies not only the end of a decade since the inception of the International Biological Programme, but also the end of its participation. During the planning phase, the exploration and conservation of plant gene pools was seen as highly appropriate for inclusion in a programme on 'The Biological Basis of Productivity and Human Welfare' (IBP, 1965). It was then envisaged that it would be executed by committees and working groups in member countries, in line with other IBP activities (IBP, 1966). This was soon found impracticable. Outside the few institutions – mostly in developed countries – involved in exploration and conservation, there was little appreciation and an almost complete lack of financial and logistic support. IBP itself lacked a logistic base for an organizational task on the scale required. Moreover, it became evident that there was an urgent need for the clarification and development of the scientific and methodological infrastructure, and for information on the urgency of the situation to be brought home to scientists, administrators and politicians. In these areas there was an obvious role for IBP. With the increasing contacts with FAO, and thanks to IBP's links with world science, IBP was in a position to become the scientific arm of FAO. At the same time, being with it yet not part of it, it became a catalyst for action. Officially, as a non-governmental organization affiliated with FAO, IBP was able to contribute to its activities – at crucial times even financially. What other contributions IBP has made may be discerned in the pages of this book.

The next ten years

We have spoken at length of the last decade and have mentioned some of the trends which may be expected to continue into the decade to come. It is always dangerous to try to predict the future but it may at least be worth saying how we view the prospect for the next ten years.

The scientific exploration and collection of the world's fast diminishing genetic resources is obviously of the greatest importance. To carry

9

out this objective well-trained people from all parts of the world will be needed. To attract and train the required scientific manpower will be an urgent task for the immediate future. By 1985 most of the genetic resources in their ancient centres of diversity may well have disappeared. The majority of the wild relatives of cultivated plants will have survived, though with reduced habitats, especially if the natural forest ecosystems continue to be destroyed at the present rate. Consequently, unless collected and stored in gene banks or preserved in other ways, many or most of the most valuable resources will be lost for ever.

The prospects for satisfactory storage of seeds in gene banks are on the whole very good, apart from the group known as 'recalcitrant' (see Chapter 22). Research during the next decade should be directed towards these species and to the problems of seed storage at very low temperatures. Further work is also needed on pollen storage. The problems of tissue and meristem cultures for long-term storage have already been pointed out in this chapter. Altogether, the next ten years should show significant 'breakthroughs' in the field of storage techniques.

The major problems ahead clearly are those of organization and funding. A 'global network' of genetic resources centres, as outlined in Part VI of this book, is now beginning to emerge. This must include the centres which are to be developed in the regions of genetic diversity, and the many institutions all over the world which are engaged in genetic resources work. The network must acquire cohesion and develop lines of communication and exchange. Material must be made available *to be used at this time*, not only preserved and stored for the future. This requires co-operative organization at all levels, from exploration to multiplication, distribution, evaluation, conservation and documentation, and to the fullest possible utilization.

This, after all, is the purpose of the endeavour: to provide material for those who will turn it to productive purposes. Informal links between institutions and individuals have existed for many years, and FAO's seed exchange has assisted in many instances. These links need to be greatly strengthened and opened to all users. Present-day information and retrieval methods have added a new dimension of usefulness to the introduction of plant material. By appropriate organization of data collecting, by encouraging feedbacks from plant breeders and other users, a vast amount of information can be gathered, stored and made available, which will make the application of genetic resources more meaningful, less subject to chance, hence infinitely more productive.

As the chapters which follow indicate, a good deal of the required knowledge exists or is in sight. No longer are there reasons why material in the field should be lost through neglect or inadequate effort. No longer need potential users want for material or information they require. No longer need we feel that we are failing in our responsibility towards the present and the future. What is now needed is action which will create the needed organization and provide the essential means.

References

FAO (1969). Report of The *Third Session of the FAO Panel of Experts on Plant Exploration and Introduction*. FAO, Rome.

FAO (1970). Report of The *Fourth Session of the FAO Panel of Experts on Plant Exploration and Introduction*. FAO, Rome.

FAO (1973). Report of The *Fifth Session of the FAO Panel of Experts on Plant Exploration and Introduction*. FAO, Rome.

Frankel, O. H. (1967). Guarding the plant breeder's treasury. *New Scientist*, **14**, 538–40.

Frankel, O. H. & Bennett, E. (eds) (1970). *Genetic Resources in Plants – their Exploration and Conservation. IBP Handbook*, no. **11**. Blackwell, Oxford and Edinburgh.

Harlan, H. V. & Martini, M. L. (1936). Problems and results in barley breeding. *USDA Yearbook of Agriculture 1936*, pp. 303–46.

Hawkes, J. G. & Lange, W. (1973). *European and Regional Gene Banks*. Eucarpia, Wageningen.

IBP (1965). *IBP News* no. **2**. Rome.

IBP (1966). *IBP News* no. **5**. London.

Miller, J. (1973). Genetic erosion: crop plants threatened by government neglect. *Science*, **182**, 1231–3.

Whyte, R. O. & Julén, G. (eds) (1963). Proceedings of a technical meeting on plant exploration and introduction. *Genetica agraria*, **17**.

PART I

Genetic variation in plant populations

2. Population structure and the effects of breeding system

S. K. JAIN

Any theory of sampling strategies for genetic conservation, as discussed by Marshall and Brown (Chapter 4), must be founded upon an extensive knowledge of the patterns of genetic variability in populations. Decisions about sampling frequency, sample size, etc., depend on certain assumptions about the between- and within-population variances, level of heterozygosity, and the allelic frequencies at individual polymorphic loci. Recent biochemical studies have raised high hopes for the comparative analyses of variation patterns in different groups of plants and animals with an aim to explore relationships between the parameters of variation and the population structure of a species. A review of these studies showed, however, that many different aspects of the reproductive biology, numerical abundance and habitat diversity will have to be studied before we could undertake such comparisons (Cavalli-Sforza & Bodmer, 1971; Selander & Kaufman, 1973; Jain, 1974). Similarly, in discussing the nature of material for germplasm exploration and conservation, Frankel & Bennett (1970) commented on the diversity of life cycle, mode of reproduction, breeding system and gene flow, and various ecological aspects. In particular, plant biologists have widely recognized the central role of breeding systems, a key parameter of the genetic systems, in terms of some general models of the evolutionary fate of obligate outbreeders, inbreeders and apomicts. Domesticated plants and their wild and weed relatives represent a highly diverse group of materials for a discussion of the role of breeding systems in evolution. In this chapter, I shall begin a survey of this diversity by first defining (1) what we mean by population structure of a species, then (2) review briefly our knowledge of population structure of the wild and weed-like plant species, (3) consider the consequences of variations in the breeding system, (4) correlate some genetic variation patterns with the variables of population structure, and finally, (5) discuss implications in the theory of genetic conservation.

Although a majority of population studies cited below had involved non-domesticated species, the basic principles and observations on genetic variation should hold equally for the wild and weed relatives of crop species.

Population structure: some basic concepts

Biogeographers study the distribution area or range of a species in terms of the spatial patterns of its members in relation to the characteristics of habitat subdivision, local density, mobility, and the dynamics of habitat changes. Evolutionists, on the other hand, are primarily interested in the adaptive features of intraspecies variation in space and time. We may therefore define population structure of a species as the totality of ecological and genetic relationships among the member individuals as well as the subdivisions of a biological species. The subdivisions, often identified on different geographical scales of study, may co-evolve as a result of some potential or actual gene exchange but also diverge under the local forces of evolutionary change. The extremes of population structures are represented by the species with one large and continuous random mating population (perhaps hypothetical) and those with one or more small discontinuous populations (isolates). The geographical subdivision, primarily as a consequence of physical barriers (e.g., islands, mountains, hills, lakes, marshes), and biotic subdivision resulting from the breeding structure and variables of gene flow, should both be recognized. The genetic structure of a population further specifies such measures of genetic diversity as the allelic frequencies at polymorphic loci, genotypic associations among loci, and genetic differentiation among subpopulations. Table 2.1 lists the genetic and ecological variables along with some of the parameters to be utilized in describing the kinds of species in terms of the kinds of population structures. It should be noted that this requires a multivariate approach rather than the conventional univariate comparisons of species groups such as inbreeders versus outbreeders, diploids versus polyploids, etc. Moreover, population structure of a species varies also over time so that historical and chance events would further influence the predictability of evolutionary patterns.

Some concepts from elementary theory of population genetics will help illustrate the interplay among the variables of population structure. A species may be distributed largely along a linear habitat such as river banks, seashore, margins of cultivated fields, etc. or in a two-dimensional habitat. Distribution may be discontinuous in the form of distinct colonies that have two possible modes of gene flow: each colony exchanges a certain proportion of pollen or seed with all the others, or only the neighboring colonies interchange genes (so-called 'island model' and 'stepping-stone model', respectively). All individuals

Table 2.1. *Factors affecting the population structure of a species*

Variate	Kinds of species (examples of extremes)	Parameters
1. Mode of reproduction	Sexual, asexual, outbreeder, inbreeder	Degree of sexuality, sex ratio; rate of out-crossing
2. Pattern of distribution	Cosmopolitan, widespread, endemic, relict	Kind and variety of habitats and communities
3. Inter- and intraspecies hybridization	Superspecies; semi-species; 'ring of races'; etc.	Rates of natural hybridization; rates of gene flow in natural, domestic or disturbed environments
4. Karyotype variation and recombination system	Dysploids, polyploids, structural hybrids; 'aberrant' systems of karyotype regulation	Chromosome number, homology, structural variations, meiotic characteristics
5. Generation length, life cycle components, survivorship in a community and population density	Perennial, biennial, annual; early successional versus climax species; common or rare	Reproductive rates; overlap between generations; intrinsic rate of population growth; population size
6. Local dispersion and gene flow	Continuous, random, versus colonial, non-random; populations subdivided to various degrees	Interdeme and intrademe gene flow; deme size; deme persistence
7. Genetic variation	Polytypic, monotypic; polymorphisms widespread or localized	Percent heterozygosis, number of alleles/locus, genotypic frequencies at linked loci
8. Environmental heterogeneity	Species with niche specificity; high or low rates of gene flow	Variation in space and time; patchiness of habitat; habitat selection
9. Phenotypic plasticity	Highly plastic in reproductive patterns	Genetic versus non-genetic variances

within a colony might be one panmictic group, or be structured into different breeding groups. Moreover, the actual number living in an area may be much larger than the effective breeding unit defined in terms of the number of progenitors of the following generation. Effective population size (N_e) is useful particularly in relation to the role of random drift since it represents the sampling process during the generation turnover. In the case of continuously distributed species, we need to define the concept of neighborhood as 'the population of a region from which the parents of individuals born near the center may be treated as if drawn at random' (Wright, 1969). Here, gene flow is treated as a function of distance in a continuum such that an effective neighborhood represents a group of individuals intermating with a reasonably high probability. Estimates of neighborhood size allow meaningful comparisons among species differing in their distribution pattern, gene flow, and demographic features such as generation length, overlap between generations, and density. As pointed out by Levin & Kerster (1971), we may also consider the neighborhood area while comparing the scales of spatial differentiation. Fig. 2.1 shows diagrammatically the population structure of a species in terms of subdivision, gene flow and fluctuations in population numbers in different parts of its range. Andrewartha & Birch (1954) provided useful insight into the problems of numerical abundance by showing how mean and variance in density might be correlated. Population biologists have recently become very much interested in the evolutionary properties of marginal populations as they presumably involve a greater role of selection and isolation as well as random drift. Theory tells us that the effective neighborhood size is relatively smaller in species with the following characteristics: linear habitat, low density, stepping-stone model of highly local gene dispersal, inbreeding, fluctuations in N_e over time, unequal sex ratios and large variance of progeny number (reproductive fitness) among individuals. Species having higher colonizing ability of temporary habitats frequently go through bottlenecks of population size and perhaps many of them have high reproductive rates (*r*-strategists) so that they have low N_e values in spite of their apparent abundance in certain years or locations. There are several good sources of literature reviews on this subject (Baker & Stebbins, 1965; MacArthur & Wilson, 1967; Wright, 1969).

We may conclude this section by giving a simple example from a recent computer simulation study (Wu & Jain, unpublished). Consider, for example, a species with mixed selfing (*s*) and random mating (*t*)

($s + t = 1$), subdivided into many partially isolated demes each of average effective size N_e, and the rate of gene flow between demes given by m, the fraction of individuals or seeds exchanged each generation. We can express the joint effects of drift and inbreeding on genotypic frequencies in terms of the fixation index F_e which is 0 for panmictic large populations (genotypic proportions as given by the Hardy–

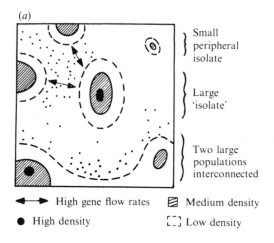

(*a*)

} Small peripheral isolate

} Large 'isolate'

} Two large populations interconnected

◀▶ High gene flow rates   ▤ Medium density

● High density   ⌐⌐ Low density

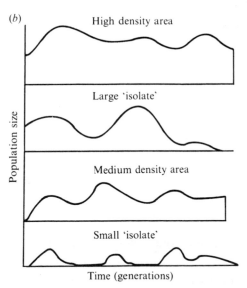

(*b*)

High density area

Large 'isolate'

Medium density area

Small 'isolate'

Population size

Time (generations)

Fig. 2.1. (*a*) Schematic relationships among different populations of a species with varying degrees of isolation; (*b*) changes in population numbers as related to the overall density which in turn depends upon habitat suitability.

19

Table 2.2. *Expected values of* F_e

s	$N_e = 25$ $m = 0.01$	0.05	$N_e = 50$ 0.01	0.05	$N_e = 100$ 0.01	0.05	Single large population
0.85	0.871	0.672	0.828	0.650	0.788	0.638	0.739
0.90	0.908	0.719	0.869	0.713	0.842	0.695	0.818
0.95	0.929	0.769	0.914	0.761	0.900	0.756	0.905
0.99	0.955	0.811	0.952	0.810	0.948	0.809	0.980

($F_e = 1 - H/2pq$, at equilibrium or steady state where H = level of heterozygosity; p, q are allelic frequencies.)

Weinberg principle) and $F_e = 1$ for completely homozygous populations. Table 2.2 gives the values of F_e at steady state for several combinations of m, s and N_e. Two points to note are: (1) drift tends to increase homozygosity with decreasing N_e and m, more so with the low values of s (say, 15 per cent outcrossing, $t = 0.15$); (2) however, with high rate of selfing (e.g., $s = 0.99$), migration and subdivision allow more heterozygosis than expected in a single large population. As noted by Cavalli-Sforza & Bodmer (1971), subdivision *per se* does not decrease the total effective population size. Maynard Smith (1970) explored the role of neutral mutations in maintaining genetic variation under subdivided structure. He concluded that the outcome depends on the ratio m/ur and N_e where u = rate of mutation, r = number of subpopulations, m and N_e as defined earlier. All other things being equal, neutral mutation theory predicts more numerous polymorphisms in species with larger N_e. Furthermore, under certain multigenic, epistatic systems, the effects of subdivided population structure on the retention of variability within the entire species become significantly large due to the increased number of and wider conditions for genotypic frequency equilibria (Karlin, personal communication). In a sense this matches well with what Wright (1940) has shown to be the optimal population structure for evolutionary change.

Population structure of plant species under domestication

In many familiar crop genera, the genetic resources are known to exist in various stages of domestication (defined broadly as evolution in man-influenced habitats). These may range from almost-wild species of primary habitats or open, naturally disturbed habitats to the weed-crop complexes adapted to the cultivated fields, dumpheaps, etc. The evolutionary forces under domestication are identical in general theoretical framework to those applying to the natural communities. Population structures in the past as well as present times must have played an important role in the processes of selective shifts, origin of diversity, and changes in breeding system, etc. Various competing hypotheses as to where and how domestication began in prehistory suggest that (1) certain plants were pre-adapted for cultivation in their ability to colonize naturally the disturbed habitats, (2) certain areas like inter-mountain valleys, or tropical lowlands near rivers or streams were most likely the sites of original domestication. Various authors have pointed out the role of cultural-ethnological factors that must have also influenced the population dynamics and genetic diversity of species involved. There are very few definite clues to the nature or intensity of past evolutionary forces in any more specific terms (see Ucko & Dimbleby (1969)).

What about our present-day knowledge on this issue? It appears from a literature survey that many plant collecting expeditions with only anecdotal and historical notes do not provide even a coarse description of the species distribution, subdivision, population numbers, and even sampling procedures, although, undoubtedly, such information might be available in unpublished field notebooks. There are several exceptions, however. Rick (1963), for instance, reported estimates of population number for a set of Galapagos tomato populations; habitat structure was studied in detail by Oka & Morishima (1967) in rice, Stebbins & Daly (1961) in *Helianthus*, Harding (unpublished) in *Lupinus*, Brown (unpublished) in *Limnanthes*, among others. A recent survey of endangered crop resources (Frankel, 1973) and several recent reports in *Plant Genetic Resources Newsletter* (e.g. Maggs, 1973; Erroux, 1972) have added valuable information on the population structures. Fryxell (1967) discussed the nature and role of disjunct species distributions in the evolution of *Gossypium*. Ladizinsky & Zohary (1968) described the ecological features of *Avena* distribution with emphasis on certain features of dispersal and population subdivision. Although, clearly we cannot expect most plant collectors to spend the required resources in this area, a plea could still be made for getting some minimal description of habitat, plant distribution, etc. in order to evaluate the sampling process of gene pools, and subsequent results on genetic variation.

Variations in breeding systems and their genetic consequences

Breeding systems in plants are widely discussed as a key feature of plant population biology. Although we speak of modal classes, viz outbreeders, inbreeders, and apomicts, it is apparent that in most species there is a great deal of variability in the mode of reproduction. Fryxell's (1957) review is a classical summary in this area. Jain & Marshall (1968) further reviewed several examples to establish the extent of variations in the outcrossing rates encountered in crop plants. A wide variety of genetic and environmental factors influence the breeding structure. Many specific genetic factors are known which regulate the degree of outbreeding, as exemplified by the occurrence of homostyly in *Primula vulgaris*, male sterility in various species, and self-sterility in *Nicotiana*. Clausen (1954), Stebbins (1950), and Harlan & Celarier (1961) among others, discussed examples of species with variable levels of apomixis. There are good reasons to speculate on the adaptive factors regulating the rate of asexual reproduction or outbreeding, assuming that the optimal population structure relies on a mixed mating system (Wright, 1940). Levin (1968) has argued for such an adaptive control of homogamy in *Lithospermum* populations.

In predominantly selfing species, the average low rate of outbreeding may however be associated with a wide potential range or high variance over seasons and localities. Table 2.3 gives a summary of outcrossing rate estimates in several annual species. Stebbins (1957) refers to the occasional outbursts of outcrossing in such species which may correspond to the values in higher range. Theoretical work using computer models (Wu & Jain, unpublished data) showed that such variations in the rate of outcrossing (t) allowed a greater level of heterozygosity within populations. For very high values of s, subdivision tends to lower it. Both of these sources of variation in heterozygosity are of great importance in relation to the maintenance of genetic variation in inbreeders. A recent study by Brown & Marshall (1974) in *Bromus mollis* provided evidence for balance between inbreeding and heterozygote advantage in local variation pattern.

In many species, a major change under domestication is increased selfing. A classic example is found in tomato (Rick, 1950). We may also note that theory predicts genes for increased inbreeding to spread in populations unless checked by homozygote disadvantage. Yet, wide variations in the rate of outcrossing indicates that a mixed mating system might be favored by evolutionary processes such

Table 2.3. *Outcrossing rates (t) estimated using gene markers (number of independent estimates given in the cells)*

Species	Class boundaries											$\bar{t}$	s_t
	0–0.009	0.01–	0.02–	0.03–	0.04–	0.05–	0.06–	0.07–	0.08–	0.09–	0.10–0.15		
Avena barbata	16	7	5	2	3	1	0	0	1	1	1	0.024	0.028
A. fatua	7	2	2	4	3	1	1	1	0	0	0	0.028	0.022
A. sterilis	2	1	1	2	1	1	2	0	0	1	0	0.040	0.030
Hordeum vulgare	8	6	2	1	2	3*	2*	2*	0	2*	2*	0.039	0.037
Bromus mollis	6	5	5	2	2	1	1	0	1	2	1	0.035	0.033
B. rubens	6†	0	0	0	0	0	0	0	0	0	0	0	0
Trifolium hirtum	0	2	0	3	1	2	1	0	0	1	0	0.045	0.024
Medicago polymorpha	0	0	0	1	0	0	1	1	0	2	1	0.082	0.031

* Populations carrying a male sterility gene (e.g. CC xiv).
† No heterozygote among a total of about 200 families from six sites.

that polymorphisms for breeding system variations are frequently maintained.

The primary consequences of inbreeding are increased homozygosis and a greater chance for selection among homozygotes (free variability) to change the gene frequencies rapidly. Table 2.4 lists a number of comparisons between the inbreeders and outbreeders derived largely on the basis of these consequences. Newly arising mutants are likely to be lost quickly since a majority of them are presumably unfavorable recessives. Different homozygous gene combinations are expected to be locally predominant in patchy environments. It follows that inbreeding species should have greater interpopulation variability and less intra-population variation than the comparable outbreeders (often congeneric species). The inbreeders will have a greater colonizing ability, not necessarily due to greater dispersal rates but on account of a higher probability of reproduction under very low densities or lack of pollinators. Ideas on a greater role of intergenotypic competition, linkage and gene interactions in building up co-adaptive gene complexes, in inbreeding species, are also derived from some simple theoretical models (Jain, 1969*b*). Therefore, the inbreeders might either possess a poorer adaptive flexibility or respond essentially through a greater individual plastic response (or a greater population turnover under the conditions of *r*-selection).

Accordingly, genetic variation analyses should relate to at least two measures. Genetic distance, computed in terms of variance among the gene frequency arrays of paired populations, may show different patterns such that there is a greater interpopulation variation typical of inbreeders. Simple theory also predicts greater heterozygosity per individual in outbreeders (Fig. 2.2). Curves A, B in Fig. 2.2 relate to the presence of associative overdominance, that is heterozygotes at different loci interact favorably, resulting in an excess of multiple heterozygotes. It should be recognized, however, that since the various parameters of population structure are highly variable due to both genetic and non-genetic factors, such predictions at best are probability statements. Most evolutionary discussions on the subject have not emphasized this fact. Let us look at some experimental studies next.

Table 2.4. *Some deductions about the genetics of inbreeding populations*

Characteristics	Inbreeders	Outbreeders
Genetic variation		
(a) Between populations*	More; greater local adaptedness	Less
(b) Within populations*	Less; fewer loci polymorphic; less heterozygosis	
Phenotypic plasticity (per individual)	Greater?	Less?
Role of intergenotypic competition/facilitation	Greater	Smaller
Co-adaptive gene combinations (role of epistasis, linkage)	Greater	Smaller
Population structure	Small, colonial populations	Large, continuous populations
Reproductive barriers among		
(a) Conspecific populations	Same by species definition	
(b) Species	Stronger; associated with more frequent sympatry	Weaker; associated with allopatry
Colonizing ability	Greater; especially due to high r or success in long range dispersal	Less than inbreeders
Response to selection in new environment	Few biotypes or microspecies with strong gene associations; greater mortality risks and population turnover	Many; weaker gene associations. Smaller turnover

* Population (sampled units) may be on different scales; besides spatial identity, we must think in terms of neighborhood or interbreeding units.

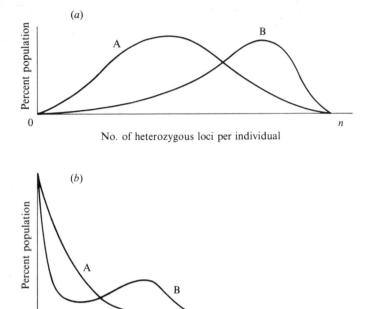

Fig. 2.2. Theoretical curves showing the distribution and heterozygosity at *n* randomly chosen loci in (*a*) outbreeders, and (*b*) inbreeders. A and B represent cases of no interaction versus associative overdominance, the latter resulting in an excess of multiple heterozygotes.

Variation in plant populations

Grant (1971), in his most recent book *Plant Speciation*, provided an excellent review of variation patterns and processes of speciation. Species groups show several general patterns ranging from those of outbreeding woody genera with interfertile species (*Ceanothus* pattern), annual herbs with varied breeding systems and sterility or incompatibility barriers (*Madia* pattern), to the annual herbaceous inbreeders like *Festuca* or *Avena*. In addition, we have examples of stable hybrid complexes involving one or more of the following: vegetative reproduction, agamospermy, permanent structural hybridity, amphiploidy, etc. There is ample evidence from the preliminary variation studies within each of these categories to suggest that most local populations have genetic variability, that inbreeders are less variable locally than the outbreeders and that geographical patterns of variation follow several scales in terms of distances and habitat structure. As noted by Allard (1970) and others, however, local differentiation, and for that matter

26

any predicted pattern might not be characteristic exclusively of one or the other type of reproductive system.

Detailed quantitative analyses of variation have been forthcoming only in recent years and certainly comparative studies of related species are very few. Katznelson (1969) compared the outbreeding and inbreeding groups of *Trifolium* species using a large array of measurement characters. His findings agreed with the theoretical predictions, as noted in Table 2.4. Isozyme variation in *Leavenworthia* species (Solbrig, 1972) also showed a fairly good fit to the expectation of less variation in inbreeders. However, notable exceptions were reported in many of these studies, besides certain universal problems of appropriate sampling procedures and genetic analyses. Hillel, Feldman & Simchen (1973) compared variation in wild populations of a selfer, *Triticum longissimum*, with those of a closely related outcrossing species, *T. speltoides*. They found greater total and within-family variability in *T. longissimum*. Schroeder & Harding (personal communication) found a lack of correlation between the estimates of outcrossing rates and the coefficients of genetic variation in several *Lupinus* species and races. Similar results showing presence or absence of such correlations have been cited at a recent symposium on *The Genetics of Colonizing Species* (Baker & Stebbins, 1965) and in monographs on crop plants like rice, sorghum, cotton and forage crops. It appears that further careful surveys are required in most instances.

In recent years, we have studied several Californian grassland species in some detail. Members of the genera *Avena*, *Bromus*, *Medicago*, and *Trifolium* were chosen for comparative variation studies as they have several features in common, namely, recent history of colonization, autogamy, annual life cycle, and a more or less overlapping range, but they vary somewhat in rates of outcrossing (Table 2.3), relative local abundance, and population structure.

Two species of *Avena*, *A. barbata* and *A. fatua*, occur very commonly in the grassland communities as well as in ruderal areas. Several features of their comparative ecogenetics were summarized in an earlier review (Jain, 1969*a*). The most striking difference in their genetic variation pattern is that most local populations of *A. fatua* are polymorphic whereas variation in *A. barbata* is confined to the central and north coastal regions. Over such large regions as the Central Valley and Sierra foothills as well as Southern California, *A. barbata* shows monomorphism. It is still hypothetical whether very small differences in the outcrossing rates and founder events could account for the observed

Table 2.5. *Variation in* Avena barbata *populations*

Region	Variation pattern for markers *b* and *ls*	Isozyme variation			
		$\overline{PI}$	σ^2_{PI}	PLP	Range of PLP
California					
Warm summer	Monomorphic	0.037	0.0008	0.068	0–0.270
Cool summer	Polymorphic	0.064	0.0036	0.157	0–0.289
Cool summer	Polymorphic	0.101	0.0100	0.207	0.010–0.434
Iberia					
Warm summer	Monomorphic	0.048	0.0028	0.118	0–0.294
Cool summer	Monomorphic	0.116	0.0120	0.181	0.029–0.353
Cool summer	Polymorphic	0.218	0.0028	0.412	0.294–0.529

PLP = % loci polymorphic in a given sample; $\overline{PI}$ = polymorphism index, defined as $\frac{1}{n}\sum_{}^{n}\frac{m}{m-1}\sum_{}^{m}p_i p_j$ (i ≠ j); p_i, p_j are allelic frequencies, m = number of alleles, n = number of loci surveyed.

patterns. Differences due to ploidy level, introgression with the culti-vated *A. sativa*, or historical events of colonization, are some of the other untested hypotheses. In fact, besides the broad regional pattern, *A. barbata* shows a remarkable patchiness in local distribution of genotypes within the polymorphic sites. A detailed survey of the Hop-land Station sites, where polymorphic and monomorphic areas are intermixed, provided some evidence for the local microclimatic in-fluences of summer fog and diurnal temperature. Variation at one of the polymorphic sites at Hopland was observed in a detailed analysis of phenotypic frequencies using small quadrats within a 30-foot long transect. Many such local surveys undertaken to explore clinal variation consistently showed a great amount of patchiness in gene frequency patterns. Similar studies with Iberian and New South Wales collections show that *A. barbata* does differ strikingly from *A. fatua* or *A. sterilis* in genetic structure and, therefore, presumably in adaptive strategies. A summary of regional patterns for the Californian and Iberian regions is given in Table 2.5. Note that isozyme variation is greater in Iberian samples and that within each region, different localities vary widely in the percent polymorphic loci per population (PLP).

The main conclusion of interest here is that within the same species, estimates of the amount of variation may vary widely, depending upon the area sampled, geographical scale of sampling, etc., presumably due to the complex interrelationships between the genetic, ecological as well as historical variables. There are several additional points of interest, not unique by any means, but well illustrated by these *Avena* studies. Two of the loci in *A. barbata*, namely, *H/h* for lemma hairiness and *Np/np* for node hairiness, showed the dominant alleles *H* and *Np* to be rather rare and highly localized in California. In contrast, none of our Iberian collections had allele *H* whereas allele *Np* was even fixed in as many as 6 per cent of the samples. Thus, individual loci can also vary widely, due to both adaptive or non-adaptive reasons, in the geographical distribution of alleles. This, of course, adds a great deal to the variation and, therefore, uncertainty about the expected allelic frequency distribution at any *specific locus*, in any *individual population*. If samples of each region are further grouped in detail for the class of habitats, from small roadside stands to the large and stable grassland populations, we still find a very wide range of values of polymorphism indices within each habitat class. Small samples received from the USDA cereals collection and from Canadian oat collections gave, as expected, lower estimates than our own collections from Iberia and California. In fact our studies suggest that *A. barbata* and *A. fatua* in California enjoy numerical abundance and enrichment of genetic diversity to the point of perhaps constituting a secondary center of diversity. There are also localized opportunities for what Mayr (1963) has called 'genetic revolutions' due to the population subdivision and bottlenecks in size. A study of *A. fatua* populations in two prune orchards, for instance, recently brought out evidence for a joint role of local selection and drift in producing highly mosaic patterns (Jain & Rai, 1974). Demographic and gene-flow data, however, allowed in this case a more detailed analysis than would be possible otherwise.

Bromus mollis and *B. rubens*, like *Avena*, are highly autogamous grassland annuals introduced from the Mediterranean region, but the two species exhibit strikingly different patterns of distribution. *B. mollis* is widespread with no colonial or apparently subdivided structure and with polymorphism as general as in *A. fatua*. *B. rubens*, in contrast, is much less variable and occurs in dense patches but within certain localized habitats of central California grasslands. A broad range of samples in *B. mollis* showed no north–south clines (Moraes, 1972), whereas Allard's (1965) discussion of sampling procedures based on *Bromus* had assumed regular clines over large areas. It appears that numerous complex environmental gradients and high phenotypic plasticity characteristic of these species often yield highly irregular variation patterns (both genetic and phenotypic criteria). Evidence for

Table 2.6. *Hierarchical variation pattern in* Bromus mollis

Source of variation	DF	Plant height MSS	Panicle length MSS
Among sites	7	5736.95	39.48
Among families within sites	184	1353.45	9.34
Among replicates within families	384	377.70	3.55
Within families (error)	2880	74.69	1.42

DF, degree of freedom; MSS, mean sum of squares.

plasticity comes from estimates of genotype–environment interaction, low heritability, etc. For two of the characters, we did find a hierarchical pattern of variation (Table 2.6), as expected on the principle of distance–differentiation colinearity. Thus, geographic variation in these two bromes also involves a great many complexities depending on the sampling, population parameters, and characters used in such analyses.

Katznelson (1969) reported on a hierarchical pattern of variation in several *Trifolium* species. As noted earlier, his findings also provided a comparison between outbreeders and inbreeders. One of the *Trifolium* species, *T. hirtum*, was introduced nearly 30 years ago into California. This species has rapidly colonized new roadside areas in several counties of California. Studies on polymorphisms showed that interpopulation variation is no greater than the amount of intra-population variation. Outcrossing rates are much higher than in Western Australia (Bailey, personal communication). A striking consequence of increased gene flow and founder effects is seen in the data on poly-morphisms in the colonies of sizes varying in the range of 10 to 60 plants. It appears that as new colonies are founded and increase in size, polymorphism is increased, presumably due to insect pollinators carry-ing pollen between the neighboring colonies.

In contrast to *Bromus* and *Trifolium*, however, *Medicago polymorpha* populations in many areas show still another pattern, viz high local patchiness in the distribution of gene frequencies, in spite of high out-crossing rates but perhaps a far more restricted *net* gene flow. The distribution pattern suggests strong habitat preference and thus the species probably has restricted colonizing ability. A study by McWilliam, Schroeder, Marshall & Oram (1970) in Australian populations of *Phalaris tuberosa* showed lack of inter-population differentiation which could be due to high gene flow, slower generation turnover, or lack of habitat diversity. Now there are numerous plant population studies in progress which involve joint analyses of genetic variation and population structure.

Table 2.7 summarizes our findings, admittedly using preliminary estimates, since there are many unknowns, and statistical problems associated with any similar set of 'few' estimates are indeed massive. While outcrossing rates ($\bar{t}$) and heterozygosity levels ($\bar{H}$) may be most directly correlated, it is apparent that given a common feature of high autogamy, *each species has a unique population structure.* We may conclude this survey by saying that variation patterns, like all other biological patterns are usually imperfect but, hopefully, comparative analyses provide after a long and patient search some evidence of regularity and explanations as to how or why they occur.

Implications in genetic conservation

We must now ask if there is one *modus operandi* or many different ones in the evolution of cultivated plants. Can we associate certain kinds of population structure and/or reproductive modes with the patterns of intra- as well as interspecific genetic variation? As pointed out earlier, no matter what specific theories are accepted about the places or dates of origin, it does appear that crop species and their wild or weed relatives have had certain features in common: relatively short evolutionary time (by the standard of most other plant species), selection for features of economic interest to man, control of reproductive cycle, phenology and to a certain degree the regulation of population numbers; local isolation and presumably disruptive selection; small size or isolation effects compensated by various levels of hybridization (cf. Harlan, 1970, on differentiation–hybridization cycles). All of this adds up to an optimum evolutionary model for maintaining species-wide diversity. Breeding systems are important only in relation to a highly local scale of heterozygosis or polymorphism whereas geographical variation is determined by the totality of population parameters. It is probable that centers or microcenters of diversity are either related to antiquity (centers of origin), or an optimum rate of evolutionary process (e.g., subdivision). Both natural and artificial selection would have, until recently, promoted this genetic enrichment. We may perhaps actually observe this process in places like California roadsides which seem to be on the way to becoming 'new secondary centers of diversity' for many introduced species including wild oats, rye, and sunflower. In view of

Table 2.7. *Population structure and variation patterns: a summary*

| Species* | Distribution | Gene flow | | | Population regulation (density response) | Patchiness | $\bar{H}$ | $\sigma^2_{\mathrm{B}}/\sigma^2_{\mathrm{w}}$† |
		$\bar{t}$	Seed dispersal	$\bar{N}_e$				
A. ba	Varied	0.024	Short range	Small? (20–200)	Plastic	High	< 0.02	Variable
A. fa	Varied	0.028	Short range	Small?	Less plastic	Low	> 0.02	Variable
A. st	Varied	0.040	Short range	?	Less plastic	Low	> 0.02	Somewhat lower than A. ba
B. mo	Varied	0.035	?	Larger	Plastic	Low	< 0.02	Variable
B. ru	Restricted	0	?	?	Plastic	High?	≪ 0.02	Variable
T. hi	Rapidly colonizing new areas	0.045	Short range	Varied	Mortal	Low	> 0.02	Low
M. po	Restricted	0.082	Short range	Varied	Mortal	High	> 0.02	Low

* Species names, as in Table 2.3.
† σ^2_{B} = between-families component of genetic variation; σ^2_{w} = within-families component of genetic variation.
Data on *Avena barbata*, adapted after Rai (1972); data on *Bromus mollis*, adapted after Moraes (1972).

the common features noted above, the variation patterns in crop genera may turn out to be simpler and fewer in relation to the goals of genetic conservation. Moreover, it is likely that preliminary fieldwork on the population structure in each case can provide useful clues to these patterns.

Marshall and Brown (Chapter 4) have dealt with the question of optimum sampling strategies for genetic conservation. Here we shall consider only briefly several courses of action: Allard (1970) had suggested a sampling plan, using *A. fatua* as an example, which involves collecting seeds of 200 plants from each location, locations being more or less spaced regularly along a rectangular grid. This scheme assumes regular gene frequency distribution, e.g., clines. Marshall and Brown, on the other hand, recommend for similar situations samples of 60–100 individuals per site with greater coverage of area and many more samples. Still another point of view was put forth by Qualset (Chapter 5) who argued that in order to sample gene combinations in the presence of linkage disequilibrium particularly for research purposes, larger samples, in the range of 500 plants, would be necessary. Clearly, these alternative strategies and possibly others rely heavily on the sampling considerations of the variety of patterns and their probabilistic nature. Any sampling scheme would have to adapt to the kind of species being collected. In inbreeding species with relatively localized patchiness of variation, samples should be drawn with a scheme of scattered sampling points rather than quadrats. This would also be the case with vegatatively propagated species. On the other hand, species with high rates of gene flow, say, through seed dispersal, would approach the population structure of an outbreeding species. Of course, a sampling strategy would also have to take note of such factors as size and location of areas, reproductive system and longevity. It should be apparent from the foregoing discussions of population structures and variation analyses that a good preliminary description of population subdivision, genetic distances among distant subpopulations, and so on, would be critical in planning sampling procedures. If the matters of urgency, expediency and absolute resource limitations are not prohibitive against a two-step sequential method, it would seem a highly desirable procedure to

(1) plan a coarse grid during the first collecting season such as to draw few but larger samples and make some detailed field and laboratory studies;

(2) plan second or subsequent rounds of sampling on a finer scale in areas of particular interest (large amount of variation or novel variation);

(3) emphasize ecological characteristics of a species distribution and ensure sampling of, say, peripheral or disjunct populations;

(4) in all these programs, work on several species with an inter-disciplinary and co-ordinated teamwork for studying the community features and parallel variation in several co-existing species of interest.

This procedure vis-à-vis the proposal of Marshall and Brown has some merit only in situations where we can afford two or more sampling seasons, where we need to know more about the field biology of variation, or when we are interested in the conservation of resources *in situ* as genetic reserves (Chapter 30). Note that Marshall and Brown discuss at some length the situations where some prior information on the variation pattern is available. Then, for example, larger samples at polymorphic sites are recommended. The proposed method will help us in getting this information along with the germplasm materials.

Our recent work in collecting *Limnanthes* species (a genus recognized for its unique oil resource as a substitute for sperm whale oil) in California was based on the above procedure. We feel that two years of sampling and concurrent genetic studies would systematically facilitate both genetic conservation and utilization of gene pools in future breeding work. The keynotes in this proposed method are: sequential decision-making, larger-than-minimum sample size for including rare alleles or gene combinations, and a good deal of teamwork for research. Genetic conservation has a great deal of urgency but hopefully, use of better sampling strategies and research-orientedness can prove of very great assistance.

References

Allard, R. W. (1965). Genetic systems associated with colonizing ability in predominantly self-pollinated species. In *The Genetics of Colonizing Species* (eds H. G. Baker and G. L. Stebbins), pp. 50–78. Academic Press, New York.

Allard, R. W. (1970). Population structure and sampling methods. In *Genetic Resources in Plants – their Exploration and Conservation* (eds O. H. Frankel and E. Bennett), pp. 97–108. Blackwell, Oxford and Edinburgh.

Andrewartha, H. G. & Birch, L. C. (1954). *The Distribution and Abundance of Animals*. University of Chicago Press, Chicago.

Baker, H. G. & Stebbins, G. L. (eds) (1965). *The Genetics of Colonizing Species*. Academic Press, New York.

Brown, A. H. D. & Marshall, D. R. (1974). The maintenance of alcohol dehydrogenase polymorphism in *Bromus mollis* L. *Aust. J. Biol. Sci.*, **27**, 545–59.

Cavalli-Sforza, L. L. & Bodmer, W. F. (1971). *The Genetics of Human Populations.* W. H. Freeman, San Francisco.

Clausen, J. (1954). Partial apomixis as an equilibrium system in evolution. *Proc. IX Intern. Congr. Genet., Caryologia,* vol. VI, *Suppl.,* pp. 469–79.

Erroux, J. (1972). Les Blés indigènes Nord-Africains. *Plant Genet. Res. Newsl.* **28,** 17–23.

Frankel, O. H. (ed.) (1973). *A Survey of Crop Genetic Resources in their Centres of Diversity.* FAO/IBP, Rome.

Frankel, O. H. & Bennett, E. (1970). Genetic resources. In *Genetic Resources in Plants – their Exploration and Conservation* (eds O. H. Frankel and E. Bennett), *IBP Handbook,* no. **11,** pp. 7–17. Blackwell, Oxford and Edinburgh.

Fryxell, P. A. (1957). Mode of reproduction of higher plants. *Bot. Rev.,* **23,** 135–233.

Fryxell, P. A. (1967). The interpretation of disjunct distributions. *Taxon,* **16,** 316–24.

Grant, V. (1971). *Plant Speciation.* Columbia Univ. Press, New York.

Harlan, J. R. (1970). Evolution of cultivated plants. In *Genetic Resources in Plants – their Exploration and Conservation* (eds O. H. Frankel and E. Bennett), pp. 7–17. Blackwell, Oxford and Edinburgh.

Harlan, J. R. & Celarier, R. P. (1961). Apomixis and species formation in the Bothriochloeae King. *Recent Adv. Bot.,* **8,** 706–10.

Hillel, J., Feldman, M. W. & Simchen, G. (1973). Mating systems and population structure in two closely related species of the wheat group. I. Variation between and within populations. *Heredity,* **30,** 141–67.

Jain, S. K. (1969a). Comparative ecogenetics of *Avena fatua* and *A. barbata* occurring in central California. *Evol. Biol.,* **3,** 73–118.

Jain, S. K. (1969b). Epistasis and linkage in inbreeding populations. *Japan J. Genet.,* **44,** *Suppl.* **1,** 135–43.

Jain, S. K. (1974). The genetic structure of plant populations. Paper presented at the Symposium on Evol. Biol. (ed. R. K. Selander), *First Intern. Congr. Syst. Evol. Biol. Boulder* (in press).

Jain, S. K. & Marshall, D. R. (1968). Simulation of models involving mixed selfing and random mating. I. Stochastic variation in outcrossing and selection parameters. *Heredity,* **23,** 411–2.

Jain, S. K. & Rai, K. N. (1974). Population biology of *Avena.* IV. Polymorphism in small populations of *Avena fatua. Theor. appl. Genet.,* **44,** 7–11.

Katznelson, J. (1969). Population studies and selection in berseem clover (*Trifolium alexandrinum* L.) and the closely related taxa. *Third Ann. Rep. Volcani Inst. Agric. Res., Israel,* 65 pp.

Ladizinsky, G. & Zohary, D. (1968). Genetic relationships between diploid and tetraploid species in series *Eubarbatae* of *Avena. Can. J. Genet. Cytol.* **10,** 68–81.

Levin, D. A. (1968). The breeding system of *Lithospermum caroliniense:* adaptation and counteradaptation. *Amer. Natur.,* **102,** 427–41.

Levin, D. A. & Kerster, H. W. (1971). Neighborhood structure in plants under diverse reproductive methods. *Amer. Natur.,* **105,** 345–54.

35

MacArthur, R. H. & Wilson, E. O. (1967). *The Theory of Island Biogeography*. Princeton Univ. Press, Princeton.

Maggs, D. H. (1973). Genetic resources in pistachio. *Plant Genet. Res. Newsl.*, **29**, 7–15.

Mayr, E. (1963). *Animal Species and Evolution*. Harvard Univ. Press, Cambridge, Mass.

McWilliam, J. R., Schroeder, H. E., Marshall, D. R. & Oram, R. N. (1971). Genetic stability of Australian Phalaris (*Phalaris tuberosa* L.) under domestication. *Austr. J. Agric. Res.*, **22**, 895–908.

Moraes, C. F. (1972). Ecogenetic studies on variation and population structure in soft chess, *Bromus mollis* L. Ph.D. Thesis, University of California, Davis.

Oka, H. I. & Morishima, H. (1967). Variations in the breeding system of a wild rice, *Oryza perennis*. *Evolution*, **21**, 249–58.

Rai, K. N. (1972). Ecogenetic studies on the patterns of differentiation in natural populations of slender wild oat, *Avena barbata* Brot. Ph.D. Thesis, University of California, Davis.

Rick, C. M. (1950). Pollination relations of *Lycopersicon esculentum* in native and foreign regions. *Evolution*, **4**, 110–22.

Rick, C. M. (1963). Biosystematic studies on Galapagos tomatoes. *Occ. Papers Calif. Acad. Sci.*, **44**, 59–77.

Selander, R. K. & Kaufman, D. W. (1973). Genetic variability and strategies of adaptation in animals. *Proc. natn. Acad. Sci. USA*, **70**, 1875–7.

Smith, J. M. (1970). Population size, polymorphism, and the rate of non-Darwinian evolution. *Amer. Natur.*, **104**, 231–7.

Solbrig, O. T. (1972). Breeding system and genetic variation in *Leavenworthia*. *Evolution*, **26**, 155–60.

Stebbins, G. L. (1950). *Variation and Evolution in Plants*. Columbia Univ. Press, New York.

Stebbins, G. L. (1957). Self-fertilization and population variability in the higher plants. *Amer. Natur.*, **91**, 337–54.

Stebbins, G. L. & Daly, K. (1961). Changes in the variation pattern of a hybrid population over an eight-year period. *Evolution*, **15**, 60–71.

Ucko, P. J. & Dimbleby, G. W. (eds) (1969). *The Domestication and Exploitation of Plants and Animals*. Duckworth, London.

Wright, S. (1940). Breeding structure of populations in relation to speciation. *Amer. Natur.*, **74**, 232–48.

Wright, S. (1969). *Evolution and the Genetics of Populations*, vol. II, *The Theory of Gene Frequencies*. Univ. of Chicago Press, Chicago.

3. Population structure and the effects of isolation and selection

A. D. BRADSHAW

When we consider the effects of isolation and selection on population structure, particularly in relation to crop plants, we are really concerned with two things: the effects that isolation and selection have on the genetic constitution of individual populations considered by themselves, and the effects they have on the genetic constitution of groups of related populations. Both of these are crucial in any consideration of crop genetic resources, since the essential question is whether one population, or very many different populations, will be an effective collection of all the important genes of any particular crop.

Levels of divergence

Essentially the effects of selection and isolation in natural situations are to cause populations to diverge progressively in gene frequency from one another, at the same time causing changes in the genetic structure of the individual populations. There are two levels in this divergence:

Level 1. The populations remain polymorphic for the same alleles but have them at different frequencies. The frequencies can be so different that the populations have quite different mean phenotypes. But the important thing is that no allele is absent from one population and present in another. In practical terms this means that it should be possible by selection to obtain the same product from any population although it is obviously going to be quicker from the population that has its alleles nearest to the frequency desired in the product. If a crop species was ever in this state then it could be represented by only one population. This is highly unlikely; but we do need to know whether it applies to parts of a crop species, so that the crop could be represented by only a few samples collected on a regional basis.

Level 2. The populations have diverged so far that certain alleles have gone to fixation (become homozygous) in different directions in different populations. If this level is reached it will not be possible to extract every product from any population because the alleles will just not be there.

37

The species would then be represented by as many samples as there are different populations, if we are to preserve all the variability in the species.

Level 1 – divergence in gene frequency

The effects of selection

The part played by selection in this process of divergence is obvious, as any calculation of change of gene frequency in a population due to selection shows, for example that by Falconer (1961, p. 33). The Illinois corn experiment shows the same in practice: it also shows that the different selection lines can be taken right outside the range of the original population, probably because of the release of variability from complexes of linked genes (Leng, 1962). This in turn shows how difficult it is to know what is actually stored in a population without carrying out selection experiments. The power of selection and the rates of change in these two examples are considerable.

Originally we did not think that in natural situations selection could be so powerful and as a result we have not, perhaps, believed that this sort of divergence was possible. But recent evidence shows that selection can indeed be very powerful (Bradshaw, 1972). Perhaps the best example is that of Akemine and Kikuchi using rice populations, since it shows what happens when the same original population is selected in different directions in different parts of Japan (Allard & Hansche, 1964). Three generations of selection were enough to cause marked divergence in heading dates which implies that the populations have developed different gene contents. Plant breeders are also familiar with the way in which existing crop varieties can change extremely quickly when moved into new environments. Thus the yield of Bottnia meadow fescue changed by a factor of more than two after one generation in south Sweden (Sylven, 1937) (Table 3.1): Bottnia meadow fescue is not able to flower in the daylength of south Sweden; the change was therefore due to selection of the few genotypes which were appropriately adapted. The yield of red clover in Tennessee declined by 10 per cent for each generation it was multiplied in the Pacific North-West USA (Beard & Hollowell, 1952) as adaptation to the disease and climate of Tennessee was lost. The variety of cotton U4 became diversified into a number of locally adapted varieties in a few years after its development and distribution in Africa (Hutchinson, Silow & Stephens, 1947). In an industrial area near Liverpool we have recently been able to show that the copper pollution from a copper smelter has caused the evolution of copper tolerance in populations of the turf grass *Agrostis stolonifera* in only a few years (Wu & Bradshaw, 1972) (Fig. 3.1). The ability of a crop plant to change under the influence of natural selection is being used as a positive plant breeding tool in several crops, e.g. the potato (Simmonds, 1966).

Table 3.1. *Changes in the yield of Bottnia meadow fescue* (Festuca pratensis) *bred in Lulea, north Sweden for northern environments, after growth for one generation in Svalof, south Sweden* (*Sylven, 1937*)

Yield in Lulea	Yield	%
After one generation in Lulea	58.9	100
After one generation in Svalof	55.9	95
Yield in Svalof	Yield	%
After one generation in Lulea	44.4	100
After one generation in Svalof	109.1	245

It is therefore not surprising to find that in virtually all species which occupy a range of habitats there is a corresponding range of different populations each adapted to its own environment. This was first demonstrated by Turesson but for our purposes the best examples are those in *Achillea millefolium* and *Potentilla glandulosa* demonstrated by Clausen, Keck & Hiesey (1940; 1948), since they showed the ranges of morphological variabilities in each population. Although neighbouring populations which differ in mean phenotype have overlapping ranges of variation, most populations not only differ in mean phenotype but also have ranges that do not overlap. They also showed that the populations are markedly different in their physiological adaptation.

Any character of a plant can be changed by selection including those that may not be immediately visible but which may be very important in the ultimate adaptation of crop varieties to their environment. A good example of this is the previously unsuspected variation in adaptation to soil nutrients found in the well investigated herbage plants, *Lolium perenne* and *Trifolium repens*, where populations from different soil conditions show quite different responses to calcium, phosphorus and other nutrients in sand and soil culture (Crossley & Bradshaw, 1968; Snaydon & Bradshaw, 1969).

In all this work on natural plant populations there is little sign that anything other than selection determines the characteristics of each

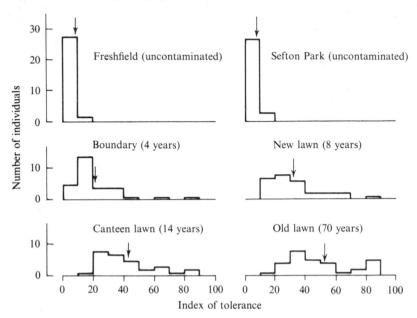

Fig. 3.1. Distribution of copper tolerance in populations of creeping bent grass (*Agrostis stolonifera*) which have recently become subject to copper pollution (Wu & Bradshaw, 1972).

population. The other possible factor, except isolation and migration which I shall discuss later, is genetic drift. This is a complex problem, and there is very little direct evidence available. From theoretical arguments plant populations appear nearly always to be too large for drift to have any effects, since the selection coefficient which will hold genetic drift in check is approximately $1/(4N)$ (where N is the population size). There are one or two instances where the characteristics of certain populations do not seem to be explicable in terms of selection (Lammerink, 1968; Palmer, 1972). But these are not common, and whether they have been determined by drift is not clear. They could be due to a form of drift, founder effect, arising from the smallness of the original population which colonised the areas. In the one case in plants, *Linanthus parryae*, where drift was originally thought to be operating, it has subsequently been shown that it could not have had the effects suggested (Epling, Lewis & Ball, 1960).

We have recently been able to look at the genetic structure of populations more precisely by their allozyme frequencies; it is interesting that the same conclusions are emerging. Variation in allozyme frequency in *Avena barbata* is only explicable as a result of selection (Clegg & Allard, 1972).

The effects of isolation

The earlier work seems to suggest that differentiation only occurs between populations which are widely separated, by distances of at least a few kilometres. It is easy to think this if one thinks that populations closer than this will tend to be kept together in gene frequency by gene flow due to migration of pollen and seed (Fig. 3.2), with the gene flow overwhelming the differentiating effects of selection.

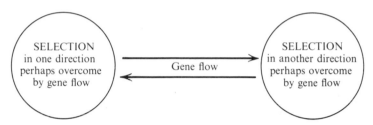

Fig. 3.2. A simple model of the effect of gene flow in preventing the evolutionary divergence of neighbouring populations under the influence of selection (see text).

But it is now clear that this is not true at all. Populations only 15 metres apart in the wind-pollinated pasture grass *Agrostis stolonifera* can be completely different with no phenotypes in common; and the pattern of differentiation over a small area of cliff follows faithfully the pattern of the environment (Aston & Bradshaw, 1966) (Fig. 3.3). Where the environment changes extremely sharply on the edge of a zinc mine, populations of the grass *Anthoxanthum odoratum* only three metres apart are quite different in their metal tolerance, again with no phenotypes in common (Antonovics & Bradshaw, 1970). Similar microgeographic differentiation has been recently reported for allozymes in *Avena barbata* (Hamrick & Allard, 1972).

In these situations it is difficult to understand why gene flow, by movement of pollen particularly, does not overwhelm the effects of selection, and therefore keep the gene frequencies of adjacent populations similar. But this would only be true if selection pressures were low and rates of gene flow high. In practice gene flow, even in a wind-pollinated plant such as *Lolium perenne*, is much less than one might expect, as low as 5 per cent over 10 m (Griffiths, 1950); and selection pressures can be very high, as we have seen. If we assume appropriate selection pressures and rates of gene flow we can arrive at a model of population differentiation which closely resembles that found in natural situations (Jain & Bradshaw, 1966) (Fig. 3.4).

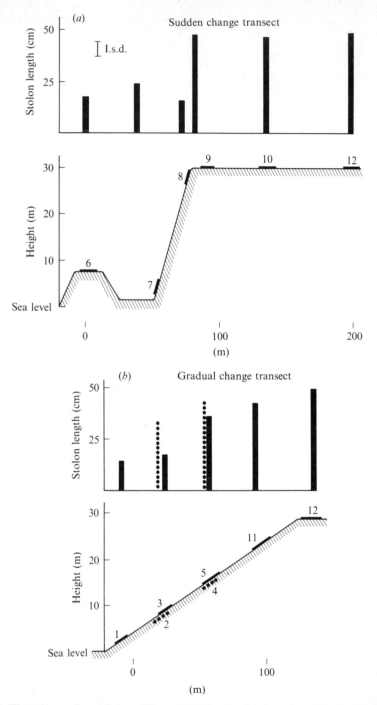

Fig. 3.3. The pattern of population differentiation in creeping bent grass (*Agrostis stoloni-fera*) in a small area of exposed sea cliffs (Aston & Bradshaw, 1966). (*a*) The pattern of differentiation in six populations in a transect across a sharp environmental boundary. (*b*) The pattern of differentiation in five populations in a transect across a nearby gradual environmental change, and in two populations (shown by dots) in a stream two metres away from the transect.

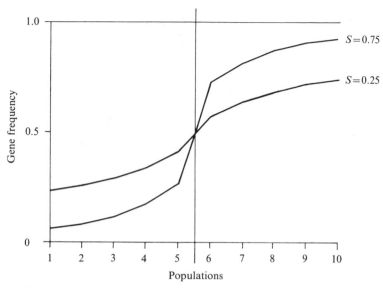

Fig. 3.4. Calculated patterns of evolutionary differentiation for a single pair of alleles in a linear set of ten adjacent populations between which there is a normal amount of gene flow: in half the populations selection is acting in favour of one allele, in the other half of the populations it is acting against the allele (Jain & Bradshaw, 1966). If gene flow is assumed to be similar to *Lolium perenne*, the populations would be four metres apart.

Such a model assumes that there is an interplay between selection and gene flow, so that the population differentiation is a product of a dynamic equilibrium. This has been very clearly shown in the copper-tolerant population on a small copper mine in North Wales (McNeilly, 1968). In this case not only could the gene flow be demonstrated, but because it was polarised due to the topography of the site the pattern of differentiation was different on the windward and leeward sides of the mine.

Selection can then readily overcome quite high levels of gene flow arising because of lack of isolation. But this does not mean that the gene flow has no effects. It is constantly increasing the variability of the population even when selection is working in the opposite direction: it will also tend to increase heterozygosity (Antonovics, 1968). This is well shown by the occurrence, at much higher than background levels, of copper-tolerant individuals in normal populations of the wind-pollinated, self-incompatible pasture grass *Agrostis tenuis* several kilometres away from a tolerant population growing on a copper mine (Khan, 1969) (Fig. 3.5). An extension of the population differentiation model of Jain & Bradshaw (1966) to 100 populations on either side of an environmental boundary (J. T. Gleaves, unpublished) confirms that

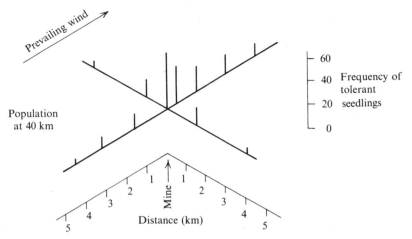

Fig. 3.5. The occurrence due to gene flow of copper-tolerant seedlings in normal populations of bent grass (*Agrostis tenuis*) growing on normal soil at various distances away
from a large copper mine colonised by a copper-tolerant population.

unadapted alleles are maintained by gene flow at a considerable distance
from the boundary.

Gene flow due to lack of isolation therefore has important effects on
the structure of populations in maintaining variability, even if it does not
prevent population divergence. It is acting along with other mechanisms
such as linkage, heterozygote advantage, and outbreeding mechanisms
which are also important in maintaining variability. In this sort of
situation plant collections made in any one of the populations connected
by gene flow would ensure that all genes would be represented in the
sample taken, providing that the sample was sufficiently large.

Level 2 – divergence in gene occurrence

All that we have so far considered has been within level 1 of the process
of divergence; the populations have the same alleles but they are at
different frequencies in different populations. We must now consider
level 2, where alleles have reached fixation in different directions in
different populations.

At some point, due to sheer distance or topography (such as the
presence of a mountain range), total isolation between two populations
must occur. Under these circumstances allele fixation can occur very
easily by selection, on the simple hypothesis that if there is directional
selection, frequency will change until it becomes fixed.

The result of selection where there is complete isolation will be distinct populations or groups of populations, such as the closely related but geographically isolated species of *Scalesia* on the Galapagos islands (Carlquist, 1965). This differentiation could of course equally come about through genetic drift. But as we have argued already it is unlikely to be very important unless the population was very small at least at one period of time after it became isolated. Perhaps the most likely time when this could occur is when a species first evolves to colonise a new habitat. However in the evolution of metal tolerance, which we have been able to examine carefully, it has been found that a large number of different individuals form the initial populations (Wu & Bradshaw, unpublished): so in this case at least the founder effect is not operative.

The fixation will be increased by inbreeding since this will reduce not only the amount of variability stored in the heterozygous condition but also the amount of gene flow. An excellent example of this is found in *Avena barbata* populations in California that we have already considered, and in the populations described by Jain in this book (Chapter 2). It can also be helped, where geographical isolation is not complete, by the development of isolation mechanisms by the plant itself. These seem to be much more common than we might expect, even between populations within species, such as the breeding barriers found by Kruckeberg (1957) in populations of *Streptanthus glandulosus* in the Bay area of California around San Francisco, causing incompatibility in crosses between populations only a few kilometres apart.

The fixation will be hindered by heterozygote advantage and disruptive selection. In self-fertilising species the evolution of ecologically co-adapted genotypes which make different demands on the environment, found in barley composite-cross populations by Allard & Adams (1969), is a further powerful force preventing fixation by generating frequency dependent selection.

Divergence is aided by mutation: if it is at a very low rate, it may cause some populations, but not others, to possess an allele, despite selection for this allele in all populations. A very clear-cut example where the process can be observed is the occurrence of warfarin resistance in rats where the single gene conferring resistance has appeared only in a few geographically separated foci (Drummond, 1966). Similarly the single gene for blackarm resistance in Upland cotton was found in native material only in a single focus in Central America despite natural selection for resistance over a much wider range (Knight & Hutchinson, 1952).

The level at which total isolation occurs, is difficult to define, or even to discover. At the same time, the factors determining total isolation and subsequent fixation of different alleles in different populations are many. So our diagram of selection and gene flow has to be much more complicated (Fig. 3.6) than it was previously (Fig. 3.2). In any one situation it may be difficult to discover which factors have been operating and at what intensity. As a result it may be impossible to predict the geographic scale on which gene fixation occurs.

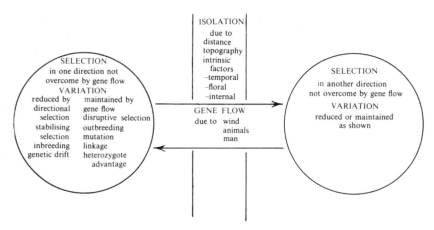

Fig. 3.6. A more complex and more realistic model of the interplay of gene flow, selection, and other factors, indicating that evolutionary divergence of neighbouring populations under the influence of natural selection can occur (see text).

The situation in cultivated crops

The crop materials with which we are concerned are the primitive cultivars, grown by peasant cultivators, now so rapidly being replaced by modern improved varieties. In these, there are special conditions operating which will affect the level at which total isolation occurs and populations arise possessing different alleles. Even although growing conditions may be primitive there has been sufficient amelioration and uniformity of cultural methods that selection can be uniform over quite wide areas. This will act against divergence. But equally, different cultural traditions can select for different genotypes in regions that are environmentally similar: this seems to have occurred in the differentiation of *Gossypium hirsutum* where quite different types of cotton have evolved in Central America in association with different ethnic groups rather than different environments (Hutchinson, 1951). In maize, very

46

distinctive varieties are maintained side by side in a state of uniformity by Indians in Central America (Mangelsdorf, 1965). But even where there is uniformity of cultural techniques this cannot overcome basic differences in environment due to altitude, exposure, rainfall or soil, all of which will have strong selective effects.

In most traditional cultures it was common practice for individual farmers to retain their own seed, and if their crop failed to get seed from neighbours: this prevents long distance movement of genotypes. But equally, in times of disturbance or migration, it was normal for groups to take their seed with them: this has led to some spectacular movements of genotypes across the world. Normal processes of trading have done the same. However after such movements the isolation and distinctness of the material has not necessarily broken down.

From many experiments we know that gene flow is reduced in out-breeding crops because all crops are normally grown in compact blocks, and pollen distribution is leptokurtic. Many crop plants have become self-fertilising, which markedly reduces gene flow. Localised differentiation can therefore readily occur. But investigations of inbreeding crops (Allard, Jain & Workman, 1968) have shown that inbreeding is rarely complete. The opportunity for gene flow therefore remains in inbreeders as well as outbreeders. This is important since while the integrity of a crop variety is unlikely to be upset by excessive gene flow, sufficient can occur to contribute new alleles, which if favoured by selection will spread through the populations affected. There are numerous examples of crops which, when moved into new areas, have picked up alleles from related native or weed species, for instance alleles for blackarm resistance in Upland cotton in Africa (Knight & Hutchinson, 1952) and for various characters in sorghum (Doggett, 1965).

Under traditional conditions of cultivation, as a rule a crop is sown from a bulk sample of the previous year's crop. Thus even a plot of 1/10 hectare of a small grain cereal will be established from seed derived from a minimum of 5000 plants, but usually from many more. As a result genetic drift and gene fixation are unlikely to occur, unless at any time for some special reason the progeny of a single plant has been retained and multiplied. However, there are many crops, from maize to melons, in which the whole of a peasant farmer's crop could be provided by seed from a single plant. And since in these sorts of crops the particular merits of individual plants will be easily visible, such selection is likely to be commonplace. In these circumstances a considerable degree of gene fixation could occur.

47

From all this it appears that differentiation is likely to be as localised in crop plants as it is in wild plants, although the exact state must depend on the properties of the individual crop. In wild plants, as we have seen, the amount of localisation is greater than we had previously assumed; I believe that this will be found to be true in crop plants if the situation is examined in detail.

The unfortunate point which is crucial to the whole of this problem is that the situation has not ever been looked at in sufficient detail to give a clear geographical picture of the genetical structure of any crop still in a primitive unimproved state. But there does seem to be enough good indirect evidence to make it advisable to assume that the scale of differentiation is very small.

This then raises considerable problems for genetic conservation. To some extent it is possible to produce by selection the same character in genetically different populations by using different genes combining to give the same effect. This will be aided by stored genetic variation as in the Illinois corn experiment. But the fact remains that populations of crop plants and their wild relatives do differ in important genes and that many crucial advances in plant breeding have depended on discovering particular populations in which a desired allele occurred.

In this case, should we sample all existing populations of a crop in order to ensure we have all possible alleles? Since we cannot do this we must adopt some way which provides reasonable samples of as many populations as possible in the way proposed by Marshall and Brown (Chapter 4). We must endeavour to sample all environments in the manner already suggested by Allard (1970).

But to cut down what will still be an excessive number of populations, we will have to adopt a system which avoids sampling populations which, because of similar selective or genetic histories, are likely to possess the same genes. There is therefore no point in selecting a set of populations from the same environment, unless they are likely to have had very different genetic histories. There is however value in selecting populations from different environments even if they could have had a common origin, because of the overriding effect of selection.

But this will still provide too many populations to sample. The only solution seems to be to adopt a two stage collecting process, as proposed by Bennett (1970). The first stage is to collect on a wide but very systematic basis in a way which covers all major environments, in every geographic region where the crop occurs. This recognises that environment is the major determinant in the occurrence of individual genes, but that historical or mutational accidents may have caused a gene to appear in an environment in one geographical region but not in the same environment in another region.

The second stage would be a very much more detailed collection of the areas which appear to contain genes of particular interest. This recognises that differentiation to level 2, the point at which populations possess different genes, occurs on a small scale, even smaller than previously expected. This also assumes that it will be possible to decide which areas are likely to be of interest. Although the decision is difficult, experience should suggest where genes of particular importance should occur. In this way the development of a rational plan for plant collection on both an extensive and an intensive basis will be possible for each crop plant.

References

Allard, R. W. (1970). Population structure and sampling methods. In *Genetic Resources in Plants – their Exploration and Conservation*, (eds O. H. Frankel and E. Bennett), *IBP Handbook*, no. **11**, pp. 97–107. Blackwell, Oxford and Edinburgh.

Allard, R. W. & Adams, J. (1969). Population studies in predominantly self-pollinating species. XIII. Intergenotypic competition and population structure in barley and wheat. *Amer. Nat.*, **103**, 621–45.

Allard, R. W. & Hansche, P. E. (1964). Some parameters of population variability and their implications in plant breeding. *Adv. Agron.*, **16**, 281–325.

Allard, R. W., Jain, S. K. & Workman, P. L. (1968). The genetics of inbreeding populations. *Adv. Genet.*, **14**, 55–132.

Antonovics, J. (1968). Evolution in closely adjacent plant populations. VI. Manifold effects of gene flow. *Heredity, London*, **23**, 507–24.

Antonovics, J. & Bradshaw, A. D. (1970). Evolution in closely adjacent plant populations. VII. Clinal patterns at a mine boundary. *Heredity, London*, **24**, 349–62.

Aston, J. L. & Bradshaw, A. D. (1966). Evolution in closely adjacent plant populations. II. *Agrostis stolonifera* in maritime habitats. *Heredity, London*, **21**, 649–64.

Beard, D. F. & Hollowell, E. A. (1952). The effect on performance when seed of forage crop varieties is grown under different environmental conditions. *Proc. 6th Int. Grassl. Cong.*, 860–6.

Bennett, E. (1970). Tactics of plant exploration. In *Genetic Resources in Plants – their Exploration and Conservation* (eds O. H. Frankel and E. Bennett), *IBP Handbook*, no. **11**, pp. 157–80. Blackwell, Oxford and Edinburgh.

Bradshaw, A. D. (1972). Some of the evolutionary consequences of being a plant. *Evolutionary Biology*, **5**, 25–47.

Carlquist, S. (1965). *Island Life*. Natural History Press, New York.

Clausen, J., Keck, D. D. & Hiesey, W. M. (1940). Experimental studies on the nature of species. I. The effect of varied environments on western North American plants. *Carnegie Inst. Washington, Publ.* **520**.

Clausen, J., Keck, D. D. & Hiesey, W. M. (1948). Experimental studies on the nature of species. III. Environmental responses of climatic races of *Achillea*. *Carnegie Inst. Washington, Publ.* **581**.

Clegg, M. T. & Allard, R. W. (1972). Patterns of genetic differentiation in the slender wild oat species *Avena barbata*. *Proc. natn. Acad. Sci. USA*, **69**, 1820–6.

Crossley, G. K. & Bradshaw, A. D. (1968). Differences in response to mineral nutrients in populations of ryegrass, *Lolium perenne* L. and orchard grass *Dactylis glomerata* L. *Crop Sci.*, **8**, 383–7.

Doggett, H. (1965). The development of the cultivated sorghums. In *Essays on Crop Plant Evolution* (ed. J. B. Hutchinson), pp. 50–69. Cambridge Univ. Press, London.

Drummond, D. C. (1966). Rats resistant to warfarin. *New Scientist*, **30**, 771–2.

Epling, C., Lewis, H. & Ball, F. M. (1960). The breeding group and seed storage: a study in population dynamics. *Evolution*, **14**, 238–55.

Falconer, D. S. (1961). *Introduction to Quantitative Genetics*. Oliver and Boyd, Edinburgh.

Griffiths, D. J. (1950). The liability of seed crops of perennial ryegrass (*Lolium perenne*) to contamination by wind borne pollen. *J. agric. Sci.*, **40**, 277–84.

Hamrick, J. L. & Allard, R. W. (1972). Microgeographical variation in allozyme frequencies in *Avena barbata*. *Proc. natn. Acad. Sci. USA*, **69**, 2100–4.

Hutchinson, J. B. (1951). Intra specific differentiation in *Gossypium hirsutum*. *Heredity, London*, **5**, 161–93.

Hutchinson, J. B., Silow, R. A. & Stephens, S. G. (1947). *The evolution of Gossypium*. Oxford Univ. Press, London.

Jain, S. K. & Bradshaw, A. D. (1966). Evolutionary divergence among adjacent plant populations. I. The evidence and its theoretical analysis. *Heredity, London*, **21**, 407–41.

Khan, M. S. I. (1969). The process of evolution of heavy metal tolerance in *Agrostis tenuis* and other grasses. M.Sc. thesis, University of Wales.

Knight, R. L. & Hutchinson, J. B. (1952). The evolution of blackarm resistance in cotton. *Genetics*, **50**, 36–58.

Kruckeberg, A. R. (1957). Variation in fertility of hybrids between populations of the serpentine species, *Streptanthus glandulosus*. *Evolution*, **11**, 185–211.

Lammerink, J. (1968). Genetic variability in commencement of flowering in *Medicago lupulina* in the South Island of New Zealand. *New Zealand J. Bot.*, **6**, 33–42.

Leng, E. R. (1962). Results of long-term selection for chemical composition in maize and their significance in evaluating breeding systems. *Z. PflZücht.*, **47**, 67–91.

McNeilly, T. (1968). Evolution in closely adjacent plant populations. III. *Agrostis tenuis* on a small copper mine. *Heredity, London,* **23**, 99–108.

Mangelsdorf, P. C. (1965). The evolution of maize. In *Essays on Crop Plant Evolution* (ed. J. B. Hutchinson), pp. 23–49. Cambridge Univ. Press, London.

Palmer, T. P. (1972). Variation in flowering time among and within populations of *Trifolium arvense* L. in New Zealand. *New Zealand J. Bot.,* **10**, 59–68.

Simmonds, N. W. (1966). Studies of the tetraploid potatoes. III. Progress in the experimental re-creation of the *Tuberosum* Group. *J. Linn. Soc. (Bot.).,* **59**, 279–88.

Snaydon, R. W. & Bradshaw, A. D. (1969). Differences between natural populations of *Trifolium repens* L. in response to mineral nutrients. II. Calcium, magnesium and potassium. *J. appl. Ecol.,* **6**, 185–202.

Sylven, N. (1937). The influence on climatic conditions on type composition. *Imp. Bureau Plant Genetics, Herbage Bull.,* **21**, 8.

Wu, Lin & Bradshaw, A. D. (1972). Aerial pollution and the rapid evolution of copper tolerance. *Nature, London,* **238**, 167–9.

4. Optimum sampling strategies in genetic conservation

D. R. MARSHALL & A. H. D. BROWN

There are definite limits to the numbers of samples which can be handled effectively in programmes for the conservation and utilisation of crop genetic resources. These limits are imposed by the financial and personnel resources available to carry out each stage in the process:

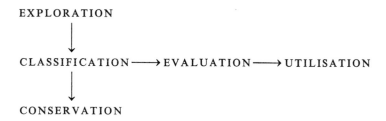

In most species there is one *major limiting factor* in the above process which determines what the upper sample limit will be. This major limiting factor varies from species to species. For example, in the case of the primitive land races of the temperate cereals in the Mediterranean basin which are rapidly approaching extinction, the major limiting factor is our capacity to collect the material before it is lost for ever. By contrast, in the case of vegetatively reproduced crops such as yams, sweet potatoes and manioc, where considerable variation still exists in the field but is difficult to conserve, the major limiting factor is conservation. Finally, in the case of the hexaploid weed relatives of cultivated oats (e.g. *Avena fatua* and *A. sterilis*) where ample material still exists in the field and there are no particular conservation problems, the major limiting factor is the breeder's capacity to evaluate and utilise the collected materials.

Obviously, by committing more resources to conservation and breeding programmes, it will be possible to raise these limits substantially. Further, by the judicious allocation of resources it will also be possible to change the major limiting factor in any species from, say, exploration to conservation. However, limits to the growth of collections will always exist, because the total resources which can be devoted to the exploration, conservation and utilisation of crop germplasm will always be

53

finite. Consequently in practice we should aim to conserve sufficient stocks of each species to saturate the plant breeder's capacity to evaluate and utilise the conserved germplasm both now and in the foreseeable future. That is, we should aim to make evaluation and utilisation the major limiting factors in all species.

The existence of crop specific limits and major limiting factors to the growth of germplasm collections highlights the need for co-ordinated and systematic planning of conservation programmes. There is no point in collecting material if it cannot be adequately conserved, nor is there any point in conserving material which cannot be evaluated and utilised. Moreover, unnecessary effort in any one species further reduces the resources which can be devoted to other species. The need for careful planning is nowhere more evident than in the formulation of sampling procedures for species where exploration is the major limiting factor to the growth of collections. These include species which are severely threatened by extinction as well as those which occur in remote areas or difficult terrain such that they are likely to be sampled only once. In these cases the plant explorer carries an immense responsibility, for it is his decisions which determine to what extent the gene resources of such species will be available for the use of future generations.

The development of efficient exploration programmes for a particular species requires decisions, and hence information, at two levels. At the first level we have decisions concerning the regions or areas of the world to be explored. Objective decisions at this level require information on (i) where significant gene pools of the species occur and where they are most threatened by extinction and (ii) the areas covered by previous exploration missions and their relative effectiveness as measured by the material held in existing collections. The need for such information has been repeatedly discussed in the literature in recent years (Bennett, 1965; Frankel, 1967; 1970*a*; Frankel & Bennett, 1970). Moreover, an extensive survey of crop genetic resources in their centres of diversity has recently been completed (Frankel, 1973, see also Chapter 37) and the Crop Ecology and Genetic Resources Unit of FAO has initiated a survey of existing collections, and this work is continuing. Consequently, these points will not be considered further here.

At the second level we have decisions concerning sampling procedures within the selected areas. These are:

(*a*) the number of plants to sample per site;

(*b*) the total number of sites to sample;

(*c*) the distribution of sampling sites within each area.

54

Objective decisions at this level, and hence the full definition of optimum sampling strategies, require a knowledge of the kinds and amounts of genetic variation in the target species and the apportionment of this variation among plants within populations, among populations within regions, and among regions within the selected areas. Unfortunately, quantitative information on the distribution of genic variation within and between populations of cultivated plants and their weed relatives is not available. Nor is it likely to become available. There is insufficient time to undertake population surveys of all the species and regions of interest prior to the preparation of exploration plans, particularly when the target populations are in jeopardy. Therefore in formulating sampling procedures for genetic conservation there is no practical alternative to using the information available from species which have been studied in some detail and extrapolating from those to the species where basic information is lacking. Much of the information on distribution of genic variation in nature in plant species has been summarised by Jain (Chapter 2) and Bradshaw (Chapter 3). Our purpose here is to formulate, as far as possible with current information, a quantitative sampling theory for genetic conservation which permits the collection of the maximum amount of genetically useful variability in the target species while keeping the number of samples within the practical limits discussed above. However, before we can consider the problem of optimal strategies in detail we need to define (i) an appropriate measure of genetic diversity – the parameter we wish to maximise, and (ii) what we regard as genetically 'useful' variability.

Measures of genetic diversity

There are a variety of measures which may be used to characterise the level of genetic diversity in a species and the apportionment of this variation within and between individuals, populations and regions. These measures fall into two classes.

(i) Measures based on genetic variance in quantitative characters. These measures are commonly used in population biology and plant breeding and have the advantage that they are familiar to all scientists in these fields.

(ii) Measures based on allelic diversity at loci governing qualitative characters. Such measures have seen increasing application in recent years due to the development of gel electrophoretic techniques to study allelic frequencies at single loci. The most commonly used measures, which are analogous to those used by ecologists to measure species diversity (Hurlbert, 1971; Lewontin, 1972) are:

(*a*) Total number of alleles in the population.

(*b*) The proportion of heterozygotes that would be produced if the population were random mated. If the frequency on the *i*th allele at a locus in p_1, the heterozygosity (*H*) is given by:

$$H = \sum_{i,j=1}^{k} p_i p_j \quad (i \neq j)$$

$$= 1 - \sum_{i=1}^{k} p_i^2$$

where k is the number of alleles.

(*c*) The Shannon–Weaver information function

$$H' = - \sum_{i=1}^{k} p_i \log_2 p_i$$

p_i and k were defined previously.

Obviously, measure (*a*) depends only on number of alleles in the population (allelic richness) while measures (*b*) and (*c*) are functions of the frequencies (allelic evenness) as well as the number of the alleles in the populations.

We reasoned that any measure of genetic diversity chosen for the present purposes should meet two criteria. First, it should be a direct measure of genic diversity. This criterion precludes the use of all measures of diversity based on variance analyses of quantitative characters (class i above). Such parameters measure only that portion of the genic variability which is expressed phenotypically. The proportion of expressed variability varies markedly with the character under consideration and the genetic background and environment in which it is expressed. Consequently, measures of genetic diversity based on variance in quantitative characters may be unreliable indicators of the diversity in a population at the level of the individual gene.

Second, it should be a function predominantly of the number of different alleles in the population. Genetic conservationists are not primarily interested in preserving a representative sample of the target species, where representative is defined in as many senses as possible. Rather, they are interested in preserving at least one copy of each of the different alleles in the target species (Bennett, 1970*b*). An accurate representation in the sample of the allelic *frequencies* in the population is of much less interest and requires sampling on a far more extensive scale. Consequently, in the context of sampling for genetic conservation, the property of allelic richness is considerably more important than the property of allelic evenness. This criterion, therefore, precludes all measures of diversity which are functions of both the number and frequency of alleles in the population. Thus, the average number of alleles per locus provides the simplest and least ambiguous measure of genetic diversity for the purposes of exploration and conservation, consistent with the above criteria.

It may be argued that the number of alleles per locus (n_a) has the disadvantage that it focuses attention on single genes, whereas the aim of conservation is to collect and conserve adaptive gene complexes. As emphasised by Brock (1971), in many circumstances induced mutations may represent a more efficient source of single gene variability than gene pools conserved from nature. Yet it is extremely unlikely that populations treated with artificial mutagens can replace natural gene pools as a source of co-adapted gene complexes which are of fundamental importance in the adaptation of populations to their environment (Dobzhansky, 1970). However, this disadvantage is more apparent than real. Firstly, co-adapted gene complexes can be regarded as 'alleles' at a 'super gene' and in the context of sampling strategies are no different in principle from alleles at a locus. Secondly, in population studies, it is impossible to study single loci independently of closely linked loci. It follows that where we use experimental data from single loci to draw inferences about collecting strategies, these inferences refer to the linked segments marked by the individual loci rather than the marker gene themselves. Thus, while we frame our discussion in terms of alleles at single loci for simplicity, the same conclusions hold for co-adapted gene complexes, which are the real target of genetic conservation.

'Useful' genetic variability

Since each gene consists of hundreds of nucleotides, each capable of base substitutions and with additional permutations possible through sequence rearrangements, additions and deletions, the potential number of allelic states at a single locus is virtually infinite. The number of different allelic combinations which may exist at several loci is even greater. Obviously, only a small fraction of these potential alleles or allelic combinations exist in a species at any one time. However, while the actual numbers of variants found in populations are small in relation to the total which may exist, they are nevertheless extremely large. In fact, as emphasised by Allard (1970*a*), most species probably contain millions, if not hundreds of millions, of different genotypes. It might be argued that we should collect and preserve all existing variants. Yet, this is clearly impractical. Consequently, we need to define the class or classes of variants which we regard as potentially most useful and to which we will assign greater priority. As a first step towards this end we will consider the distribution of alleles or allelic combinations in populations in both theory and practice.

Theoretical distribution of alleles in populations

Neutral alleles. Kimura & Crow (1964; 1970) have formulated the expected number of neutral alleles (n_a), with allelic frequencies lying between p and q ($0 < p < q < 1$), in an equilibrium population of effective size, N_e, as:

$$n_a = \int_p^q \phi(x)\, dx = \theta \int_p^q (1-x)^{\theta-1} x^{-1} dx$$

where $\phi(x)\,dx$ is the expected number of alleles with frequencies lying between x and $x + dx$ ($0 < x < 1$); u is the mutation rate to a novel new allele, and $\theta = 4N_e u$. In this model, the shape of the allelic profile is determined by the value of θ. Fig. 4.1 depicts the allelic profiles for θ equal to 0.5, 1.0 and 2.0. When θ is small (i.e. in relatively small populations) most alleles are either very common or very rare. As the population size or mutation rate increases, a greater proportion of alleles occur at intermediate (0.05 to 0.30) frequencies. The curves also emphasise that in large populations a virtually infinite number of alleles occur at very low frequencies.

To find the expected number of alleles with frequencies greater than 0.50, 0.10, 0.05 and 0.01, $\phi(x)$ was integrated for each value of θ. The calculated values of n_a are given in Table 4.1. The important point of note is that populations usually contain only two to four alleles per locus in intermediate to high frequency (> 0.05); the other alleles are relatively rare.

It may be objected that the assumption of neutrality is unrealistic. Consequently, it is also desirable to consider deleterious variants on the one hand and alleles favoured by balancing selection on the other. The distributions of deleterious and neutral alleles will differ in that selection will reduce the number of common and intermediate variants and increase the number of relatively rare alleles. The distribution of

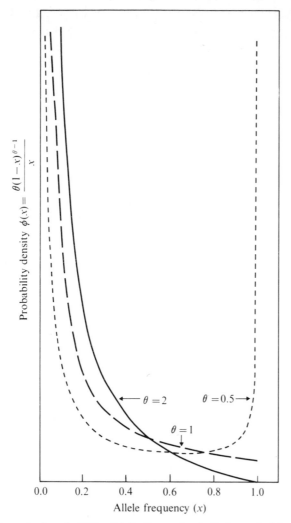

Fig. 4.1. Theoretical probability distributions of selectively neutral isoallele frequencies for various values of $\theta = 4 N_e u$. See text for further details.

neutral and selectively maintained alleles will also differ. As a case in point we will consider alleles maintained by over-dominance.

Over-dominant alleles. To study the effect of over-dominance on the allelic profile of a population we will use Kimura & Crow's (1970) model in which all heterozygotes have unit fitness and all homozygotes have relative fitness of $(1-s)$. Under this model

$$\phi(x) = Cx^{-1} \exp\left[-\sigma(x-f)^2 - \theta x\right]$$

59

where $\sigma = 2N_e s$, $1 - f$ is the probability that an individual chosen at random is heterozygous at the locus in question and C is such that

$$\int_0^1 x\phi(x)\, dx = 1.$$

Table 4.1. *Expected number of neutral alleles (n_a) in various frequency classes*

Model	$\theta = 0.5$	$\theta = 1.0$	$\theta = 2.0$
$\phi(x)$	$(2x\sqrt{(1-x)})^{-1}$	x^{-1}	$2(1-x)x^{-1}$
$\int\phi(x)$	$\log_e[(1-\sqrt{(1-x)})/\sqrt{x}]$	$\log_e x$	$2[\log_e x - x]$
Frequency class	Expected number of alleles (n_a)		
> 0.50	0.85	0.69	0.39
> 0.10	1.83	2.30	2.81
> 0.05	2.19	3.00	4.09
> 0.01	3.00	4.61	7.23

Table 4.2. *Effective number of over-dominant alleles (n_e) and expected number of alleles (n_a) with frequency > 0.05, assuming constant selection differential s against homozygotes*

Selection differential	$\theta = 0.5$	$\theta = 1.0$	$\theta = 2.0$
	Effective number of alleles (n_e)		
$s = 0$	1.5	2.0	3.0
$s = 0.01$	9.0	14.0	21.0
$s = 0.10$	25.0	36.0	54.0
	Actual number of alleles (n_a)		
$s = 0$	2.2	3.0	4.1
$s = 0.01$	9.0	8.0	7.0
$s = 0.10$	3.0	2.0	0

Kimura & Crow (1970) have tabulated values of f for various values of s, N_e and u. Let us assume $u = 10^{-5}$, $s = 0.01$ or 0.1 and that θ takes values 0.5, 1.0 and 2.0 as before. The values of f^{-1}, the 'effective' number of alleles (n_e) or the number of equally frequent alleles required to give the observed level of heterozygosity, are shown in Table 4.2. It will be noted that over-dominance ($s > 0$) substantially increases the effective number of alleles over neutrality ($s = 0$).

60

Table 4.3. *Number of alleles per variable locus in populations of* Avena barbata, Drosophila willistoni, *and* Homo sapiens

| Allelic frequencies | *Avena barbata*[1] | *Drosophila willistoni*[2] | | *Homo sapiens*[3] |
		Island	Continental	
> 0.50	1.00	0.97	1.00	0.93
> 0.10	1.84	1.39	1.43	1.91
> 0.05	1.71	1.71	1.98	2.21
> 0.01	2.00	2.80	2.92	2.61
Number of loci	5	16	16	14

[1] Clegg, M. T. & Allard, R. W. (1972). *Proc. natn. Acad. Sci. USA*, **69**, 1820–4.
[2] Ayala, F. J., Powell, J. R. & Dobzhansky, T. (1971). *Proc. natn. Acad. Sci. USA*, **68**, 2480–3.
[3] Lewontin, R. C. (1972). *Evol. Biol.*, **7**, 381–98.

Approximate values of C can be obtained as $C = \theta \exp[\sigma f^2]$, $\phi(x)$ plotted and the actual number (n_a) of common and intermediate alleles ($0.05 < P < 1$) estimated graphically as the area under the curve. The estimates obtained in this way are given in Table 4.2. It is clear that weak over-dominance leads to a marked increase in the number of intermediate and common alleles in the population. However, under this model strong over-dominance leads to maintenance of large numbers of alleles all at relatively low frequencies.

Distribution of alleles in natural populations

In Table 4.3, the above theoretical distributions can be compared with some published experimental data. It will be noted that on the average there are approximately two detectable alleles at each variable locus with frequency greater than 0.05.

The impact of a breeding system of predominant self-pollination is displayed by the *A. barbata* results. Under close inbreeding, it is more difficult to maintain large numbers of alleles in populations at low frequencies. Consequently, in *A. barbata* there were usually only two alleles per polymorphic locus and those occurred at intermediate frequencies. By contrast, populations of *D. willistoni* have a greater number of alleles in the class $0.01 < P < 0.05$. These presumably represent deleterious recessives and constitute the mutational load. This load seems to be less in human populations. However, many very rare variants have been found in human populations when the sample size has been sufficiently large. For example, Hopkinson & Harris (1971) have detected 2.6 rare variants per locus for 12 genes specifying soluble enzymes in samples of 3000–13000 gametes. This represents a total average frequency of rare heterozygotes of 0.0017 or average frequency for rare alleles of 0.0008.

Assignment of priorities

From the above considerations it is obvious that the alleles in a population can be *arbitrarily* divided into those which are common ($P > 0.05$; usually less than four alleles) and those which are rare ($p < 0.05$; many alleles). In addition, any particular allele can be categorised as to whether it is *widespread* and occurs in many of the populations in the target area or *local* and restricted to one, or a few adjacent populations. These subdivisions yield four possible classes of alleles based on frequency and distribution. As it is possible to collect only a fraction of the many variants in a species, the critical question is: should we give priority to any particular class of alleles?

Considering first the common alleles or variants of widest occurrence; this class is the easiest to sample (indeed they are probably often present in modern agricultural varieties) and will be inevitably included in the sample regardless of strategy. Second, there is the class of widely occurring but locally rare alleles. From probability considerations, it follows that the sampling of this class depends only on the *total* number of plants taken from the target area, rather than how these are distributed between and within sampling sites. Consequently, the number of such alleles recovered is also largely independent of sampling strategy.

The third class of alleles, those which are *locally common*, are of greatest importance as regards sampling strategy. By contrast, the final class of alleles, which are both rare and restricted to a few populations, do not merit the same attention. We reach this conclusion from two arguments. First, the great majority of common alleles or allelic combinations (whether widespread or local) presumably represent adaptive variants maintained in populations by some form of balancing selection (Dobzhansky, 1970). Consequently, common variants are likely to be of far greater interest to plant breeders than rare variants which presumably represent either newly arising mutants or recombinations, or deleterious genes or gene combinations maintained in the population by a balance between mutation, migration or recombination and selection. Second, rare variants are obviously much more difficult to collect than their common counterparts. Indeed, artificially induced mutation or in the case of gene blocks, artificially enhanced recombination, coupled with an effective selection screen, could be a far more efficient means of obtaining rare variants than randomly searching for them in natural populations. Consequently, the aim of plant exploration can be defined as *the collection of at least one copy of each variant occurring in the target populations with frequency greater than 0.05*. It follows that the most appropriate measure of genetic conservation is not simply the average number of alleles per locus, but the average number with population frequency greater than 0.05.

Definition of optimum sampling strategies

In discussing sampling strategies we will first consider a basic procedure for use in crops where we have no specific information on the distribution of variation in nature. We will then consider more sophisticated procedures for use where such information is available.

The basic strategy

Number of plants per site

By the term 'site' we wish to designate that area from which one bulked sample will be drawn and one set of ecological recordings kept in a single collecting instance.

The delineation of the most appropriate sampling area is usually relatively simple in annual crop plants. Populations of these species are harvested in bulk and a portion is used to sow the following crop. The high degree of mixing which occurs during harvesting and sowing each year ensures that all fields planted from a single seed source will not vary in genetic structure. Obviously, some genetic differentiation may take place by selective plant losses during the growth of the crop. However, in most circumstances, this will be minimal. Consequently, in annual crops the sampling target will be the individual field or farm if each farmer uses different seed stocks or a group of fields or farms if farmers use a common seed stock.

In wild or weed species the definition of the appropriate sampling unit is much more difficult. First, such species are often continuously distributed in nature and there are no artificial boundaries defining discrete, relatively homogeneous populations as there are in crop plants. Thus, the plant explorer must decide where in the continuum he will

63

collect and the total size of the area to be sampled (e.g. 1000, 10000 or 100000 m²). Secondly, as emphasised by Bradshaw (Chapter 3), gene flow through pollen and seed dispersal is extremely limited in wild plants, particularly in comparison with cultivated annuals, and as a result, natural populations of wild species often show marked genetic differentiation over distances as small as a few metres. In these circumstances, the explorer must also decide whether to collect a single random sample from the chosen area or to sample the differentiated subpopulations separately. A number of populational and ecological factors will influence the explorer's decisions on these matters. The most important populational factor is the size of the interbreeding unit which is a function of the number and density of plants, the mating system and the level of pollen and seed dispersal. The most important ecological factor is the degree of environmental heterogeneity for such variables as soil type, aspect, slope, moisture regime, and associated flora. Consequently, the delineation of the appropriate sampling area is best left to the explorer at the time of collection since he is in a position to evaluate the factors involved.

Once an appropriate target population has been defined we are interested in obtaining as representative a sample of the common alleles in the population as possible. However, since excessive sampling of any one site limits the explorer's opportunities to discover and sample other, perhaps more interesting, sites the samples should be kept as small as possible. Therefore, it seems logical to define the optimum sample size per site to be the number of plants required to obtain, with 95 per cent certainty, all the alleles at a random locus occurring in the target population with frequency greater than 0.05. This definition is very similar to that of Oka (1969) based on '95 per cent of the genes distributed in the population with a frequency of 5 per cent'.

If complete information were available on the distribution of alleles in the target species, it would be possible to define an optimum sample size for each population. However, in species where such information is lacking and which are of greatest interest here, this is obviously impossible. The simplest and safest practical alternative for such species is to define, using the theoretical and experimental data on the distribution of alleles in populations discussed earlier (Fig. 4.1, Tables 4.1–4.3), a sample size which ensures that we meet our objective in the most common circumstances, and to apply this sample size to all populations sampled. This we will now proceed to do.

Sampling theory of allelic profiles. Consider a population in which two alleles (A_1, A_2) occur with frequencies p_1 and p_2, respectively. The probability that a random sample of n gametes contains at least one copy of each allele $(P[A_1^+, A_2^+])$ is given by:

$$P[A_1^+, A_2^+] = 1 - (1-p_1)^n - (1-p_2)^n + (1-p_1-p_2)^n.$$

If $p_1 = 0.95$ and $p_2 = 0.05$ then a sample of 59 gametes is required to obtain at least one copy of each allele with 95 per cent certainty.

For a larger number of alleles, the exact probability expression becomes more cumbersome (Moran, 1968). For example, for four alleles:

$$P[A_1^+, A_2^+, A_3^+, A_4^+] = 1 - \sum_{i=1}^{4} (1-p_i)^n + \sum_{i,j=1}^{4} (1-p_i-p_j)^n$$

$$- \sum_{i,j,k=1}^{4} (1-p_i-p_j-p_k)^n + (1 - \sum_{i=1}^{4} p_i)^n; \quad (k > j > i).$$

The behaviour of this and analogous expressions for increasing values of n was studied for five types of allelic profiles (Table 4.4). At one extreme, the first profile simulates the case of four alleles maintained by strong over-dominance. The other extreme is the fifth profile, obtained when three alleles are held in low but significant frequency under a mutation-selection balance. The intermediate cases are derived from the three profiles drawn in Fig. 4.1.

Table 4.4 shows the number of gametes (n) which must be sampled to be 95 per cent certain of obtaining at least one copy of each allele, and the probability, P, of achieving this objective given that $n = 100$. The results show that a surprisingly small random sample could be claimed as the 'minimum representative sample' under our criterion. It is clear from Table 4.4 that the value of n for which P achieves 0.95 is heavily dependent on the actual frequency of the rarest allele of interest to the collector. In general, alleles which are actively maintained in the population by balancing selection at intermediate frequencies require smaller samples than rarer deleterious recessives. Nevertheless, even in the limiting case of twenty alleles with frequency 0.05, a random sample of about 120 gametes will include with 95 per cent certainty one copy of each allele. Consequently, it seems safe to conclude that a random sample of 50–100 would be more than adequate under most circumstances.

This conclusion is further strengthened by the fact that in practice we do not sample individual gametes or even seeds, rather we sample single heads, panicles, other fruiting bodies or vegetative propagules. Individual heads or panicles will represent a number of different

Table 4.4. *Sample sizes (n) required to be 95 per cent certain of obtaining at least one copy of each common allele (frequency > 0.05) and probabilities (P) of achieving this objective if n = 100 for five contrasting models of allelic profiles*

| | Balanced polymorphism 1 | Neutral | | | Mutation/ selection 5 |
Model		$\theta = 0.5$ 2	$\theta = 1.0$ 3	$\theta = 2.0$ 4	
Allele					
A_1	0.25	0.76	0.63	0.49	0.80
A_2	0.25	0.20	0.23	0.22	0.05
A_3	0.25	—	0.09	0.12	0.05
A_4	0.25	—	—	0.07	0.05
Remainder	—	0.04	0.05	0.10	0.05
Sample size (n)	19	15	37	43	80
Probability (P)	1.00	1.00	1.00	0.99	0.98

gametes in most crop species, particularly if they are outbreeding. Vegetative propagules carry at least the diploid complement.

It would also be instructive to know how many different alleles per locus we might expect to include on average in a random sample of say 50 or 100 gametes. Unfortunately, a general answer to this question is only available for the case of neutral loci. Ewens (1972) has shown that in a random sample of n gametes one would expect to capture

$$E(k) \simeq \frac{\theta}{\theta} + \frac{\theta}{\theta+1} + \dots + \frac{\theta}{\theta+n-1}$$

neutral alleles where $\theta = 4N_e u$. The sampling variance of k is

$$\text{var}(k) = E(k) - \left[\frac{\theta^2}{\theta^2} + \frac{\theta^2}{(\theta+1)^2} + \dots + \frac{\theta^2}{(\theta+n-1)^2} \right].$$

He has tabulated these moments for many values of θ and n. For illustrative purposes we will consider only two cases. When $\theta = 1$, the expected number of alleles in a random sample of 50 gametes is 4.5, with variance 2.9; whereas in a sample of 100 gametes, the expected number is 5.2 and the variance is 3.6. Thus on average one expects to collect some rare alleles additional to the 3.0 common alleles (Table 4.1) we define as the conservation target.

Biased sampling procedures. Bennett (1970a) has argued in favour of enriching the random sample with a biased sample of rare phenotypic variants if these occur in the target population. Such a procedure has the obvious advantage that it allows the explorer to collect more of the observable variants in the population. Nevertheless, it suffers from a number of serious disadvantages. First, biased sampling will be far more time consuming than simple random sampling because of the need to search through large numbers of plants for the rare off-types. Consequently, by devoting time to the collection of rare types at one site, the explorer reduces his chances of sampling the common variants at other sites. Further, as we have argued before, despite the few spectacular cases of rare phenotypically distinguishable variants which later proved to be of great value (e.g. the opaque mutant in maize), common, currently adaptive alleles are more likely to be valuable to the plant breeder than rare variants. Second, there is the danger of collecting unrecognised disease specimens which could cause the whole collection to be destroyed by quarantine authorities. Third, if the random and biased samples are bulked, this will upset the correlation between that sample and its location and thus severely reduce any value the sample possesses for population research. On the other hand, if the samples are kept separate, this will greatly increase the effort which must be devoted to recording, multiplication and screening of the material. We conclude therefore that sampling of deviates should play only a minor role, if any, in genetic conservation.

In some instances an explorer may wish to sample only a small number of populations and can afford to spend more than the minimum amount of time at each site. In such situations, we would suggest that he should increase the size of the random sample at those sites which are highly polymorphic for morphological marker loci or highly variable for quantitative characters. Recent studies in *Avena fatua* and *A. barbata* indicate that populations which are highly polymorphic at morphological marker loci also tend to be highly polymorphic for those loci without observable phenotypic effects (Marshall & Allard, 1970). If this finding holds in general, it means that taking a larger random sample at sites highly polymorphic for observable variants offers a more efficient means of increasing the total variation in a collection than just collecting rare morphological variants at each site.

Number of sites

Since each site offers the prospect of sampling a different set of alleles, where we have no information on the distribution of variation in nature, the optimum number of sites to sample is the maximum possible. The number of populations which can be sampled will be determined by such factors as the length of the collecting season, relative abundance of the target species, roughness of the terrain, etc. These factors place a strict upper limit on the number of samples which can be collected in a single season. Under ideal conditions this limit may be as high as 1000 samples, although in most circumstances it will be considerably less (Bennett, 1970*a*; Allard, 1970*a*).

Distribution of sites within the target area

As emphasised by Bennett (1965; 1970*b*) and Bradshaw (Chapter 3) it has been found repeatedly that the pattern of genetic differentiation within species is strongly correlated with environmental heterogeneity. Consequently, in species where there is no reliable information on the distribution of variation in nature, the explorer should aim to sample as broad a range of environments as possible.

There are many ways of achieving this objective. We will discuss two contrasting sampling strategies to illustrate the issues involved. In the first case the sampling sites are over-dispersed, that is, distributed more or less evenly over the target area (Fig. 4.2*a*). This procedure will capture the maximum amount of genetic variation associated with broad geographic differences in edaphic and climatic factors. In the second case the sampling sites are clustered in, say, groups of five, and the groups are more or less evenly distributed over the target area (Fig. 4.2*b*). This procedure will capture a significant amount of the variability associated with both geographic and *microgeographic* differences in environmental factors.

Each of these strategies has its advantages and disadvantages when considered in relation to a particular species or group of species. The first is the preferred procedure when sampling annual crop species. As noted earlier, such species are harvested in bulk and a portion is used to sow the following crop, often in a different field. The high degree of mixing associated with the harvesting and sowing procedures and the constant migration from field to field, virtually precludes the development of differentiated populations on a microgeographic scale. Thus, there is little point in collecting a number of populations within a region in such species.

The second strategy is likely to be the preferred procedure in wild or weed species. These species often grow on non-arable as well as arable land and as a result, occupy a greater range of habitats than

(a)

(b)

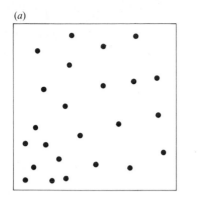
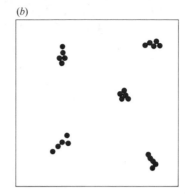

Fig. 4.2. Diagramatic representations of (a) evenly dispersed and (b) clustered sampling patterns.

crop plants. On the other hand, gene flow through pollen and seed dispersal is usually relatively small in wild species. As a result, a considerable portion of the total variability in such species has developed in response to, and is maintained by, heterogeneity in local environments. The clustering of samples has a number of other benefits in wild species. First, it reduces time spent in travelling between sites and permits the collection of a greater total number of samples in a given time. Second, it forces the explorer to search consciously for markedly different habitats within a region and avoids the possibility, which exists when only one sample is taken per region, that the explorer will unconsciously select, say, all lush or all arid sites. Third, it increases the value of the collections for studies of population structures and the adaptive significance of observable polymorphisms.

The procedures illustrated in Fig. 4.2 represent extremes and many intermediate strategies are possible. In many species it is conceivable that one strategy may be preferred in one area and an alternative strategy preferred in other areas. However, whatever strategy is used it is obviously important that the plant explorer assemble the fullest information possible, on the climate, soil and vegetation and their local variations, in the area to be explored before deciding on the distribution of sampling sites.

Modifications to the basic procedure

In a number of species which are targets for conservation we have at least some information on the distribution of variation in nature (Frankel & Bennett, 1970). This information can allow the explorer to develop sampling procedures which are more efficient than the basic exploration strategy discussed above. For example, if it is known that the target species shows little, or no, inter-populational differentiation (e.g. *Phalaris tuberosa* in south-eastern Australia; McWilliam, Schroeder, Marshall & Oram, 1971) then the optimum strategy would be to collect a large number of individuals from a few populations. Alternatively, if it is known that the target species is highly differentiated and the majority of populations contain one, or a few, homozygous genotypes (e.g. *Trifolium subterraneum* in Western Australia; Gladstones, 1966), then the optimum strategy would be to collect a few individuals from a large number of populations. Consequently, we will now consider what modifications can be made to the basic strategy when something is known of the distribution of genetic variation in the target species.

Number of plants per site and number of sites

The critical problem here is to determine, given information on levels of variation within and between populations of the target species, the optimal allocation of sampling resources within and between sites so that the explorer recovers the maximum amount of variability. Oka (1969) has presented a useful formulation of this problem. He estimated the fraction of the total genetic variation in the target area captured by a particular sampling procedure, G, as

$$G = 1 - \{(1-P) + P(1-p)^n\}^N,$$

where $1 =$ total genetic variation in the target area,

$\quad\quad P =$ proportion of total variation represented by population or sampling site,

$\quad\quad p =$ proportion of genetic variation per population or site represented by an individual plant (or seed),

$\quad\quad N =$ number of populations or sites sampled,

$\quad\quad n =$ number of plants (or seeds) sampled per site.

Oka described the effect of increasing n and N on G, assuming values of P and p suggested by experimental data on various kinds of target species (modern and primitive cultivars of rice and its wild relatives), in order to determine the values required for G to exceed 0.95. That is, he calculated the values of N and n required to ensure that the explorer captured more than 95 per cent of the total genetic variability in the target area.

However, this treatment does not immediately answer the problem of

70

Table 4.5. *Theoretical values of (i) optimal number of plants to sample per site and (ii) optimal number of sites to sample per day for a range of genetic models*

				Outbreeding species	
Population	Modern cultivars (1)	Primitive cultivars (2)	Wild relatives (3)	(4)	(5)
P	0.01	0.05	0.10	0.50	0.75
p	0.95	0.20	0.05	0.05	0.05
a/b ratio	(i) Number of plants per site (n)				
25	1	10	15	30	36
100	2	15	39	50	55
a b	(ii) Number of sites per day (N)				
25 1	18	14	12	9	8
50 0.5	10	8	7	6	6

the *optimal* allocation of sampling resources. There are in fact many combinations of N and n which give $G > 0.95$. The critical point is which of these combinations is the most efficient. To answer this question we must assume a relationship between N and n consequent upon the limited availability of sampling resources. The simplest kind of relationship is:

$$E = N(a+bn),$$

where E represents the total effort available for sampling,

   a represents that amount required for each site visited
      (i.e. transport, ecological records, site inspection, etc.), and
   b represents that amount expended to sample one plant.

This relationship assumes that as bn is reduced, there is more time to travel to and sample other sites, allowing a greater density of sampling within the target area.

   The problem now becomes to choose from values of N and n subject to the constraint E, those which maximise G, for particular values of p and P. The solutions to five cases are presented in Table 4.5. Three of these cases (examples 1–3) employ values of p and P given by Oka (1969) for modern and primitive cultivars of rice and their wild relatives. The remaining two use a much higher value of P as implied by recent studies of genic variability in cross-fertilising organisms (e.g. Lewontin, 1972; Selander & Johnson, 1972). Table 4.5(i) gives the values of n, the number of plants per site, for maximum efficiency. These estimates depend on the ratio of a/b and not on their actual values. However, the estimates of N, the number of sites sampled, does depend on ascribing

71

Table 4.6. *Number of days required to collect 95 per cent of variation* $(G > 0.95)$ *using maximally efficient sampling procedure under a range of genetic models*

				Outbreeding species	
Population	Modern cultivars (1)	Primitive cultivars (2)	Wild relatives (3)	(4)	(5)
P	0.01	0.05	0.10	0.50	0.75
p	0.95	0.20	0.05	0.05	0.05
a b		Days to collect 95% of variation			
25 1	16	5	5	1	1
50 0.5	30	8	5	1	1

actual values to a and b. For simplicity, we have used units of time, since this is the major factor limiting the explorer's activities once he is in the field. The values used are $E = 500$ minutes (i.e. it is assumed that the explorer spends 8.33 hours/day in sampling), $a = 25$ or 50 minutes (average time travelling between sites etc.) and $b = 1$ or 0.5 minutes (average time spent in collecting a single plant). Table 4.5(ii) gives the number of populations optimally sampled in a day.

It is clear from these results that partial genetic information may cause the explorer to alter markedly his sampling pattern from that previously discussed. In particular, these calculations reinforce the point that too much attention to one site sacrifices considerably the efficiency of sampling. Even when the amount of effort required per site compared with the amount to sample one plant is very high, we still find it most efficient to sample more sites per day and fewer plants per site.

Table 4.6 gives the number of days required to collect 95 per cent of the variation in a particular region, given that the population structure of the species is uniform in that area. It will be noted that in outbreeders where each population may contain 50–75 per cent of the total variation in the region as a whole, one day in the field will yield 95 per cent of the common alleles in the area. On the other hand, in plants where each population may contain as low as one per cent of the total variation in a region, the explorer must spend 30 days in the field to achieve the same objective. These calculations serve to emphasise that the greater the genetic differentiation in a species, the harder it is to sample effectively.

The above model suffers from a number of deficiencies. First, it assumes that each population contains a constant proportion of the total variation in the target species. This assumption is, of course, highly unrealistic. The level of variation generally varies quite markedly from population to population. However, deviations from this assumption do not detract from the conclusions that the optimal strategy is to take as few plants as possible per site and sample a maximum number of sites. Obviously, if the level of within-population variation varies markedly from population to population the more populations sampled the greater is the probability the explorer will sample the more variable populations. Second, it is implicit in this model that equal numbers of plants will be collected at each site. In many instances it is neither practical nor wise to do this. For example, if the genetic structure of populations varied markedly between sites, a more efficient procedure would be to collect the minimum number of plants at uniform sites and increase sample size at highly variable sites. Third, the model ignores important practical realities. In particular, it takes no account of the fact that samples need to be distributed as well as conserved. In most species the practical minimum sample size is at least 25 plants per population unless the explorer is prepared to spend a great deal of time multiplying samples prior to distribution.

Despite these deficiencies, the above analysis does provide an insight into the relative efficiency of different sampling strategies. In particular it emphasises that collecting 200–300 plants per population as recommended by Bennett (1970*a*) and Allard (1970*a*) is an extremely wasteful procedure. In most circumstances, sample size should not exceed 50 plants per population and in no circumstances is it desirable to collect more than 100 plants per population.

Distribution of sites within the target area

Information on the population structure of the target species can also be important in determining the most effective distribution of sampling sites within the area to be explored. Recent studies of the patterns of population structure and differentiation for a variety of morphological and enzymatic marker loci in the slender wild oat (*Avena barbata*) in central California (Jain, 1969; Marshall & Allard, 1970; Clegg & Allard, 1972; Hamrick & Allard, 1972) illustrates this point well. Over large areas of California, populations of this species are of two types for the loci surveyed. Either they are predominantly homozygous and monotypic for one array of alleles, or they are homozygous and monotypic for an alternate array. Further, there is a strong correlation between the occurrence of the two genotypes and the major climatic regions of California. One type occurs primarily in the arid interior of the state (Mediterranean warm summer region), while the other is limited to the more mesic coastal areas (Mediterranean cool summer region). However, highly polymorphic populations of *A. barbata*, displaying most of the allelic combinations of the loci surveyed, occur in the area of overlap between the major climatic zones. Obviously, if the aim of an exploration mission was to obtain as representative a sample of allelic combinations of these genes as possible in central Californian populations of *A. barbata*, the most efficient strategy would be to take all samples from the zone of overlap between the two major climatic regions and ignore the rest of the state.

Conclusions

It is obvious that information on the population structure of the target species can greatly improve the efficiency of sampling procedures by indicating the optimal numbers of plants per site, numbers of sites per region and the distribution of sampling sites within the target area. For estimating optimum sample sizes, only a minimum amount of information is required (e.g. estimates of the average number of alleles per locus or within and between family variances for five to ten populations). However, for determining the optimal distribution of sites within a region, substantial information is required before the optimum strategy can be defined with any degree of confidence. Nevertheless, some information is better than none, no matter how meagre. Consequently, the explorer should endeavour to collect all available information on the population structure of the target species prior to the planning of the collecting mission.

Maintenance of collected samples

There is little point in striving to collect the maximum amount of genetic variability in a species unless the material is adequately conserved to prevent the subsequent loss of the variation during the maintenance of the collection. We do not intend to consider in detail all aspects of the maintenance of germplasm collections. Rather, our purpose is to focus on one central issue which has been the subject of considerable debate in recent years – the relative merits of the maintenance of sexually propagated crops as stored seed compared with that by propagation in bulk populations or mass reservoirs (Simmonds, 1962; Frankel, 1970*b*; Allard, 1970*b*).

There are two main issues involved here:

(i) how best to conserve collections so that the maximum amount of variability is maintained for future generations?

(ii) how best to maintain collected variability for use by current plant breeders?

These represent distinct but interrelated issues. Obviously, if a particular collection is maintained in the long term as a bulk population it must be used in the short term in that form. However, if a collection is maintained in long-term storage as separate items, then these can be compounded into a bulk population or used individually as the situation demands.

Base collections *

The question we must answer here is: will maintenance as seed or living collections preserve the most variation for future generations?

Simmonds (1962) originally proposed the use of mass reservoirs for long-term conservation because he regarded 'museum collections', as he termed them, as a wasting resource, often with very high rates of attrition (see also Bennett, 1965; 1968). At the time Simmonds wrote his article this was true in many instances. However, over the last decade there have been considerable advances in the technology of seed storage and it is now possible to maintain seed of many species for periods of 20–25 years or more before regeneration is necessary, as compared to three to five years in the past. At the same time, there has also been a greater appreciation of the need for genetic conservation, and first-class seed storage facilities have been established in several countries including the USA, USSR, Japan, Turkey, Philippine Islands and Mexico, and more are planned or being developed in Europe, Asia and Latin America. The increased availability of long-term storage facilities and the greater appreciation of their function means that attrition of samples in storage is no longer the problem it was in the past. We do not wish to imply that there will be no losses during storage. As noted by Allard (1970b) completely 'static' preservation is impossible and loss of genetic variation can occur through differential survival of genotypes in storage and selection, hybridisation and genetic drift during the rejuvenation process. However, we emphasise that these losses can be controlled by the judicial management of the collection and reduced to acceptable levels such that the stored seed even after 50 years and two

* For definition of the different types of collections see Chapter 37.

cycles of regeneration would contain a significant fraction of the original variation. The same cannot be said for 'mass reservoirs'. One of the most notable features of the composite cross populations was the rapid reduction in genetic variation in the early generations (Jain, 1961; Allard & Jain, 1962). For example, in Composite Cross v (CCv), the total genetic variation for height and heading date was reduced by 30–40 per cent between generations 4 and 14. By generation 19, the depletion in genetic variation for these characters was from 50 to 70 per cent. The conclusion seems inescapable that the bulk populations, or 'mass reservoirs' as they have come to be called, are of little value in *preserving* variation, potential or expressed, for future use. It is often argued that this tremendous loss of variability inherent in the use of bulk populations can be avoided by growing the bulk populations at a number of sites. It is envisaged that a different fraction of the variation would be preserved at each of the sites. However, the need to grow the populations at different sites negates to a large degree one of their main potential advantages, that is, there are few entries to grow and maintain. To take this argument to its ultimate conclusion, we might envisage growing CCv at 20 sites to ensure maintenance of as much variability as possible. Yet, it was originally derived by intercrossing 30, apparently pure-line, parental varieties. We would expect that the original 30 lines would be far easier to maintain than 20 bulk hybrid populations.

A second line of evidence which argues against the use of bulk populations for the long-term maintenance of variability comes from a study by Clegg, Allard & Kahler (1972) of allelic frequencies at four esterase loci in CCii and CCv. It might be assumed that complex inter-allelic and intergenic interactions would maintain different alleles in different CC populations, as these populations do not have all parents in common. However, this study indicated that the same pair of four locus-complementary, gametic types became predominant in both populations at Davis, California. In other words, both populations were approaching the same end point despite their different initial compositions. If it should prove to be at all general, this finding has important implications in the present context. In particular it means that bulk populations, although they may be developed from lines from different regions of the world, could after 20–40 generations become very similar. In other words, bulk populations when grown in a common environment not only retain a small portion of the potential variability expressed on crossing a set of parents, but they also tend to retain the same spectrum of variation.

76

Consequently, we would conclude that seed stored as separate items offers a far more effective means of preserving variation collected from nature than bulk populations. Further, as emphasised by Frankel (1970*b*), collections make it possible to study individual gene pools and to relate their characteristics to the environment from which they originated and this potential is lost if most or all the populations are bulked into mass reservoirs for maintenance.

Active or working collections

The question we must answer here is: should working collections, which represent the seed of lines held in storage for immediate distribution, be in the form of individual entries or bulk populations? Again we would argue for separate entries. However, in this case our arguments are based largely on the question of flexibility of use. If seed is supplied as individual entries, the breeder can decide how many or how few he requires and he uses them as he wishes.

It should not be assumed that bulk populations have no place in plant breeding. They are obviously a valuable adjunct to breeding for the same reason as they are unsuitable for use in the long-term maintenance of germplasm. They offer an effective means of selecting a range of locally adapted genotypes from the vast numbers generated by intercrossing parental lines or populations. However, it is important to realise that the requirements of each breeder will differ along with his local environment. Therefore he should be able to determine the parental composition of a bulk population, rather than receive a preformed pool from a conservation centre, since such a pool may have lost much variation potentially useful to him.

Summary and overall recommendations

1. There are definite limits to the numbers of samples which can be handled effectively in programmes for the conservation and utilisation of crop genetic resources. Consequently, there is a need for the co-ordinated and systematic planning of conservation programmes to ensure the preservation of the maximum amount of useful genetic variability while keeping the total number of samples within these limits.

2. The number of alleles per locus is the simplest measure of genetic diversity for the purposes of exploration and conservation. Linked,

co-adaptive gene complexes can be regarded as 'alleles' of a 'super-gene' and in the present context are no different in principle from alleles at a single locus. Consideration of allelic profiles encountered in populations in theory and practice on the one hand, and the virtually infinite potential for genes to vary on the other, led us to focus on the class of alleles which are *locally common*, as critical in determining optimal sampling strategies. With this in mind, the aim of plant exploration was defined as the collection of at least one copy of each variant occurring in the target populations with a frequency greater than 0.05.

3. Where little, or no, information is available on the distribution of variation in nature, the optimal strategy is (i) to collect 50–100 individuals per site, (ii) to sample as many sites as possible within the time available, and (iii) to ensure that sampling sites represent as broad a range of environments as possible.

4. When partial information on the population structure of the target species is available it is possible to modify this basic procedure. A detailed analysis of alternative strategies indicates that, in many circumstances, even fewer than 50 plants per site, but more sites, should be sampled in order to maximise the genetic diversity in the collected sample.

5. There is no point in striving to collect the maximum amount of genetic variability in a species unless the material is adequately conserved. An important issue here is the relative merits of maintenance as stored seed compared with that by propagation in bulk populations. This question is resolved in favour of maintenance as stored seed on the basis of increased availability of first-class, long-term storage facilities and the now clearer insight into the goals of gene conservation.

References

Allard, R. W. (1970*a*). Population structure and sampling methods. In *Genetic Resources in Plants – their Exploration and Conservation* (eds O. H. Frankel and E. Bennett), *IBP Handbook*, no. **11**, pp. 97–107. Blackwell, Oxford and Edinburgh.

Allard, R. W. (1970*b*). Problems of maintenance. In *Genetic Resources in Plants – their Exploration and Conservation* (eds O. H. Frankel and E. Bennett), *IBP Handbook*, no. **11**, pp. 491–4. Blackwell, Oxford and Edinburgh.

Allard, R. W. & Jain, S. K. (1962). Population studies in predominantly self-pollinated species. II. Analysis of quantitative changes in a bulk hybrid population of barley. *Evolution*, **16**, 90–101.

Bennett, E. (1965). Plant introduction and genetic conservation: gene-cological aspects of an urgent world problem. *Scott. Pl. Breed. Stn Rec.*, pp. 27–113.

Bennett, E. (1968). *Record of the FAO/IBP Technical Conference on the Exploration, Utilization and Conservation of Plant Genetic Resources, 1967.* FAO, Rome.

Bennett, E. (1970a). Adaptation in wild and cultivated plant populations. In *Genetic Resources in Plants – their Exploration and Conservation* (eds O. H. Frankel and E. Bennett), *IBP Handbook*, no. 11, pp. 115–29. Blackwell, Oxford and Edinburgh.

Bennett, E. (1970b). Tactics in plant exploration. *Genetic Resources in Plants – their Exploration and Conservation* (eds O. H. Frankel and E. Bennett), *IBP Handbook*, no. 11, pp. 157–79. Blackwell, Oxford and Edinburgh.

Brock, R. D. (1971). The role of induced mutations in plant improvement. *Radiation Botany*, 11, 181–96.

Clegg, M. T. & Allard, R. W. (1972). Patterns of genetic differentiation in the slender wild oat species *Avena barbata. Proc. natn. Acad. Sci. USA*, 69, 1820–4.

Clegg, M. T., Allard, R. W. & Kahler, A. L. (1972). Is the gene the unit of selection? Evidence from two experimental plant populations. *Proc. natn. Acad. Sci. USA*, 69, 2474–8.

Dobzhansky, T. (1970). *Genetics of the Evolutionary Process.* Columbia Univ. Press, New York.

Ewens, W. J. (1972). The sampling theory of selectively neutral alleles. *Theor. Pop. Biol.*, 3, 87–112.

Frankel, O. H. (1967). Guarding the plant-breeder's treasury. *New Scientist*, 14, 538–40.

Frankel, O. H. (1970a). Genetic conservation of plants useful to man. *Biological Conservation*, 2, 162–9.

Frankel, O. H. (1970b). Genetic conservation in perspective. In *Genetic Resources in Plants – their Exploration and Conservation* (eds O. H. Frankel and E. Bennett), *IBP Handbook*, no. 11, pp. 469–89. Blackwell, Oxford and Edinburgh.

Frankel, O. H. (1973). *Survey of Crop Genetic Resources in their Centres of Diversity. First Report.* FAO/IBP, Rome.

Frankel, O. H. & Bennett, E. (eds) (1970). *Genetic Resources in Plants – their Exploration and Conservation, IBP Handbook*, no. 11, 554 pp. Blackwell, Oxford and Edinburgh.

Gladstones, J. S. (1966). Naturalised subterranean clover (*Trifolium subterraneum* L.) in Western Australia: the strains, their distributions, characteristics and possible origins. *Aust. J. Bot.*, 14, 329–54.

Hamrick, J. L. & Allard, R. W. (1972). Microgeographical variation in allozyme frequencies in *Avena barbata. Proc. natn. Acad. Sci. USA*, 69, 2100–4.

Hopkinson, D. A. & Harris, H. (1971). Recent work on isozymes in man. *Ann. Rev. Genet.*, 5, 5–32.

Hurlbert, S. A. (1971). The non-concept of species diversity. A critique and alternative parameters. *Ecology*, **52**, 577–86.

Jain, S. K. (1961). Studies on the breeding of self-pollinating cereals. The composite cross bulk population method. *Euphytica*, **10**, 315–24.

Jain, S. K. (1969). Comparative ecogenetics of *Avena fatua* and *A. barbata* occurring in central California. *Evol. Biol.*, **3**, 73–118.

Kimura, M. & Crow, J. F. (1964). The number of alleles that can be maintained in a finite population. *Genetics*, **49**, 725–38.

Kimura, M. & Crow, J. F. (1970). *An Introduction to Population Genetics Theory*. Harper and Row, New York, 591 pp.

Lewontin, R. C. (1972). The apportionment of human diversity. *Evol. Biol.*, **7**, 381–98.

Marshall, D. R. & Allard, R. W. (1970). Isozyme polymorphisms in natural populations of *Avena fatua* and *A. barbata*. *Heredity*, **25**, 373–82.

McWilliam, J. R., Schroeder, H. E., Marshall, D. R. & Oram, R. N. (1971). Genetic stability of Australian phalaris (*Phalaris tuberosa* L.) under domestication. *Aust. J. agric. Res.*, **22**, 895–908.

Moran, P. A. P. (1968). *An Introduction to Probability Theory*. Oxford Univ. Press, London.

Oka, Hiko-ichi (1969). A note on the design of germplasm presentation work in grain crops. *SABRAO Newsletter*, **1**, 127–34.

Selander, R. K. & Johnson, W. C. (1972). Genetic variation among vertebrate species. *Proc. XVII Int. Congr. Zool.*, pp. 1–31.

Simmonds, N. W. (1962). Variability in crop plants, its use and conservation. *Biol. Rev.*, **37**, 442–65.

5. Sampling germplasm in a center of diversity: an example of disease resistance in Ethiopian barley

C. O. QUALSET

For some cultivated species, very large world collections have been amassed. By virtue of their size alone we might think that a sufficient genetic sample is on hand. However, examination of collection records reveals four important aspects. First, large gaps exist so that only a few or no samples have been collected in some geographical regions. For the most part, these regions have been identified and efforts are being made to obtain collections. Second, usually roadside and marketplace samples are obtained, and large regions have not been sampled because of their relative inaccessibility. Roadside samples might not represent the variability found in the species in the region. Third, the number of plants taken from local populations has been very small (often 10 or fewer), or if large samples were taken they have not been preserved as populations in such a way as to prevent loss of genetic variability. With small samples it is not possible to determine the nature and extent of genetic variability within the populations and comparisons of populations cannot be made. This is especially critical for rare genes and for physiological or resistance factors that cannot be identified without special tests. Fourth, often insufficient records of environmental features have been made at the collection sites. Not all collections are equally deficient in these four categories, and in fact for a few collections sufficient records and samples were taken to allow study from an ecogenetic point of view. Guidelines were developed at a previous FAO/IBP Conference (Frankel & Bennett, 1970) for obtaining germplasm samples that are useful for study of evolutionary and ecological aspects of cultivated plants and no doubt future explorations will be more cognizant of the importance of adequate sampling.

It is the objective here to illustrate the investigation of the evolution of characters and character complexes in a cultivated species. Recommendations will be developed regarding sampling for adequate

81

germplasm preservation and for the study of evolutionary processes in cultivated plants.

This analysis will be illustrated with a typical and recurring scenario in agricultural production: A new malady is shown to be due to an adverse interaction of host and pathogen; the only feasible control of the pathogen is through use of genetic resistance; a search for resistance is successful; the resistance is transferred to agriculturally useful forms and further crop losses are averted, at least temporarily.

The barley yellow dwarf virus disease

In 1951 a widespread epiphytotic occurred in barley in California. Plants showed extreme yellowing and dwarfing resulting in severe yield losses. Oswald & Houston (1953) subsequently found that this previously unknown disease was caused by an aphid-transmitted virus which they called the barley yellow dwarf virus (BYDV). It was later determined that this same virus affected oats and wheat and is harbored in many native grasses. The virus is carried to seedling small-grain crops in the early fall or spring depending on conditions favorable for aphid flights. In retrospect it was established that this virus disease had occurred in various areas of the United States for many years (Bruehl, 1961); this was emphasized by a national outbreak of BYDV in 1959 which caused large yield reductions in oats (Murphy, 1959). BYDV has been found in many areas of the world and causes significant yield losses in some years. It was concluded quite early that insecticidal control of the insect vector would not be effective because momentary feeding by the insect would transmit the virus to healthy plants even in the presence of an insecticide. Resistance seemed to offer the only possible way to control this disease effectively.

Resistance was found in the world collection of barley (Schaller, Rasmusson & Qualset, 1963; Qualset & Schaller, 1969). Resistant types were not immune and showed limited symptom expression. The virus can be recovered from the resistant types (Jones & Catherall, 1970). Highly resistant types in the world collection were found only from Ethiopia. Genetic analyses of 16 varieties showed that resistance is conferred by one gene on chromosome 3 (Rasmusson & Schaller, 1959; Damsteegt & Bruehl, 1964; Schaller, Qualset & Rutger, 1964). Strains of the virus have been identified, but the most highly resistant varieties have been resistant up to now at several locations (Arny & Jedlinski, 1966; Hayes, Catherall & Jones, 1971). Some resistant varieties have less tolerance than others. These types have sometimes been scored resistant in one location and susceptible in another. The resistance was easily transferred by backcrossing as demonstrated by Schaller, Chim, Prato & Isom (1970). Comparisons of resistant and susceptible isogenic lines of the variety California Mariout in the areas of production have shown a rather consistent 15–20 per cent yield advantage for the resistant type. BYDV caused yield losses much greater than anticipated on the basis of visual symptom expression in production fields.

Apparently it is unlikely that additional sources of resistance will be found in geographic regions other than Ethiopia. Only a single gene for resistance has been identified. This is disturbing because if genetic changes in the virus result in virulence to this source of resistance we would have no further sources for the protection of barley. Will more extensive sampling and genetic tests reveal additional genes for resistance? A detailed study of all available barleys from Ethiopia in the US Department of Agriculture world collection was conducted to determine the extent of sampling of barleys in Ethiopia and to attempt to discover factors that may have led to the evolution of resistance in this region.

Ethiopian barley

When barley was first observed by agronomists and botanists in Ethiopia it was apparent that a great deal of genetic variability was present. N. I. Vavilov and H. V. Harlan were impressed by the vast range in ecological conditions and the uniqueness of the barley during their collection trips there in the 1920s. It was first thought that Ethiopia was a center of origin of cultivated barley, but Schiemann (1951) suggested that two-rowed and six-rowed barleys were introduced to Ethiopia via the Syria–Egypt or Yemen–Eritrea routes. Subsequent diversity may have arisen through mutation and probably extensive hybridization (H. V. Harlan, 1957) at the cool, high elevation sites. Ethiopia is now considered an 'accumulation center' (Schiemann, 1951) or 'center of concentration' (Ward, 1962) for genetic diversity rather than the region of origin of cultivated barley (J. R. Harlan, 1971).

Orlov (1929) pointed out the distinctiveness of Ethiopian barley as 'a marked ecological type' strongly differing from the types of Asia and Europe. He described Ethiopian barleys as having a long tillering period, large number of tillers, long and narrow leaves of milky green color, a trapezoid-shaped ligule, open-flowering, and large grains. Vavilov (1960) mentioned these same characters and pointed out that crosses of Eastern Asiatic and Eurasian barleys with Ethiopian types 'display a considerable degree of sterility (through seed infertility), by this trait alone revealing a certain degree of genetic isolation'. The sterility mentioned by Vavilov has not been investigated further, but it may be similar to the genetically controlled floral sensitivity mechanism that causes seed failure during manual hybridization in certain Ethiopian barleys (Qualset & Schaller, 1968). Both Orlov (1929) and H. V. Harlan (1957) implied that hybridization may be extensive in this usually highly autogamous species. Anderson (1961) found a high degree of segregation among tetraploid wheat progenies obtained from a field in Ethiopia, which also suggests that hybridization may be very important in the establishment and maintenance of genetic diversity in Ethiopian cereals.

In the study by Qualset & Schaller (1963), phenotypic frequencies of several morphological characters were obtained from about 650 Ethiopian barleys (Table 5.1). The frequencies can be compared with the world-wide distribution of characters as described by Ward (1962). The frequencies in Table 5.1 for Ethiopia are based on a larger sample than was available to Ward and confirmed his conclusions concerning the distinctness of Ethiopian barleys for grain color, deficiens (absence of lateral florets on the spike), irregulare (lateral florets irregularly present or absent), and short-haired rachilla. Giessen, Hoffmann & Schottenloher (1956) reported an analysis of collections made in Ethiopia in 1937–38 by K. Troll and R. Schottenloher. They recognized 12 regions in Ethiopia based on geographic, geologic, climatic and agricultural features and collected barley from ten of these regions. They identified 170 different morphological types which were classified as 68 varieties, based on 'systematically important' characters. Several characteristics were believed important from a plant breeding point of view: tillering, straw strength, kernel size, disease resistance (mildew and dwarf rust), and high-protein content. Qualset & Suneson (1966) constructed a breeding population which was developed to incorporate many of these characteristics of Ethiopian barley into adapted genotypes. Additionally, it is of considerable interest that an Ethiopian barley has been found with high-lysine and high-protein content (Munck, Karlsson & Hagberg, 1971).

Table 5.1. *Frequencies of characters observed in Ethiopian barley compared to world-wide distribution*

Character	Description	Symbol	No. Ethiopian collections	Frequency, % Ethiopia	Frequency, % World*
Spike type	6-rowed	6	430	66.0	71.0
	2-rowed	2	33	5.1	25.1
	Deficiens	D	124	19.1	2.0
	Irregulare	I	64	9.8	1.2
Grain color	White	W	268	41.2	55.1
	Blue	Bl	177	27.2	36.7
	Black	B	129	19.8	5.7
	Purple	P	77	11.8	2.5
Rachilla hair	Long	Lg	175	26.9	60.9
	Short	Sh	476	73.1	39.1
Glume awn†	Long	LG	374	57.6	55.2
	Short	SG	275	42.4	44.8
Spike density	Lax	Lx	601	92.6	—
	Dense	D	48	7.4	—
Caryopsis	Covered	C	572	87.9	87.5
	Naked	N	79	12.1	12.5
Heading time	Early	E	400	63.4	—
	Midseason	M	149	23.6	—
	Late	L	82	13.0	—
BYDV reaction	Susceptible	S	511	78.5	98.0
	Resistant	R	140	21.5	2.0

* From Ward (1962), except BYDV reaction from Schaller *et al.* (1963).
† Ratio of glume awn to glume; LG ratio ≥ 1, SG ratio ≤ 1.

Origin of BYDV resistance

Schaller *et al.* (1963) found that about 21 per cent of the barleys from Ethiopia were resistant to BYDV (Table 5.1). Stewart (1963, personal communication) has observed BYDV in Ethiopia. Major collections of Ethiopian barley were made by Harlan and by Vavilov. Subsequently other explorers provided samples for the USDA World Collection. The BYDV reaction of nearly every entry was determined and the frequency of resistance was obtained (Table 5.2). The overall frequency of 17.5 per cent is similar, but somewhat lower, than Schaller *et al.* (1963) observed previously. Only one sizable collection, that of Troll and Schottenloher, differs considerably from the mean frequency of resistant types. This is probably because a somewhat different sampling route was taken by these explorers than the others listed in Table 5.2.

Table 5.2. *Frequencies of BYDV-resistant Ethiopian barleys collected at various times*

Collector and year		CI numbers included	Number of collections	Number resistant	Percentage resistant
H. V. Harlan	1923	3204–4224	97	26	26.8
N. I. Vavilov	1927	5785–5876 ⎫ * 9954–9957 ⎭	93	12	12.9
G. R. Giglioli	1937	6378–6391	14	2	15.4
K. Troll, R. Schottenloher	1937–8	12148–12229	82	7	8.5
H. E. Myers	1945	7190–7239†	49	14	28.6
W. A. Archer	1951	9584–9852 9967–9984	255	62	24.3
J. R. Harlan	1961	11701–11721	21	3	14.3
I. E. Siegenthaler	1961	11723–11745 12038–12055	42	9	21.4
E. L. Smith, C. E. H. Thomas	1963–4	12513–13098	396	49	12.4
Total			1049	184	17.5

* Presented to J. G. Dickson by N. I. Vavilov.
† Presented to H. E. Myers by the Egyptian Ministry of Agriculture.

There is no systematic change in frequency of resistance during the 40-year period in which collections have been made and it is apparent that resistance has been present in Ethiopian barley for quite some time. However, because no migration to other regions in the world has been observed it is suspected that the resistance may be the result of a relatively recent mutation.

Elevations were recorded at the collection sites for the collections made by Harlan in 1923 and Smith and Thomas 1963–64. When the frequency of resistance was determined for various altitudes (Table 5.3) a very striking cline was observed. The samples from high elevations had a much higher frequency of resistance than the lower areas. Ecological features of the high elevation sites apparently are more favorable for the virus and the aphid vector, resulting in natural selection for resistance. Alternatively, the resistance may have become established at high elevations by chance alone. Direct observations of the incidence of the disease and its insect vectors should be made to distinguish between the two possibilities.

Collection records indicating the approximate place of collection of the barley samples were available from the World Collection records

Table 5.3. *Frequency of collections resistant to BYDV from various elevations in Ethiopia**

Elevation (ft)	Number of collections	Percentage resistant†
< 7000	78	3.8
7–8000	129	8.5
8–9000	87	12.6
9–10000	69	17.4
10–11000	33	33.3
> 12000	17	64.7

* Total of 413 collections; 22 from H. V. Harlan made in 1923 and 391 from E. L. Smith and C. E. H. Thomas, 1963–64.
† $\chi^2_{(5)}$ for independence = 56.2, $P < 0.005$.

for the H. V. Harlan collection and the Smith–Thomas collection. From the records of the Smith–Thomas collection it is possible to locate the origin of resistant types. Collections from adjacent areas were grouped and the proportion of resistant collections is given in Fig. 5.1. It is obvious that resistance is not restricted to one region in Ethiopia. Indeed, one would expect some movement of this gene throughout the barley populations of Ethiopia. However, one, and possibly two, centers of concentration of resistant types can be tentatively identified. One center is located in the regions north of Addis Ababa toward Fiche and north-east to Debre Birhan, a second center being 200 km south of Addis Ababa near Adaba. Obviously there are many gaps in the areas of collection and the samples were of inadequate size to locate a definite center of origin of resistance.

Also plotted on the map are the approximate sites where H. V. Harlan obtained samples in 1923. His route was Addis Ababa–Debre Birhan–Dese–Debre Tabor–Lake Tana. For his collections only resistant types are plotted. The number of samples collected was rather small and the records of origin somewhat obscure. The Debre Birhan area was certainly the center of concentration of resistance along his route. His collections also revealed a few resistant types north of Debre Birhan to Dese and also in the Lake Tana region.

From the Smith–Thomas and Harlan collections, the following tentative conclusions can be drawn. Since BYDV resistance is found at high frequency in the highlands north of Addis Ababa, this is probably the center of origin of the resistance. Resistance is found at low frequency throughout the barley-growing areas of Ethiopia, probably as

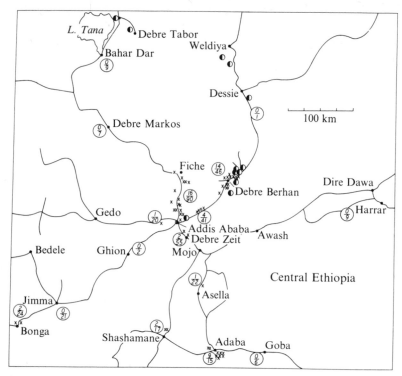

Fig. 5.1. Central Ethiopia. Sites of collection of barley resistant to BYDV indicated by ×
for the Smith–Thomas collection. Fractions indicate number of resistant per number of
samples from the Smith–Thomas collection. ◑ indicates site of collection of resistant
barley by H. V. Harlan.

a result of migration from the center of origin. It was pointed out earlier
that only a single gene for resistance had been found. With the present
information on the localization of resistance it is obvious that resistant
collections from various regions should be intercrossed and progenies
studied to determine if they have the same genes for resistance. Thus,
knowledge of the geographic origin of germplasm collections can be
very useful in attacking long-range plant breeding problems.

Evolution of character complexes

We are virtually certain that the mutation for BYDV resistance occurred
in Ethiopia and, since the disease is present there, we have reason to
believe that there has been natural selection for resistance, particularly
in the centers of concentration of resistance. It is well known from the
early work of Anderson (1939), Clausen & Hiesey (1960), Bal, Suneson

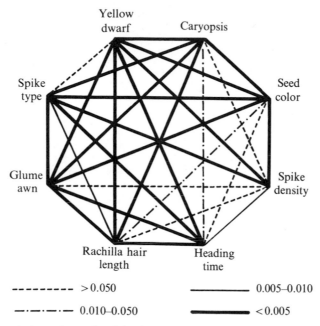

Fig. 5.2. Association polygon for eight characters in a collection of 650 Ethiopian barleys. The type of line connecting two characters indicates the probability of independent association.

& Ramage (1959), Ward (1962) and others, that characters, and therefore genes, do not occur at random in populations. This is true for linked and unlinked genes. Qualset & Schaller (1963) showed that when all of the Ethiopian barleys are taken as a group from the USDA World Collection very few character combinations were present at frequencies predicted by independent distribution. The characters studied included heading time, seed and spike characters, and BYDV reaction. Fig. 5.2 is an association polygon that shows non-random distribution of characters, taken two at a time. The probability of independent distribution exceeds 5 per cent for only 5 of 28 combinations. Of particular interest are the associations of six of the seven characters with BYDV resistance. Resistance is associated with white and purple grain color, short-haired rachilla, long glume awn, lax spike, covered caryopsis, and late flowering time (Table 5.4). Natural selection can be expected for time of flowering, and it is likely that the late maturing types would be found at high elevations as was found for BYDV resistance (Table 5.3). For the other characters there are no known direct selective advantages for the various morphological forms.

89

Table 5.4. *Characters positively associated with resistance to BYDV in Ethiopian barley*

Character	Number of resistant collections	Deviation from independent distributions, number of collections
Spike type		
6-rowed	97	4.5 NS*
Irregular	16	2.2 NS
Grain color		
White	77	19.4**
Purple	22	5.4**
Short rachilla hair	134	31.6**
Long glume awn	77	17.7**
Lax spike	138	8.4**
Covered caryopsis	134	11.0**
Heading time		
Medium	61	29.8**
Late	56	39.1**

* NS, $P > 0.05$
** $P < 0.01$, contingency χ^2 for independence.

It is expected that natural or human selection is important for disease resistance. Thus a study of associations of resistances to a group of diseases should provide information concerning the evolution of adaptive gene complexes. Ethiopian barley is well suited to such analysis because most of the known barley diseases occur in Ethiopia (Stewart, 1963). A study was co-ordinated in which investigators in the USA, Canada, and Sweden surveyed more than 650 Ethiopian barleys for resistance to nine diseases (Qualset & Moseman, 1966 and unpublished). Table 5.5 shows the frequencies of resistant collections to each of the diseases. Resistance was found to all diseases, but Ethiopian barley was not generally resistant to any disease. The probabilities for independent distribution of resistance among these diseases was investigated and the results for associations with BYDV resistance are indicated in Table 5.5. Significant departures from independence were found for 10 of the 16 combinations. Of the 10 significant associations, six were resistant–resistant and four were resistant–susceptible. When pairwise associations for all nine diseases are examined (using data from only one culture for each disease) 23 of the 36 possible associations were nonsignificant. In the remaining 13 cases, nine were significant resistant–resistant and four were significant resistant–susceptible associations. There is a slight excess of resistant–resistant combinations which would

Table 5.5. *Resistance to nine virus and fungal diseases in Ethiopian barley and analyses for associations with BYDV resistance*

Disease	Race or fungal culture	Percentage of collections resistant	Association with BYDV resistance[a]
Powdery mildew[b]	59.11 + 59.21	22.2	+ **
	Cr 3	8.0	+ **
	A 6	2.9	−
	D 1	3.8	− **
	B 8	2.3	− *
	C 10	8.3	+ *
Leaf rust[c]		3.9	−
Net blotch[d]	102	7.0	−
	526	21.0	−
	Bulk	10.9	+
Septoria[e]		9.6	+
Scald[f]		31.6	+ *
Spot blotch[g]		3.8	− **
Loose smut[h]	58–12	31.2	+ **
	49–70	28.7	+ **
BSMV[i]		6.0	− **
BYDV		21.7	

[a] + Association of resistance of one disease with resistance to BYDV; − association of susceptibility of one disease with resistance to BYDV; * and ** probability for independence less than 0.05 and 0.01, respectively.

Causal organisms
[b] *Erysiphe graminis* DC. f. sp. *hordei* Marchal.; [c] *Puccinia hordei* Otth.; [d] *Pyrenophora teres* Drechs.; [e] *Septoria passerinii* Sacc.; [f] *Rhynchosporium secalis* (Oudem.) J. J. Davis; [g] *Helminthosporium sorokinianum* Sacc.; [h] *Ustilago nuda* (Jens.) Rostr.; [i] Barley stripe mosaic virus.

be expected with natural selection for multiple resistance. This trend is not as strong as might be expected but it must be remembered that the cultures of the organisms used were not taken from Ethiopia. It is important to examine Ethiopian barleys for resistance to Ethiopian forms of the various pathogens before a full understanding of the evolution of patterns of variability in a cultivated species can be realized.

The non-random associations of morphological characters, and possibly disease resistance factors, with BYDV resistance were found when Ethiopian barley was studied as one group and we have already seen that there is geographic specificity for BYDV resistance and other characters as H. V. Harlan (1957) pointed out. With the above reservation in mind, we can consider the question of how such character complexes might have arisen. Two alternatives are obvious: (1) The associated characters represent the phenotype of the prototype from which the mutant was derived. If this is the case, the present predominant associated characters suggest that the susceptible prototype was a late maturing type with white grain, short-haired rachilla, long glume awn, lax spike, and covered caryopsis. After a mutation for resistance occurred the prototype combination of characters would remain relatively unchanged for a long period of time because of predominant self-pollination. Linkages among some genes would tend to slow the diffusion of the gene for resistance to random distribution. Associations of this type are relics from the original mutational event. Qualset & Schaller (1963) termed these *residual associations*. (2) Alternatively, the present associations with BYDV resistance do not reflect the non-mutated prototype but represent types that were selectively favored during the cultivation of barley because of environmental requirements or preference for certain types by man. Character combinations that arise through selection were termed *adaptive associations* previously (Qualset & Schaller, 1963). Some of the characters studied may *a priori* be considered neutral and others to have selective value under some circumstances. Additional collections and genetic analyses are necessary to evaluate these two hypotheses and it is likely that the observed non-random associations are both residual and adaptive.

Discussion

The forces of evolution that mold the genetic structure of populations should be considered when evaluating agricultural species as well as non-cultivated species. It is clear from this study that the materials as they exist in World Collections are inadequate to study evolutionary problems except on the broadest scale. The World Collections already contain much of the useful germplasm for breeding, but the sampling procedures for ecogenetic analysis must be more systematic throughout regions. There are two aspects of sampling populations of cultivated plants that deserve emphasis. The first deals with the problem of germplasm conservation and the second with evolutionary aspects (cultural and biological) of germplasm in its natural state. Germplasm conservation is concerned primarily with capturing and preserving genes from populations that are in danger of extinction because of new agricultural practices. Marshall and Brown have considered this problem in detail in Chapter 4. To be reasonably sure ($P \geqslant 0.95$) of obtaining at least one copy of each allele at a locus that occurs at a frequency greater than 0.05 in a sample from a population, they showed that a sample size of about 50 individuals is adequate. Bennett (1970) and Allard (1970) have provided detailed recommendations for designing sampling strategies and documenting ecological conditions at the sample sites.

To study evolutionary aspects of crop plants, as illustrated with Ethiopian barley here, emphasis should be placed on frequencies of combinations of characters (genes) as well as individual gene frequencies. Frequencies of character combinations can be quite low because with random distribution, their expected frequencies are obtained as products of the gene frequencies. For example, the probability of observing an individual in a population that has alleles from each of two loci with frequencies of 0.05 is 0.0025. To ensure that interesting, but rare, combinations of characters are obtained in samples taken from agricultural populations, the sample size should be five to ten times larger than that recommended by Marshall and Brown, for example, 500 individuals rather than 50. This extends the time required for sampling and the resources needed for maintenance and evaluation of the collections. This recommendation is not practical in general application because detailed study of character combinations need not be done for all populations. The strategy chosen depends upon the crop species and amount of variability present, but it is strongly urged that consideration be given to both aspects of genetic resources: conservation and evolution. To do this, a field sampling strategy should be developed whereby samples are taken according to Marshall and Brown's recommendations for germplasm preservation from most of the populations and simultaneously for evolutionary study a portion of the populations, say ten per cent, should be sampled more extensively.

It should be apparent that Ethiopian barley provides a very suitable subject for study of the evolution of important characters. There can be no doubt that Ethiopian barleys provide genes that are useful for barley breeding. In fact, one of the BYDV resistant collections, Benton (CI 1227), was introduced directly for cultivation in the USA (Foote, 1966). Interestingly, this conclusion was not reached by J. R. Harlan (1956), drawing on experience with the H. V. Harlan–C. A. Suneson composite-cross populations of barley where Ethiopian types apparently did not contribute significantly to population improvement.

Barley is only one of many species that are widely variable in Ethiopia. J. R. Harlan (1969) has examined crop resources in Ethiopia and points out that 83 cultivated species are present, with broad diversity in most of them. His plea for urgent attention to the study of genetic resources in Ethiopia was based on the following: (1) a great range of variability present, (2) a unique opportunity to study genetic systems, (3) the antiquity of seed agriculture and tropical 'vegeculture', and (4) evolution of agriculture systems can be documented because little change has occurred since prehistoric times. Other regions that still have landraces of cultivated species should be considered in this way.

Conclusions

It has been shown here that resistance to a disease is non-randomly distributed throughout the world barleys (in Ethiopia only), non-randomly distributed in its general area of origin (within Ethiopia), and that it is non-randomly distributed with regard to several morpho-physiologic characters. From the existing collections it was not possible to infer the nature of the selective forces that have led to such specificity. An important problem that could not be answered is whether linked blocks of genes of adaptive significance have evolved, as for example, linked genes for disease resistance. The identity of genes in Ethiopian barley with genes in primitive barleys from the presumed area of origin of barley in the fertile crescent of the Middle East (Harlan & Zohary, 1966) would provide evidence for the migration of barley from the primary center of origin to Ethiopia.

It is apparent that, even for a well-known crop such as barley, we have just begun to attain an understanding of the evolution of the crop and that we almost certainly have not discovered or captured all of the important genes already in existence. Cultivated plants should be viewed as part of the natural vegetation and research on the ecologic, genetic, and cultural factors that lead to their evolution should proceed in a very systematic and thorough manner. Only then will we under-stand the conditions, as in Ethiopia, that have resulted in the origin and maintenance of such vast variability. The existing germplasm collections are not adequate for this purpose, nor were they designed to answer all of these questions. Co-ordinated studies by geneticists, plant pathologists, entomologists, geographers, anthropologists, soil scientists, and others should be organized as soon as possible to examine all aspects of cultural and biological evolutionary processes.

References

Allard, R. W. (1970). Population structure and sampling methods. In *Genetic Resources in Plants – their Exploration and Conservation* (eds O. H. Frankel and E. Bennett), *IBP Handbook*, no. **11**, pp. 97–107. Blackwell, Oxford and Edinburgh.

Anderson, E. (1939). Recombination in species crosses. *Genetics*, **24**, 668–98.

Anderson, E. (1961). The analysis of variation in cultivated plants with special reference to introgression. *Euphytica*, **10**, 79–86.

Arny, D. C. & Jedlinski, H. (1966). Resistance to the yellow dwarf virus in selected barley varieties. *Plant Disease Reporter*, **50**, 380–1.

Bal, B. S., Suneson, C. A. & Ramage, R. T. (1959). Genetic shift during 30 generations of natural selection in barley. *Agron. J.*, **51**, 555–8.

Bennett, E. (1970). Tactics of plant exploration. In *Genetic Resources in Plants – their Exploration and Conservation* (eds O. H. Frankel and E. Bennett), *IBP Handbook*, no. **11**, pp. 157–79. Blackwell, Oxford and Edinburgh.

Bruehl, G. W. (1961). *Barley Yellow Dwarf. Amer. Phytopath. Soc. Mono.* no. **1**, 52 pp.

Clausen, Jens & Hiesey, W. M. (1960). The balance between coherence and variation in evolution. *Proc. natn. Acad. Sci. USA*, **46**, 494–506.

Damsteegt, V. D. & Bruehl, G. W. (1964). Inheritance of resistance in barley yellow dwarf. *Phytopathology*, **54**, 219–24.

Foote, W. H. (1966). Benton barley. *Crop Sci.*, **6**, 93.

Frankel, O. H. & Bennett, E. (1970). *Genetic Resources in Plants – their Exploration and Conservation, IBP Handbook*, no. **11**. Blackwell, Oxford and Edinburgh.

Giessen, J. E., Hoffmann, W. & Schottenloher, R. (1956). Die Gersten Äthiopiens und Erythräas. *Z. PflZücht.*, **35**, 377–440.

Harlan, H. V. (1957). *One Man's Life with Barley*, 223 pp. Exposition Press, New York.

Harlan, J. R. (1956). Distribution and utilization of natural variability in cultivated plants. *Brookhaven Symp. Biol.*, **9**, 191–208.

Harlan, J. R. (1969). Ethiopia: a center of diversity. *Econ. Bot.*, **23**, 309–14.

Harlan, J. R. (1971). On the origin of barley: a second look. In *Barley Genetics*, vol. II (ed. R. A. Nilan), pp. 45ff. Proc. 2nd Int. Barley Genetics Symp., 1969. Wash. State Univ. Press, Pullman, Washington.

Harlan, J. R. & Zohary, D. (1966). Distribution of wild wheats and barley. *Science*, **153**, 1074–80.

Hayes, J. D., Catherall, P. L. & Jones, A. T. (1971). Problems encountered in developing BYDV tolerant European cultivars of barley. In *Barley Genetics*, vol. II (ed. R. A. Nilan), pp. 493–9, Proc. 2nd Int. Barley Genetics Symp., 1969. Wash. State Univ. Press, Pullman, Washington.

Jones, A. T. & Catherall, P. L. (1970). The relationship between growth rate and the expression of tolerance to barley yellow dwarf virus in barley. *Ann. appl. Biol.*, **65**, 137–45.

Munck, L., Karlsson, K-E. & Hagberg, A. (1971). Selection and characterization of a high-protein, high-lysine variety from the world barley collection. In *Barley Genetics*, vol. II (ed. R. A. Nilan), pp. 544–58, Proc. 2nd Int. Barley Genetics Symp., 1969. Wash. State Univ. Press, Pullman, Washington.

Murphy, H. C. (1959). The epidemic of barley yellow dwarf on oats in 1959. *Pl. Dis. Rep. Suppl.* **262**.

Orlov, A. A. (1929). The barleys of Abyssinia and Eritrea. *Bull. Appl. Bot.*, **20**, 283–342.

Oswald, J. W. & Houston, B. R. (1953). The yellow-dwarf virus disease of cereal crops. *Phytopathology*, **43**, 128–36.

Qualset, C. O. & Moseman, J. G. (1966). *Disease reactions of 654 barley introductions from Ethiopia*. US Dept. Agric., Crops Research Division, CR–65–66.

Qualset, C. O. & Schaller, C. W. (1963). Nonrandom association of characters in a collection of Ethiopian barley. *Amer. Soc. Agron. Abst., 1963 Annual meetings*, p. 88.

Qualset, C. O. & Schaller, C. W. (1968). Genetical and developmental analysis of sterility in barley conditioned by floral sensitivity. *Z. PflZücht.*, **60**, 247–59.

Qualset, C. O. & Schaller, C. W. (1969). Additional sources of resistance to the barley yellow dwarf virus in barley. *Crop Sci.*, **9**, 104–5.

Qualset, C. O. & Suneson, C. A. (1966). A barley gene-pool for use in breeding for resistance to the barley yellow dwarf virus disease. *Crop Sci.*, **6**, 302.

Rasmusson, D. C. & Schaller, C. W. (1959). The inheritance of resistance in barley to the yellow-dwarf virus. *Agron. J.*, **51**, 661–4.

Schaller, C. W., Chim, C. I., Prato, J. D. & Isom, W. H. (1970). CM67 and Atlas 68 – two new yellow-dwarf-resistant barley varieties. *Calif. Agr.*, **24** (4), 4–6.

Schaller, C. W., Qualset, C. O. & Rutger, J. N. (1964). Inheritance and linkage of the *Yd2* gene conditioning resistance to the barley yellow dwarf virus disease in barley. *Crop Sci.*, **4**, 544–8.

Schaller, C. W., Rasmusson, D. C. & Qualset, C. O. (1963). Sources of resistance to the yellow-dwarf virus in barley. *Crop Sci.*, **3**, 342–4.

Schiemann, Elisabeth (1951). New results on the history of cultivated cereals. *Heredity*, **5**, 305–20.

Stewart, R. B. (1963). Personal communication. Unpublished host-pathogen index for Ethiopia.

Vavilov, N. I. (1960). *World resources of cereals, leguminous seed crops and flax, and their utilization in plant breeding*, 442 pp. Acad. Sci. USSR: Moscow 1957.

Ward, D. J. (1962). Some evolutionary aspects of certain morphologic characters in a world collection of barleys. *US Dept. Agric. Tech. Bull.*, **1276**, 112 pp.

PART II

Exploration

Part IIA. Exploration targets

6. Genetic resources survey as a basis for exploration

O. H. FRANKEL

History and purposes of the survey of genetic resources

The first survey of genetic resources in their centres of diversity was conducted in 1971/2 and published early in 1973 (Frankel, 1973). This survey had a long history. It was first suggested in the early proposals for the International Biological Programme in 1964, and was included among the recommendations of various FAO meetings (FAO, 1969*a*, *b*; 1970), elaborated by Frankel & Bennett (1970) and finally organized and completed in 1971/2.

In the conception of this survey the emphasis was on *action*. It had become widely known and appreciated that the 'genetic treasuries' (Frankel, 1967) in the centres of diversity were not only immeasurably valuable, but rapidly vanishing assets. But the rate of salvage remained frustratingly low. Hence the survey was conceived as an inventory and guide for the specialists directly involved – plant collectors, plant breeders, evolutionists, agronomists – and also as a call to action to governments and administrators concerned with the management and conservation of natural resources and with the efficiency and productivity of agricultural systems. The purpose of the survey was to produce detailed evidence of existing genetic resources and the extent to which they are in jeopardy. It was particularly concerned with high-lighting present emergency situations and those likely to develop in the coming years. Indeed, the survey heightened the sense of urgency conveyed in the recommendations for a world-wide programme of genetic conservation, which were unanimously adopted by the United Nations Conference on the Human Environment at Stockholm in June 1972.

In spite of its limitations this survey is the first attempt at an inventory of genetic resources existing in the field, as distinct from those available in institutional collections. An account of the survey and its main

99

findings deserves a place in a book concerned with the exploration and conservation of genetic resources, especially since the survey report will have only limited circulation.

Objectives and limitations

To make a comprehensive and exact survey would have taken a long time and substantial financial resources; and by the time the survey was finished, much of the material it covered might have been eroded or have disappeared. The initial need was for signposts to danger spots rather than for precise maps which could only be produced by exhaustive exploration on the spot. The guidelines for the survey are contained in the Report of the Fourth Session of the FAO Panel of Experts on Plant Exploration and Introduction, held at FAO in Rome in April 1970 (FAO, 1970):

In order to conserve the remaining primitive cultivated material and related wild species, it is necessary to survey their geographical distribution, the degree of threat which each faces and as far as possible to assess the variation present. The Panel believes that a great deal of valuable information exists which should be collected and analysed.

This survey should be as comprehensive as possible. Considering the magnitude of the task and, in many cases, the need for urgent salvage, an order of priorities is required in conducting the survey, taking into consideration the importance of the crop, the degree of urgency, availability of information, and other factors. Of necessity, this survey would be somewhat superficial and incomplete to begin with in view of the need for rapid action, but further information can be added as it comes to hand, and incorporated in a data handling system. An enormous amount of information is available in publications and unpublished reports, in notebooks and in people's minds. There are plant explorers, ecologists, taxonomists and evolutionists with a broad knowledge of regions and habitats of importance for particular species.

Highest priority was to go to the major food crops, but less important crops were to be included as opportunity arose, especially if they were particularly threatened and/or information was readily available.

The survey was planned and financed jointly by FAO and IBP, with FAO as the executing agency. It was conducted by experts in or near the centres of diversity, and by specially commissioned consultants covering larger regions. In view of the growing urgency of the task the target for completion was set at the end of 1972. Limitations of time and finance restricted the scope of the enquiry and are responsible for some of the gaps – such as the omission of cassava, of fruit species in Iran, or of all vegetable crops: and for deficiencies due to the limited number of experts who could be consulted.

The main emphasis was placed on primitive (or traditional) cultivated material which in general is more immediately threatened than wild or weed material. This has had the result that some seriously eroded progenitors and other relatives of cereal, pulse, fruit and vegetable crops in the Mediterranean region have not yet been included, and there are similar gaps elsewhere, such as potatoes and other tubers in the High Andes, beans in tropical America, and *Euchlaena* in Mexico.

Some difficulties of delimitation occurred which caused apparent inconsistencies of inclusion or exclusion. In principle the distinctions between the three main categories of genetic resources – wild and weed species, primitive (or traditional) cultivars, and advanced (or bred) cultivars – are clear enough.

But in practice one finds intergrades which pose some problems of delimitation. In some fruit species the distinction between 'wild' and 'primitive' is difficult, and probably irrelevant, since wild material may be as valuable a resource as a primitive cultivar. At the other end of the scale there were difficulties of delimitation between primitive populations and locally selected cultivars. Some cultivars of cereals have the appearance of perfect uniformity, although indubitably indigenous and hence heterogeneous, in an area where plant breeding is as yet in its infancy. Problems were also created by relatively recent introductions – of some 30 or 40 years' standing – which had become introgressed by local material and now exhibit some local differentiation. Such material may be of value to local plant breeders, but is of wider interest if this is the only form in which the indigenous material has persisted.

Difficulties arising from insufficient information were emphasized by several contributors. In some instances this is due to the sporadic and inaccessible distribution of residual primitive varieties; in other instances to the lack of information on the varietal distribution in a country or region.

Although the Soviet Union is an area of enormous wealth in genetic resources it has not been included in this initial survey. Extensive surveys and collecting expeditions already conducted by the Vavilov Institute of Plant Industry in Leningrad have accumulated such a wealth of information, and of plant collections, that that Institute alone has the information, and the capacity, to deal with any emergency situations within its territory. Both information and material are made freely available by the Institute. Unfortunately, at present little information is available about another extremely important region, the People's Republic of China. It is hoped that as a result of freer communications, information gathered by Chinese scientists will become generally available before long.

Existing collections

Since it was the purpose of this survey to designate priority targets for the salvage of threatened genetic resources, the status of related material in existing collections is of direct relevance. Some contributors to the survey made reference to material in collections and botanic gardens, and several have emphasized the need to avoid duplication in further collections.

Plausible as such observations are, there are good reasons why every effort should be made to collect and conserve as much as possible of the genetic diversity which persists in the field before it is lost:

(i) Relatively little is known about the holdings of primitive cultivars in many of the larger collections, but published records indicate that with some notable exceptions they are neither large nor representative, and few have accessible, let alone computerized, records.

(ii) Many collection entries, and especially many older ones, are not representative of the populations from which they were taken but are derived from a few plants, and frequently depleted by subsequent selection. Unfortunately, many of the 'world collections' are similarly constituted. While a small and biased sample is better than none at all, there is good reason for seeking a more adequate representation while there is time.

(iii) Many collections have suffered serious genetic erosion as a consequence of hybridization, selection, genetic drift, unsuitable growing conditions, or human error during propagation. Conditions for safe-keeping were not widely available nor even generally recognized until fairly recently (see Frankel, 1970).

However, there are several excellent collections of primitive material which are well known to experts in their respective crops; hence duplication of collecting efforts is unlikely to occur. Nevertheless, it is both necessary and urgent that a guide to such collections be prepared to ensure their fullest utilization by plant breeders everywhere.

Plan and execution

In the main the survey followed the list of priorities for areas and crops specified by the FAO Panel of Experts at its third meeting (FAO, 1969). All priority areas were included with the exception of Oceania, and within them nearly all specified priority crops, the gaps being due to lack of available information. Similarly the amount of detail contained in the various reports was limited by the degree of information obtainable in the time available.

The sections which follow summarize the main results contained in the published account (Frankel, 1973).

Mediterranean (wheat)

In *Greece* the indigenous wheats which still predominated in the 1930s have virtually disappeared except in remote mountain areas. Even there genetic erosion is far advanced, and any exploration for native germplasm must yield a relatively low return for effort expended. The valuable collections of indigenous wheats made by Papadakis and his predecessors Papageorgiou and Melas no longer exist in their original form, the remaining material being selections from the original populations.

Cyprus. In spite of an active wheat improvement programme based on indigenous material, Cyprus still has a considerable wealth of variation in its *durum* wheats. Careful exploration could yield valuable genetic diversity from regions subjected to hot drying winds and from a wide variety of soil types.

In the main wheat growing areas in *Sicily* modern Italian *durum* varieties prevail, but varietal mixtures are common. These, and the old native races which still prevail in the mountainous interior, are being studied by the Germ Plasm Laboratory in Bari.

Near East

Cereals. Throughout the region primitive wheats at all ploidy levels are in jeopardy, with the partial exception of Afghanistan. *Turkey* has conducted extensive collecting expeditions in recent years and proposes to collect in the remaining areas. There are sporadic remnants of both *T. durum* and *T. aestivum* in the Pontic and Taurus mountains. Indigenous varieties of barley and to a lesser extent oats are being

103

rapidly replaced by modern cultivars. In *Syria* genetic diversity is rather limited. Native *durum* varieties are safe, being needed for export, until a high-yielding semi-dwarf *durum* is available. There is growing awareness in *Iraq* of the rapid erosion of its rich resources of indigenous wheat, and there are plans for establishing a gene bank; but there is need for immediate action. In both *Iran* and *Pakistan* indigenous land races of wheat can still be found, mainly in the mountainous regions under rainfed conditions, but sporadically elsewhere. In both countries there is need for immediate action if what remains is to be salvaged. In *Afghanistan* indigenous wheats were relatively safe until recently. Now, however, Mexican wheats have entered from the south on a considerable scale.

Pulse crops in this region are not in immediate danger. Possible exceptions are lentils in *Turkey*, and lentils and lupins in *Lebanon*.

Primitive tree fruit varieties need to be conserved in *Turkey* – with first priority to apples, apricots, peaches and sweet cherries, and second to plums, pistachios, pears and grapes – and in *Iraq* and *Pakistan*, but not at present in *Syria*. No information was available from *Iran*.

India

Cereals. The once great diversity of *wheats* in India has now all but disappeared from the fields; but large and representative collections were assembled in time and a major one is being maintained by the Indian Agricultural Research Institute. Improved varieties of Indian origin are also an important resource, now much restricted in use by the advent of the new 'high-yielding varieties'. The remaining unselected indigenous wheats are confined to the hills in the north and north-east in Himachal Pradesh, in UP, and in Kashmir. The IARI collected in recent years in Kashmir, Nepal, Sikkim and Bhutan.

Until recently nearly 5500 varieties of *rice* were cultivated in India, 700 of which were improved cultivars, mostly selections from local stocks. Now these are rapidly being replaced by the new dwarf varieties, especially in the plains. The report supplies detailed information on the areas in North-East India, Orissa, the North-East Deccan Plateau, and, to a lesser extent, in Kashmir, Mysore, Tamil Nadu and Kerala, where indigenous types persist, and provides brief notes for some of them. While extensive collections have been assembled, it is clear that a good deal of material remains to be collected.

Among the wealth of variability of *maize* in India, of special interest are the primitive types of corn in the north-eastern hilly regions, in the north Himalayan region, in the forest area of Orissa, and elsewhere. Although this valuable and in some respects unique material is not in imminent danger, it should be systematically collected and conserved.

High-yielding hybrids of *sorghum* are replacing local varieties, including improved cultivars, but primitive and locally adapted culti-vars of sorghum and also of the other millets can still be found in several regions.

Four species of *Phaseolus*, *P. aureus* (mung bean), *P. mungo* (black gram), *P. aconitifolius* (moth bean), and *P. calcaratus* (rice bean), have been collected and widely used by plant breeders in India. Locally adapted indigenous material, however, is still widely available and none of these species is at present threatened with genetic erosion.

South-East Asia

Malaysia and Indonesia are the centres of variation for a large number of tropical fruits, and most of the wild relatives of tree species occur as components of the rich forest vegetation. In *Malaysia* these valuable resources are now increasingly threatened by widespread land develop-ment and by the forest regeneration programmes. Protection can be given only within adequate forest reserves. The 'mixed orchards' constitute rich reservoirs of genetic variation which are not adequately replaced by collections of selected clones.

Indonesia has an enormous wealth of local varieties of tree fruits, some of which are subject to considerable genetic erosion. In the course of the survey a great deal of information was collected on the location, the degree of genetic erosion, and the representation in botanic gardens of a large number of such varieties.

Africa

Sorghum. Of the four wild races, *arundinaceum*, *verticilliflorum*, *virgatum* and *aethiopicum*, only *aethiopicum* appears to be threatened, due to extreme overgrazing as a result of the resettlement of people evacuated from Wadi Halfa (now Lake Nasser). Primitive land races are to be found throughout the sorghum belts of Africa. The American hybrid sorghums are adapted only in South Africa, where the impact has been considerable. In Upper Volta, Sudan, Nigeria, Niger and Senegal, there is progressive genetic erosion, and in the next few years there may be drastic changes. In the words of Dr J. R. Harlan, 'the time to collect is *now*, before the germplasm disappears'.

Millets. The survey shows the areas where wild, weed and cultivated *pearl millets* are found. There is no risk of erosion since there is no plant breeding, nor any crop which could replace this most important millet. The finger millet (*Eleusine coracana*) has its centre of variation in Ethiopia and Uganda, and *Eragrostis* (tef) is confined to Ethiopia. *Digitaria exilis* (fonio) and *Brachiaria deflexa* have lesser claims on collecting efforts.

African rice. Two groups of African rices appear to be of interest to plant breeders, mainly for the improvement of Asian rice: the perennial *Oryza longistaminata*, and the wild, weed and cultivated annuals, *O. barthii*, *O. stapfii*, and *O. glaberrima*. The cultivation of *O. glaberrima* has been declining rapidly, and thorough and systematic sampling should be made without delay.

Barley. Ethiopia is the richest source of barley germplasm in Africa and one of the richest in the world. Areas accessible by the few main roads have been explored by several collecting expeditions, but a great deal of valuable germplasm remains to be collected. North African barley has been well collected, except perhaps in the Atlas Mountains.

Wheat. The existing collections, even if combined, are not regarded as an adequate representation of African wheat germplasm. Ethiopia is the richest source, and genetic erosion – as a result of the spreading of Kenyan and Mexican wheats – is accelerating. In the North African countries erosion has gone much further, and salvage of remnants is urgent. The Atlas Mountains and Sahara oases are of interest. Information on material in existing collections is lacking.

Yams. The two principal cultivated African species, *Dioscorea rotundata* and *D. cayenensis*, have an enormous diversity at the village level, with extreme local specialization found in ecological niches, food habits and even cultural and ritual associations going back for centuries or even millennia. This variation is now threatened by 'better' cultivars, by the Asiatic *D. alata*, and by other crops – cassava, rice and maize. These circumstances prevail throughout West Africa, but the greatest concentration of types can be found in the former Eastern Region of Nigeria. Much material has already been lost and there is a situation of some urgency to obtain an adequate representation for conservation.

Coffee. Wild coffee in the Ethiopian rainforest was the traditional source of seed, and a continuing source of variation. The area under forest is shrinking rapidly, and further collecting of such material, or its adequate protection in forest reserves, should receive urgent attention.

Latin America

In spite of the massive introduction of Euro-Asiatic crops, both as food and for export production, this vast area still retains an enormous variety of indigenous variability in the many major and minor crops to which it gave rise. 'Native variability', in Dr J. León's words, 'is maintained only because of the conservative approach . . . by the small farmers, especially the Indians, but this is steadily changing', due to the agricultural, social and economic transformation. Small holdings give way to large agricultural enterprises; and the destruction of forests and their replacement by plantations and pastures is a further contributory factor in the process of genetic erosion or destruction of indigenous germplasm.

Short of reproducing Dr León's introductory chapter, it is impossible here to attempt an account of the distribution of genetic resources in the regions of greatest diversity – Mexico, the Pacific slopes of Central and South America, the intermountain valleys of the Andes, and the Atlantic lowlands. I must restrict myself to a general reference to the survey report, and to brief and inadequate references to the more detailed contributions dealing with specific crops.

Zea spp. All countries in the region maintain germplasm collections of corn which, taken together, represent the wide variation of the original populations. A close study of the more than 24 000 entries in major collections led to the description of 290 races and 47 subraces. Two main centres of genetic diversity emerged: (1) The Andean region (Colombia, Ecuador, Peru, Bolivia and Chile); (2) The Meso-american region (Mexico and Guatemala). These centres have been adequately covered. However, there are large regions which have not been explored, particularly in the Amazon basin; and there are possibilities of finding new genetic complexes in the mountain region, especially when attention is paid to ethnobotanical factors.

Comparisons between populations in identical sites, but 20 years apart, demonstrate the essential stability of the components, yet dynamic changes in their proportions.

The report records the distribution of teosinte (*Zea mexicana*) and of *Tripsacum* spp. and provides information on collections in which material of these species is preserved.

Exploration targets

Phaseolus spp. A large number of samples of Meso-american species – *P. vulgaris*, *P. coccineus*, *P. coccineus* ssp. *darwinianus*, *P. acutifolius*, and *P. lunatus* – has been collected in Mexico and Guatemala, and the representation is considered adequate, as is the method of preservation.

Capsicum spp. – *Cucurbita* spp. A large number of samples of *Capsicum* spp. has been collected throughout Tropical America. However, the author of the report believes that an adequate representation of both *Capsicum* and *Cucurbita* spp. is being maintained by the rural population.

Solanum spp. The information (provided by Dr C. Ochoa) is fully covered in his chapter in this book (pp. 167–73).

Cacao. The detailed report on the status of cacao germplasm in Mexico, Guatemala, Nicaragua, Costa Rica, Colombia, Venezuela, Ecuador and Brazil, does not lend itself to presentation in summary form. Attention is drawn to acute genetic erosion of some of the ancient varieties, especially of the 'Criollo' type. Chapter 15 by Dr J. Soria gives an account of recent collecting activities.

Conclusions

In spite of the shortcomings of the survey in both coverage and depth it makes abundantly clear that an emergency situation exists in many regions and for many species. It shows where the threat is most acute, and leaves no doubt that similar situations are likely to arise before long in many other places.

As it stands, the survey provides many leads for plant collectors in general, but especially for those concerned with the salvage of material which is threatened with imminent extinction. It should be useful in planning resource conservation on a global scale, and in the allocation of support from international sources such as the United Nations Environment Fund.

This survey can be expanded and strengthened only with the support of all those who possess additional information, which it is hoped will be forthcoming as soon as possible. If the present report holds a lesson it is that *there is no time to waste.*

This chapter is based on the report on the FAO–IBP Survey of Crop Genetic Resources published by FAO. Acknowledgment is due to the contributors to the survey who are named in the original publication. This, however, is the opportunity for acknowledging the very great contribution made by Dr J. León, Chief, Crop Ecology and Genetic Resources Unit of FAO, who organized the participation in the survey, took part in the editorial work, and translated the reports written in Spanish – contributions which, owing to FAO rules, could not be duly recognized in the original publication.

References

FAO (1969*a*). Report of the *FAO/IBP Technical Conference on the Exploration, Utilization and Conservation of Plant Genetic Resources, held in Rome, September 1967*. FAO, Rome.

FAO (1969*b*). Report of the *Third session of the FAO Panel of Experts on Plant Exploration and Introduction*. FAO, Rome.

FAO (1970). Report of the *Fourth session of the FAO Panel of Experts on Plant Exploration and Introduction*. FAO, Rome.

Frankel, O. H. (1967). Guarding the plant breeder's treasury. *New Scientist*, **14**, 538–40.

Frankel, O. H. (1970). Genetic conservation in perspective. In *Genetic Resources in Plants – their Exploration and Conservation* (eds O. H. Frankel and E. Bennett), *IBP Handbook*, no. **11**, pp. 469–89. Blackwell, Oxford and Edinburgh.

Frankel, O. H. (ed.) (1973). *Survey of Crop Genetic Resources in their Centres of Diversity*. FAO–IBP, Rome.

Frankel, O. H. & Bennett, E. (1970). Genetic Resources. In *Genetic Resources in Plants – their Exploration and Conservation* (eds O. H. Frankel and E. Bennett), *IBP Handbook*, no. **11**, pp. 7–17. Blackwell, Oxford and Edinburgh.

7.   Seed crops

J. R. HARLAN

Some of the practical problems that confront plant explorers are: political climate, local co-operation, transportation, sampling, timing, shipping collections, and plant quarantine.

I shall not deal here with political climate or local co-operation, but it is obvious that if the explorer is denied a visa or barred entry to the region in which collections are to be made, there is nothing that can be done. On the other hand, local co-operation and participation can be enormously helpful. Apart from these considerations, the interaction between transportation and sampling provide most of the practical problems in the exploration of the seed crops.

For our major crops, the easily accessible material has been largely sampled. Little can now be gained by visiting the main markets or driving along the highways. What remains to be collected is mostly in remote and less accessible places. Collecting by foot, horseback, or canoe, however, is very slow and the sampling of a large region requires either repeated campaigns or massive efforts with multiple collecting teams.

One sampling problem which confronts every plant explorer concerns the strategy of sampling in very remote mountainous areas. Thus, if one is travelling along a valley road at the base of a mountain range the temptation is to collect in the markets of the towns and villages in the valley itself, rather than to spend time and effort by making side trips on horseback up into the difficult regions above. It might be thought that everything of value ought to be found in the valley markets, but it is quite possible that these might contain common varieties brought in by truck from another region. The explorer who knows his crop will be able to tell if this is so, and he may also be able to pick up information in the markets on the origin of the samples he is collecting. Yet there will always be an uneasy feeling that he is missing something of immense value if he does not go up into the mountain range itself.

Yet is it worth the time and effort necessary? There are several valleys that decend to the plain. Each one would require a number of days to collect on foot or horseback. Is it necessary to collect from each one? Negotiating for horses and guides takes considerable time. It also takes a long time to ride from valley to valley. Is it necessary to renegotiate at the mouth of each valley to get riding and pack horses? Would a horse trailer be practical here to move horses from watershed to watershed along the mountain range? It has been suggested that much could be done with a motor bike and truck to carry it. The lower trails of a valley are often negotiable by motor bike in parts of Asia and this would avoid the necessity of arranging for horses. One could ascend each valley as far as practicable and finish on foot. Trails in Ethiopia, however, tend to take the shortest distance between two points and go straight up and down the mountain sides.

The use of a helicopter has often been suggested and has been tried out a few times. The best use appears to be as a vehicle to move a small party into some advance base camp. The group collects on foot for an agreed number of days, then the helicopter returns and moves the party to another base camp. The local inhabitants may interpret this as an invasion, however, and some advance communication may be required. I have been in places where even fixed-wing planes get hit by rifle fire. Helicopters are expensive under any circumstances and frequently not available.

However one goes, there are repeated decisions to be made. Going up a valley, there are always branches to the streams and villages off to one side or the other. Are they worth visiting? Would one collect anything really new by spending an additional day or two? How much alike are the landraces in neighbouring villages? How important is it to get a really good sampling grid? Will a single pass along the main trail be adequate? Are we going to spend all season on this mountain range, or are we going to sample two or more?

There are no real answers to these questions for they have not been seriously studied. I know of no cases where different sampling procedures for crops have been compared. Experience may sometimes sharpen subjective opinions and feelings about a given situation, but real knowledge is lacking. It seems to me that grains and field crops in general tend to be somewhat mobile, being exchanged from village to village to the point that a rather coarse grid is likely to pick up all but the very rare items. Vegetables, on the other hand, seem to be more personal. One gardener in a village may have things his neighbour in the next compound may not have. Sampling procedures should probably be varied accordingly, but we have to remember that it takes a lot more time to go from gardener to gardener than to pick a few seed-heads from a field.

Villages at the limits of agriculture may be a bit special. The last settlements seen when going up a high pass and the first ones met with going down the other side may have crops and cultivars not found in the lower villages. The same situation may apply to villages at the edge of a desert. On the other hand, collecting may be thin. If you arrive before harvest you may find that the people have consumed the entire previous crop and there is nothing to collect. People at the margins of agriculture often lead a tenuous existence.

Crops and cultivars often vary with ethnic groups. Two villages side by side may have radically different arrays of material if they are inhabited by different tribes. The collector should be sensitive to such situations or he may miss useful genetic material.

Sample size and methods of taking samples are problems forever with the collector. In general, I would agree with the conclusions and suggestions presented by Marshall & Brown in Chapter 4, although I have usually taken seed from fewer plants than they have recommended. On the other hand, I would plead for highly subjective and biased samples rather than random ones. It seems to me that the charge to the collector is to get as much of the genetic diversity as he can. He should, therefore, try to get at least a few seeds of every kind that he can see. In the attempt he will, of course, get the commoner things as in a random sample but he will also get rare items that random samples miss.

A specific example is the discovery of guinea sorghums being grown by the Konso tribe of Southern Ethiopia. The guinea race is basically West African although it is grown in Malawi and elsewhere in East Africa. We have yet to find it anywhere else in Ethiopia. But guinea is found in Konsoland at a frequency of far less than one in a thousand. Random field samples would surely have missed it, although large market samples might have included it.

There is a certain snobbishness among some explorers who feel that market samples are not really sporting and are a lazy man's way if not a fraud's way to collect. Yet, if the market is truly local, and villagers bring in their produce on camel-back, donkey-back or on their heads, the marketplace is not only the easiest place to collect, but is likely to lead to the best sampling for some crops. Materials from 25 or 30 villages might be sampled at one time and the explorer is not likely to visit that many within a marketing area. One should be cautious about markets where produce is trucked in and out for fear of getting no more than standard varieties from a long distance away. Furthermore, some kinds of seeds reach the market only seasonally or not at all, and may therefore be missed.

Attitudes of the local inhabitants are often important factors. A large party in remote areas is likely to be looked upon as a show of authority from the local source of political power and may be resented. It will also overtax the resources of a village. I prefer to go with a single interpreter or even alone if possible. While this keeps a low profile there are hazards of misinterpretation: I have been run out of fields at rifle point more than once. Much more common is to be half killed by kindness. You must come in and eat; you must taste this or sample that; you are not permitted to go on until you are 'refreshed'. How much 'refreshment' can the stomach take? Well-meaning as country people are, it becomes an endurance contest, nevertheless.

Timing is critical. Collections should be made during the harvest season to obtain field information. In seed crops, one can usually obtain samples of stored seed in off-season, but information is very restricted. For wild materials, however, it is absolutely essential to be in the right place at the right time. To do this consistently is almost impossible, and the explorer frequently finds he is either too early or too late.

The best, most systematic and most complete collections should come from resident collectors who live in or near the regions to be sampled and who can more often be at the right place at the right time. If material is missed in one season, it can be collected in another. Resident collectors should know the material and terrain much better and, over a period of years, should be far more effective than the expedition-type explorer from outside. They do, however, require financial support and encouragement from outside the region.

The potential value of a collection cannot be assessed in the field. Perhaps this statement could best be illustrated by PI 178383, a wheat I collected in a remote part of Eastern Turkey in 1948. It is a miserable-looking wheat, tall, thin-stemmed, lodges badly, is suceptible to leaf rust, lacks winter hardiness yet is difficult to vernalize, and has poor baking qualities. Understandably, no one paid any attention to it for some 15 years. Suddenly, stripe rust became serious in the north-western states and PI 178383 turned out to be resistant to four races of stripe rust, 35 races of common bunt, ten races of dwarf bunt and to have good tolerance to flag smut and snow mould. The accession is now used in all wheat breeding programmes in Montana, Idaho, Washington and Oregon. The improved cultivars based on 178 383 are reducing losses by a matter of some millions of dollars per year.

For this reason, the explorer should try to sample everything; he should not confine his attention to the best-looking material. It is the plant breeder's job to improve the crop. It is the plant explorer's job to collect genetic variation. In this connection the wild and weedy races of our crops are too often ignored completely. These can be useful sources of genes, but in most crops have been so poorly collected that breeders have had little chance to use them.

The shipping of seed collections is easy compared with the handling of vegetative material. Collecting envelopes should be taken to the field but cloth sacks can usually be found or made up almost anywhere. Although seeds keep well, it pays to send collections to a receiving station by air, since otherwise, they may be consumed by weevils or damaged by being left unprotected on a dock for some months.

Quarantine restrictions have caused undue problems in some countries. No one wishes to transfer serious diseases or pests from one part of the world to another, but there is no point in assembling germplasm unless it can be evaluated and utilized. Plant introduction programmes can usually succeed only where they are carried out with the full co-operation of the plant quarantine service.

Conclusions

Without *a priori* information on the usefulness of the material, the collector is supposed to assemble a good sampling of genetic diversity, avoiding at the same time undue redundancy. Variation in crops is often closely associated with ethnic groups, elevation, rainfall and soil types. It is necessary to make judgements as to how to sample a single field, fields of a local area and the areas that should be sampled. These subjective judgements are based on minimal information, intuition, guesswork and sometimes experience, but not on sound experimental evidence. Transportation in remote areas is always slow and can be difficult.

8. Vegetatively propagated crops

J. G. HAWKES

From a practical point of view it is much easier to collect seeds than vegetative parts of the plant. Seeds are relatively small, fairly abundant, easily stored, and easily dried and packed for transmission to the introduction station or gene bank. Furthermore, they incorporate a wider range of genetic diversity, since one parent plant of an outbreeding species can produce as many genotypes as there are seeds. Thus the genetic diversity of the seeds collected from a single plant can incorporate quite a large part of that of the local gene pool.

On the other hand, vegetative propagules and storage organs, such as rhizomes, tubers, roots, bulbs, corms or cuttings, are the same genotype as the mother plant. Hence vegetative collections need to be more extensive than seed collections, in terms of the numbers of plants sampled in any one area, if a comparable range of genetic variability is to be sampled.

Another difficulty with vegetative material is that it is often very bulky and awkward to transport. Thus, to take an extreme example, the underground organs of wild *Dioscorea* species may weigh as much as several kilos each. To sample adequately one ought to make several hundred collections from each district, and hence the problems of bulk become very difficult indeed in such cases.

Great care is required, also, to keep vegetative collections alive during an expedition and during transmission to the gene bank. Small storage organs will easily dry out and die if they are not wrapped in waterproof materials. On the other hand, if they are kept too moist they will easily rot. Care must be taken, furthermore, to prevent premature sprouting. The ideal solution, though one not always realizable in practice, is to wrap the collected materials in a semipermeable material (not a plastic, such as polythene, however) and keep them in cool storage conditions such as a domestic refrigerator.

The present writer has had considerable difficulty with the small tubers of wild potatoes which often lose viability, even when they do not dry out or rot. This seems to be due to the fact that they were collected when immature or when they had already begun to sprout and develop daughter plants. Of course the ideal time to collect wild potato tubers is during dormancy, when no vegetative shoots at all appear

117

above the soil surface. However, at that time, no plants can be seen above ground level and the tubers cannot be found. A reasonable compromise is to try to collect when the plants are just dying down but are still visible, and the tubers are reasonably mature.

Quarantine and customs difficulties may also provide problems in attempts to send vegetative material from one country to another. For this reason, the collector must study the appropriate quarantine laws and must provide himself in advance, whenever possible, with the necessary import and export permits.

Yet another problem in the collection of vegetative material is the difficulty of estimating genetic variability in morphological characters, when the material to be sampled lies under the soil surface and must be dug up before it can be examined. If the fields are visited earlier in the season when flowers and vegetative characters are visible, the storage organs will not be mature and a further visit will be needed later, to make the collections. On the other hand, in a cereal field for instance, the spikes or panicles are well visible and can be collected straight away. The above difficulty of collecting vegetative materials does not arise, of course, if we are concerned with entirely random collections, since we shall then not be involved with subjective selections of materials based on morphological characters.

Another point that needs to be kept in mind when sampling vegetative materials is the question of the size of clonal populations. What may appear at first sight to be a population of plants of varying phenotype and therefore presumably of varying genotype may in fact be a single clone of a standard genotype growing under a wide range of environmental differences. It is therefore unwise to sample too closely in the field for fear of taking a range of samples from a single widespread clone.

The question then arises – why collect vegetative materials at all if they are so difficult to collect and manipulate. The answer is, simply, that no other materials are often available. Fertility may be very low (as for instance with *Dioscorea* and *Ipomoea* cultivars) so that seeds may be sparse or absent. If seeds are found they may be too few to provide a reasonable sample of the diversity within the local gene pool. Very often they will be completely lacking.

Then again, if we are interested in problems of hybridization, gene flow, introgression, etc., or material showing ranges of cytotypes of reduced fertility, the vegetative collections will provide a complete range of everything that is viable. Seed collections, on the other hand, may represent only a small part (the highly fertile one) of the range of variability. Aneuploids, odd-number polyploids, and materials with various types of sterility, whether genetical or chromosomal, will not produce seeds (or only very few) and hence can only be collected in the form of vegetative propagules.

It is worth emphasizing that even genotypes of very low fertility may be of value to the breeder and hence would be worth collecting as vegetative material. These may be used as female parents in successful crosses with highly fertile pollen parents, even though with their own pollen and in natural conditions they may be almost or completely sterile.

Sampling techniques vary to some extent between wild and cultivated material.

Wild species

These should be sampled by making at least 10–20 random collections from a population in one square kilometre. Marshall and Brown (Chapter 4) recommend from 50–100 samples per site, though with vegetative materials the time taken in sampling and the bulk of the samples when collected may limit the numbers to less than this. As was mentioned above, samples should not be taken too close together, for fear of collecting several ramets of the same clone. Sampling should be repeated over the area at a density conditioned by factors such as topography, population density, observed variability, etc. If the environment is highly varied, with steep altitude gradients or large edaphic and climatic changes, then the intervals between the populations sampled should be decreased. It should be pointed out that this sampling strategy, which seems reasonably good in practice, still awaits experimental verification.

Cultivated species

The population structure for cultivated species that are propagated vegetatively is very different from that of the corresponding wild species, and also from that of sexually reproducing species. Unfortunately, there seems to be very little knowledge of what this structure is, so sampling strategy must largely be a matter of guesswork. Under primitive agriculture single clone cultivation does not exist, and fields contain diverse mixtures of genotypes of what are generally referred to as

119

'primitive forms', and even mixtures of distinct species. However, it is extremely doubtful whether such mixtures of clones could be considered as natural populations, in any way comparable to those occurring in the wild. Variability under cultivation could be assumed to represent only a small portion of the total series of genotypes produced at various times in the past by sexual reproduction and by the slow flow of vegetative mutations. Natural selection would in any case be reducing the total number of genotypes, and this process is of course common to all plants, wild or cultivated, sexually or vegetatively propagating. With vegetative crops, it is likely, however, that a strong artificial selection is also exercised at harvest, during storage and at planting time. It is impossible to be certain of the kind and degree of this selection, but it would seem to be much more intense than that obtaining with seed-propagated crops. With potatoes, for instance, only a limited number of genotypes seems to exist in each market area or district within the centre of variability of this cultigen. Hence, one can easily collect from 50 to 200 samples in a region, with a high degree of certainty that all, or a very high proportion of the existing clones have been sampled. In other words, since the total 'selected' remains of the original population is so small it would seem that a sample of every variant can be collected without much difficulty, and random sampling then becomes an unnecessary ideal.

It could be argued that one is only looking at morphological variation, and that this then gives a false picture of the total variability, much of which would be of a physiological nature which does not show a phenotypic effect. It is worthy of note, however, that in potatoes nearly every variant that has been found as a result of screening to show a useful quality of resistance or special adaptability, can be distinguished also on morphological characters.

In practice, when one is exploring an area for cultivated material it is useful to identify each district served by a town or village market. It is then possible to sample in the district from fields, markets and stores. Such collections should also be supplemented whenever possible from garden plots, where material is grown for home consumption and is not sent out to stores or markets.

It is very unlikely that, in the collection of vegetative materials, anyone has yet made completely random collections with cultivated species, though this has probably been done with the wild ones. As was mentioned above, it may perhaps not be necessary to try to do this. In markets and in fields one almost instinctively looks for different colours, colour patterns, shapes and sizes, and it is extremely difficult to do otherwise. Perhaps this is correct, since we are here not in any way looking at natural populations, but almost certainly at the results of very intensive selections by the native cultivators, who can recognize every variant by its tuber shape, colour and colour patterning. Such combinations of morphological characters have already been related to physiological characters, such as adaptation, palatability, cooking quality, disease resistance, etc., by the farmers themselves. Thus, perhaps what we as collectors should be doing, is to collect the remnants of the population samples that the farmers have already selected for us; indeed we can do little else. And since most vegetatively propagated crops are highly heterozygous they hold an enormous reservoir of genetic variability within each clone which is released when true seed is produced. What more need we ask for?

9. Tree crops

J. T. SYKES

There are a number of problems connected with tree-crop collecting which merit special attention. The following account, based on the writer's experience at the Agricultural Research and Introduction Centre in Izmir, Turkey, draws attention to these.

Three distinct components of exploration, each of which presents problems are: (*a*) survey techniques and their interpretation, (*b*) sampling methods for wild species and primitive cultivars and (*c*) the collection and propagation of accessions.

Survey techniques

Survey procedures should be particularly effective for tree crops, since they are less susceptible to short-term fluctuation than are annual crops. Information from published Floras as well as field surveys are useful, particularly where species distribution data may be interpreted in relation to the agro-ecological factors of the surveyed area (Davies, 1966; Bunting & Kuckuck, 1970). However, such surveys are time-consuming and tend to be incomplete because certain areas are inevitably excluded.

Other sources of information are available to provide data from which distribution maps may be prepared, such as detailed statistics on regional production, which might be interpreted to determine the presence of tree-crop populations well-adapted to different climates. Thus, walnut (*Juglans regia* L.) occurs in each of the nine agricultural regions of Turkey, which suggests that native germplasm has considerable adaptability. Since 24 separate climatic types in four main regions have been recognised in Turkey (Erinc, 1949), winter-hardy and drought-tolerant genotypes from the north-eastern and south-eastern regions respectively would, therefore, merit investigation and possible collection. Again, although only two per cent of the Turkish national production of chestnuts (*Castanea sativa* Mill.) is derived from the relatively few trees which occur in the four south-eastern agricultural regions, nevertheless the low rainfall and high temperatures which prevail there in summer would indicate that potentially drought-tolerant chestnuts from these four regions might constitute a valuable genetic resource.

123

Aerial surveying techniques, in certain areas and where economically feasible, may be a supplement to ground reconnaissance. Many different techniques have been used to evaluate vegetation resources and crops, but most of these until recently have identified species with only limited success.

Images from long wavelength radiation obtained from side-looking airborne radar (SLAR) have been interpreted to evaluate vegetation resources but, as a vegetation inventory tool, this technique is more expensive than aerial photography and it has some limitations, especially over undulating topography. However, based on the recognition of different tone and texture categories, indications of differences in vegetation type and species composition have been obtained over flat terrain (Daus & Lauer, 1971). Multiband photography has proved superior to singleband aerial photography. Black and white, colour and infrared films have been used, separately or in combination, to discriminate between deciduous trees and conifers in the temperate northern hemisphere. However, these techniques have identified only one species, black oak (*Quercus velutina* Lam.) with an acceptable level of accuracy (Lauer, Benson & Hay, 1971). Very large-scale aerial photography (e.g. 1:1500) has been used in eastern Canada in the successful identification of a wide range of forest tree species.

Narrow band, multispectral sensing techniques have recently been used by Yost & Wenderoth (1971) with some greater success. Unique spectral 'signatures' of several tree species were obtained, based on leaf spectral reflectance as this relates to factors which affect pigmentation, water stress and cellular development. Colour differences were also reported when the images from four bands were superimposed to identify *Quercus kelloggii* Newb., *Abies concolor* Hilderbr., *Pinus contorta* Dougl. and *Alnus rhombifolia* Nutt. Howard (1970) has reported using multiband photography and enhanced colour in Australia to identify *Eucalyptus regnans* F.v.M., *E. obliqua* L'Hér. and *E. goniocalyx* F.v.M. Rohde & Olson (1972) used six channels in the 0.4 to 1.0 μm range in the spectral analysis of photographs taken from 500 m and higher altitudes. The data obtained were computer-processed and proved substantially better than any combination of photographic interpretations in the discrimination between conifers and deciduous trees and between pine and spruce. The following trees were successfully recognised by this method: *Quercus rubra* Duroi, *Q. alba* L., *Juglans nigra* L., *Robinia pseudoacacia* L. and *Acer saccharum* Marsh. With some justification, these workers conclude that improvements in data collection and data processing offer considerable promise in future for the rapid recognition of ecological communities and the identification of plant species.

Where and to what extent these technological advances may be used in future for mapping plant distribution will depend on cost factors and the need to obtain approval before specific areas are photographed. If undertaken at appropriate intervals, the plant resources inventory thus obtained would indicate major vegetational changes and, with greater refinement, perhaps these techniques could be used in detecting the times of flowering, fruit maturity and leaf abscission. If this information were available, the planning of expeditions in relation to the evaluation of trees and the collection of propagation material at the optimum times would be greatly facilitated.

Sampling methods

The development of a definitive, statistically valid rationale for sampling heterozygous individuals depends upon the feasibility of collecting random samples from a population. However, individual trees are frequently widely dispersed, with some occurring in inaccessible regions. Therefore, random sampling may not be possible in all areas because of such practical difficulties.

An even greater problem results from the very large number of accessions obtained by random sampling, which necessitates intensive collecting activity and, subsequently, a large area for the conservation of living specimens. It is estimated, for example, that approximately three million almond trees occur in Turkey, a large majority of which is indigenous and has originated from seed. A very considerable effort would be required to sample at random one tree in 1000 although even this sampling intensity, perhaps inadequate in some areas and populations, would yield a total of 3000 accessions. Conserving these accessions as single, unreplicated trees at a seven by seven metre spacing would require an area of 15 hectares. Similarly, the total area required for a germplasm collection of wild species, ecotypes and cultivars of pistachio has been estimated at 21 hectares (Maggs, 1973). Restraints of such magnitude in collection and conservation are considerable, even though the problem might be solved by utilising dwarfing rootstocks and/or multiple grafting techniques, coupled with adequate virus control and maintenance programmes. Thus, although no satisfactory evaluation could be undertaken, it would be technically possible to graft as many

125

as 200 scions onto one rootstock as a 'genetic resources holding operation'.

The alternative is biased sampling, which may be inevitable when collectors are few in number and their time is limited. Although in practical terms the expediency of biased sampling may be acceptable, an additional problem to resolve is which particular phenotypic characters to select as a sampling criterion. Relatively few studies of infra-specific variation have been undertaken in ornamental and tree fruit, compared with forestry species. However, quite a wide range of differences in morphological and physiological characters has neverthe-less been reported. These include pattern, rate and habit of growth, (Vaartaja, 1959; Smithberg & Weiser, 1968; Fordham, 1971), seed source in relation to quality and germination requirement (Wilcox, 1968; Flint, 1970) and vigour in potential plum rootstocks (Beakbane, 1969*a*). Additionally, other characters are important in fruit species, for example, root and shoot anatomy (Beakbane, 1953; 1969*b*), leaf structure (Beakbane, 1967) and ovule longevity in relation to effective pollination period (Williams, 1966).

However, many of these characters are not easily detectable in the field, and laboratory techniques are required for their precise evaluation. Therefore, *only* the accessions with subjectively-assessed, empirical phenotypic characters and which are relevant to current requirements are obtained by biased sampling. This procedure is limited, however, because the present objectives of breeding programmes and production systems are profoundly subject to future change. The demand for pre-cocious-flowered and easily propagated germplasm, for example, would be greatly stimulated if very intensive, high-density planting proved commercially feasible as a fruit production system (Hudson, 1971*a*, *b*).

Collection procedures

Two categories of fruit tree germplasm merit particular emphasis; these are, firstly, species existing as wild trees, and secondly, locally developed primitive cultivars (Zagaja, 1970). Collection and evaluation data in Turkey for almond and walnut in the first category and for quince in the second are given below, to illustrate some of the problems associated with their collection.

Collection of wild species

Fifty nuts were selected at random from two different mixed samples of almonds obtained from wild trees; each nut and kernel was weighed individually and the shelling percentage was calculated. These values are shown as frequency distribution diagrams in Figs 9.1a and 9.2a. Variation was considerable; from 0.4 to 1.6 g and from 0.4 to 1.4 g in kernel weight (Fig. 9.1a) and, in shelling percentage (Fig. 9.2a), from 9 to 53 per cent and from 9 to 37 per cent.

Individual trees in the Çeşme peninsula west of Izmir showed considerable variation in flowering time (Zagaja, 1970). These were surveyed and evaluated, based on the following characters: lateness of flowering (for frost-avoidance), tree yield, and health status; also non-bitterness and thin-shelled quality of the nuts harvested from each tree. Twenty almonds, randomly selected from the harvested crop of each of 59 pre-selected trees, were measured and weighed and, for each tree, the mean values of kernel weight and shelling percentage were determined. The frequency distribution diagrams of these values are shown in Figs 9.1b and 9.2b. The range of variation was from 0.6 to 2.0 g (Fig. 9.1b) and from 17 to 53 per cent (Fig. 9.2b) for kernel weight and shelling percentage respectively. Twenty-five trees were selected (cross-hatched) from which budsticks were collected for vegetative propagation and in order to conserve several plants of each of these genotypes.

However, to evaluate and collect these accessions satisfactorily, the various sites had to be visited at least three times. First, in early spring when the time and duration of flowering were determined, these field records being confirmed the following year. A second series of visits was made in summer co-incident with harvest, at which time tree yield was determined, nut quality and non-bitterness were initially assessed and nut samples were obtained for laboratory evaluation. A third visit was made to collect budsticks from selected trees. If these budsticks had been collected earlier, possibly concomitantly with nut evaluation, more time would have been taken in doing both operations simultaneously, and then in propagating an excessive number of accessions, not all of which could have been conserved subsequently.

It may be concluded that obtaining mixed or market samples may be useful for the initial, coarse-grid sampling of the range of variability in an area. Subsequently, if individual genotypes are then sampled in a non-random manner, several visits to the collecting site may be required which, together with laboratory studies, enables individual trees to be

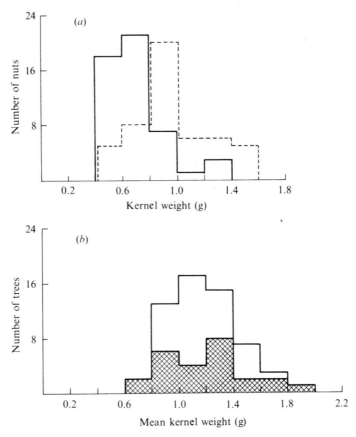

Fig. 9.1. Frequency distribution of almond kernel weight in wild populations. (*a*) Single nuts from two market samples. Solid line, market sample from Gördes; dashed line, mixed sample from Government farm. (*b*) Mean values from 59 individually sampled trees. Cross-hatching represents the 25 trees finally selected.

identified and then conserved in a programme of, generally, not less than two years' duration.

Considerable variation was also evident in the size, shape and quality of walnuts collected from several areas of western Turkey. This variation is expressed quantitatively in Fig. 9.3 in terms of the mean values, for 96 individual trees, of kernel weight (Fig. 9.3*a*) and shelling percentage (Fig. 9.3*b*). These values varied, respectively, from 1.5 to 8.5 g and from 30 to 58 per cent.

Superior trees (cross-hatched) with a large yield of high-quality nuts may be identified from these data. However, using traditional summer patch budding techniques, walnuts are much more difficult to propagate

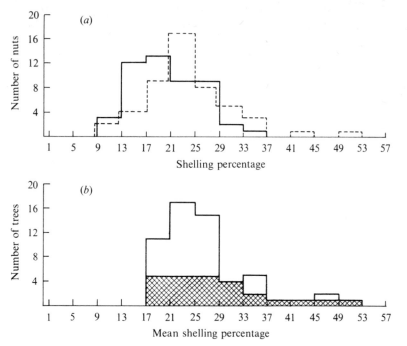

Fig. 9.2. Frequency distribution of shelling percentage (kernel weight/nut weight × 100) of almonds. (*a*) Single nuts from two market samples. Solid line, market sample from Gördes; dashed line, mixed sample from Government farm. (*b*) Mean values of samples from 59 wild trees. Cross-hatching represents the 25 trees finally selected.

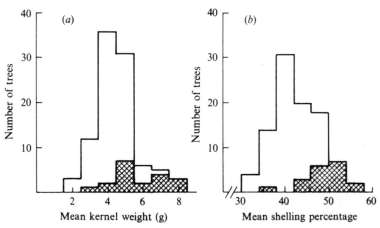

Fig. 9.3. Frequency distribution of (*a*) mean values of kernel weight and (*b*) shelling percentage of walnuts from 96 individually sampled wild trees. Cross-hatching represents 25 selected trees.

vegetatively than are almonds. When another expedition is required to collect dormant shoots as graftwood, this extends the duration of such a programme. Therefore, although field evaluation provides data as criteria for biased sampling, it involves a sustained, long-term programme of several years' duration. More intensive sampling and a detailed study of phenotypic characters and tree performance may reveal clinal or discontinuous variation in tree populations as additional information of value in planning future exploration routes.

Can a compromise be achieved between random sampling, yielding an excessive number of genotypes, and biased sampling by which some valuable genetic material may not be recovered? Perhaps, in future, two complementary sampling procedures might be considered. Coarse-grid sampling on a large scale would ensure that some collections were made from as many populations as proved practicable. The total number of accessions collected must, with our present inadequate techniques and facilities, be limited by the amount of exploration undertaken and because of a limit in the number of accessions that can, at present, be satisfactorily propagated and conserved. More intensive, fine-grid sampling could be restricted, firstly to specific areas, for example gene microcentres and populations in which variation is considerable (Harlan, 1951); secondly, to the biased sampling of populations, even single plants, which have easily detectable characters sought by the plant breeder.

Collection of primitive cultivars

High-yielding, modern cultivars increasingly threaten to replace, in Turkey and elsewhere, climatically well-adapted primitive cultivars. Collection of these locally-developed cultivars merits a high priority rating, but they may be confused with other cultivars, bred elsewhere and introduced into the country in the distant or recent past, especially when the latter have acquired local cultivar names. However, few modern cultivars of quince and pomegranate (*Punica granatum* L.) have been bred elsewhere or introduced into Turkey where these fruits are still quite widely grown.

Twenty-seven accessions of 13 supposedly different quince cultivars were collected from orchard and village sites in western Turkey. Their pomological characters were analysed and some confusion in cultivar nomenclature was evident. Eight cultivars, all of which were grown in small, discrete areas, were clearly identified. In contrast, the nomenclature of the five other more popular and widely-grown cultivars was more complex and confused. Sometimes the same genotype had acquired different cultivar names in certain areas while, conversely, quite different taxa had assumed an identical cultivar name (Sykes, 1972).

Another potential problem is occasionally encountered when primitive fruit cultivars are collected. Some cultivars are highly esteemed by their owners, whose generosity and help in Turkey is invariably overwhelming and abundant. Such very willing co-operation would be endangered if it was suspected that these cultivars might be exclusively exploited elsewhere for private commercial advantage.

If a clone comprises genetically identical progeny which originate exclusively by vegetative propagation from a single plant, collecting a minimum number of plants, theoretically only one, would recover all the genes in the plants of that clone. This definition, with the exception of mutation-induced chimeras and budsports, presupposes absolute genetical uniformity among individuals of the clone. However, primitive cultivars have been grown and vegetatively propagated for many decades, even centuries. The mutations and virus diseases that some clones or individual plants could have accumulated, both in type and amount, may be substantial. The different genotype–environment interactions and the complex quantitative effects of infection by various viruses may account for some of the within-clone differences that have been reported.

These include several physiological and other characters, such as self-compatability and growth differences in apple (Campbell & Jetzer, 1972; Stott & Campbell, 1971), differences of graft-compatability in time and between clones in pear (Coles & Campbell, 1965; Posnette & Cropley, 1962) and in the propagation ability of several species, including Leyland Cypress, × *Cupressocyparis leylandii* (Jackson and Dallimore) Dallimore (Matthews, Waller & Potts, 1960) and apple rootstocks, due to virus infection (Howard, 1972).

Whether due to genetic or other factors such as virus infection, more variation is evident in the clones of some cultivars than has previously been assumed. Accessions of each cultivar, preferably sampled from the entire area of their distribution, should be sufficient in number to allow for any possible intra-clone differences. The taxonomic identity of each cultivar should be verified by comparing the characters of several accessions grown under uniform conditions in a 'clone archive' or 'living collection'.

Collection and propagation methods

The earliest recorded plant introduction was the frankincense tree (*Boswellia carteri* Birdw.) brought as living plants in pots from East Africa to Egypt almost 35 centuries ago. Of the many woody species since introduced into cultivation most of them were, until about 150 years ago, transported as living plants. Seed was, somewhat surprisingly, not usually collected in early plant exploration and, as a practical method of introducing new species, seed collection was initially recommended only in the early nineteenth century (Whittle, 1970).

In addition to rooted living plants, seed, pollen, graftwood, budwood and various types of cuttings are all now used in plant introduction. Certainly, compared with seed, the critical conditions needed for the transport and storage of vegetative material and quarantine requirements present a formidable array of collection and introduction problems (Pansiot, 1962). Several conference and collectors' reports have noted these difficulties but, with few exceptions, these have included limited reference either to sampling intensity or to the type of material collected (Whyte, 1958; Barton, 1968; Brooks, 1968).

Seed collection

When seed is collected the characters of the trees in the population, especially those of the seed source and surrounding trees as potential pollen parents, should be carefully recorded. The precise location and its agro-ecological characters should also be noted because the re-collection of seed may be required at a later date, for example from a specific seed source – either as a single tree or a small population – which yields virus-free and relatively uniform seedling rootstocks or, alternatively, from a source which yields valuable genetic variability.

However, seed viability of many tree species is of limited duration. For example, the longevity of most tree fruit seeds rarely exceeds three years although, following low temperature storage in sealed bottles, 90 per cent germination has been reported in such seed stored for seven years (Solovieva, 1966).

Naturally, if it is required to conserve a specific genotype in a heterozygous population, this cannot be done by the collection of seeds, despite the advantage due to a small volume of material collected. Seed and pollen collection may have only a limited practicality at present for direct introduction or long-term conservation, even though the range

132

of genetic variability conserved will be very much greater than in vegetative collections. Obviously, what is urgently needed here is research on the long-term conservation of seeds which at present can be kept for only very short periods (see Roberts, Chapter 22).

Vegetative material

A range of alternative propagation materials is available for vegetative propagation, which ensures that the genetic integrity of a particular clone or genotype is preserved. Naturally-layered shoots and rooted suckers are easily collected and established but usually they occur only on the plants of wild species. In the collection of cultivars, alternatives are available, such as the various types of cuttings, or shoots used as scions and grafted or budded onto appropriate rootstocks.

A leafy, non-lignified cutting is very vulnerable to loss after collection but, paradoxically, in many species it is the type most easily and quickly rooted if wilting is prevented by automatically-controlled mist application. However, portable mist units are too small to be usefully included in expedition equipment, which restricts the collection of immature shoot cuttings to the area close to a centre where mist facilities are available.

Fully mature dormant shoots as 'hardwood' cuttings are less liable to desiccation and have a longer natural viability, even under adverse conditions, than immature leafy cuttings. Low temperature has given some success in storing dormant shoot cuttings of various ornamental species for several months (Snyder & Hess, 1956; Flint & McGuire, 1962), and dormant cuttings of fruit rootstocks (Howard & Garner, 1966) and chestnut scionwood (Jaynes, 1969) for a one-year period. Preliminary studies suggest that reduced atmospheric pressure (hypobaric storage) may also have a future application to the storage of cuttings (Burg & Burg, 1966). At present, two factors limit the usefulness of this type of cutting. First, many species and particularly rosaceous trees are not readily propagated from 'hardwood' cuttings, a proportion of which usually do not survive; second, the climatic unsuitability of the period from autumn to early spring when dormant shoots must be obtained. Often it is difficult to undertake expeditions during the winter months to collect dormant shoots, for use either as cuttings or dormant scions, and especially when this necessitates sampling trees in remote areas.

In contrast, small leafless budsticks offer many advantages. They may be collected from deciduous trees throughout a long summer period and their small size facilitates easy and inexpensive transport by air. Bud-sticks survive for at least two weeks after collection if, during an expedition, they are wrapped in plastic and kept moist and cool in an ice-box. When budding or grafting is practised, however, an additional dimension is imposed on collection programmes. Appropriate rootstocks of different species must be available in sufficient quantity at a well-organised nursery staffed with personnel competent in propagation procedures. Most seedlings are free of virus and they may be preferred as rootstocks to more uniform clonal material. The nursery site should be carefully selected since the growth characters, phenotypic expression and performance of collected genotypes may be affected by the type of rootstock used and the local climatic conditions.

Conclusions

Priorities in collecting must be established between alternative areas and, for different genera and species, between cultigens and wild material. Frequently this presents a problem when facilities are limited. Experience in Turkey would suggest that emphasis should be given to primitive cultivars, as the rate of their replacement by modern cultivars generally exceeds the rate of loss of indigenous wild species.

The collection of primitive cultivars may pose certain problems, however, related to the origin of the individual tree from which an accession is collected. For example, has it been perpetuated exclusively by vegetative propagation or may it have originated, at some time in the distant or recent past, as a seedling or as a mutation of the original primitive cultivar? Its origin may, in some instances, be confirmed if data on the age and history of the tree are available or if there is an old graft union at the base of the tree. However, even if clonal origin may be assumed, there may be morphological or physiological differences between clones of the same cultivar. The taxonomic validation of a sufficient number of accessions of the same cultivar collected from a wide area will help to resolve many apparent anomalies in cultivar nomenclature.

When wild species are collected, perhaps a realistic compromise could be achieved between random and biased sampling. Large scale, coarse-grid sampling at random of populations could be practised to obtain a general concept of the variation pattern. This could then be supplemented, where appropriate in particular areas or populations, with detailed random sampling together with biased sampling of specific and morphologically distinct genotypes.

134

Trees have a long life cycle and evolution proceeds slowly in tree populations. Therefore, the concept of natural reserves should also be considered for their conservation *in situ*. This would be particularly applicable to areas that are sparsely populated, have a high plant population density and which, as protected areas, could be satisfactorily and permanently maintained.

Both collecting techniques and propagation facilities have been developed or greatly improved during the last two decades. They now include air transport and the use of mist systems, plastics and anti-desiccants for moisture retention, and refrigeration for the short-term preservation of vegetative material. However, despite such recent advances, the optimum type of vegetative material to collect has, in many species, not yet been clearly determined. Easily and quickly collected, ideally it should be of minimal size, for example, single bud cuttings, buds or budstocks. Such a vegetative unit should have the ability to survive short-term transportation and be responsive to relatively simple and reliable propagation methods by which the collected material could be conserved.

Tree genetic resources are the essential raw material on which future advances depend. In their exploration, they must be surveyed and evaluated more thoroughly and, in most areas, collected more widely than is being done at present. This is a comprehensive and formidable challenge in tree crops and success is dependent upon the combined efforts of the collector, taxonomist, propagator, horticulturist and phytopathologist.

References

Barton, D. W. (1968). Horticultural germ plasm: its exploration and preservation. *Hort. Sci.*, **3**, 241–3.

Beakbane, A. B. (1953). Anatomical structure in relation to rootstock behaviour. *Proc. Int. hort. Cong., Lond.*, **1**, 152–8.

Beakbane, A. B. (1967). A relationship between leaf structure and growth potential in apples. *Ann. appl. Biol.*, **60**, 67–76.

Beakbane, A. B. (1969*a*). A new series of potential plum rootstocks. *Report of East Malling Research Station for 1968* (1969), pp. 81–3.

Beakbane, A. B. (1969*b*). Relationship between structure and adventitious rooting. *Combined Proc., Int. Plant Propagators' Society, Wooster, Ohio*, **19**, 192–201.

Brooks, H. J. (1968). Collecting wild fruits in the USSR. *Hort. Sci.*, **3**, 258–60.

Bunting, A. H. & Kuckuck, H. (1970). Ecological and agronomic studies related to plant exploration. In *Genetic Resources in Plants – their*

Exploration and Conservation (eds O. H. Frankel and E. Bennett), *IBP Handbook*, no. **11**, pp. 181–8. Blackwell, Oxford and Edinburgh.

Burg, S. P. & Burg, E. A. (1966). Fruit storage at subatmospheric pressures. *Science*, **153**, 314–15.

Campbell, A. I. & Jetzer, P. (1972). Clonal variations in apple. *Report of Long Ashton Research Station for 1971* (1972), pp. 37–8.

Coles, J. S. & Campbell, A. I. (1965). Compatability of 'Bristol Cross' pear with some quince rootstocks. *Report of Long Ashton Research Station for 1964* (1965), pp. 113–6.

Daus, S. J. & Lauer, D. T. (1971). SLAR imagery for evaluating wildland vegetational resources. *Proceedings of the American Society of Photogrammetry Technical Sessions and Symposium on Computational Photogrammetry*, pp. 386–92. San Francisco, California.

Davies, P. H. (ed.) (1966). *Flora of Turkey and the East Aegean Isles*. Edinburgh Univ. Press, Edinburgh.

Erinc, S. (1949). The climates of Turkey according to Thornthwaites's Classification. *Ann. Assoc. Am. Geog.*, **39**, 26–45.

Flint, H. L. (1970). Importance of seed source to propagation. *Combined Proc. Int. Plant Propagators' Society, Wooster, Ohio*, **20**, 171–9.

Flint, H. L. & McGuire, J. J. (1962). Response of rooted cuttings of several woody ornamental species to overwinter storage. *Proc. Am. Soc. hort. Sci.*, **80**, 625–9.

Fordham, A. J. (1971). Canadian hemlock variants and their propagation. *Combined Proc., Int. Plant Propagators' Society, Wooster, Ohio*, **21**, 470–5.

Harlan, J. R. (1951). Anatomy of gene centers. *Am. Natur.*, **85**, no. 821, 97–103.

Howard, B. H. & Garner, R. J. (1966). Prolonged cold storage of rooted and unrooted hardwood cuttings of apple, pear and quince rootstocks. *Report of East Malling Research Station for 1965* (1966), pp. 80–2.

Howard, B. H. (1972). Depressing effects of virus infection on adventitious root production in apple hardwood cuttings. *J. Hort. Sci.*, **47**, 255–8.

Howard, J. G. (1970). Multiband concepts of forested land units. *Proceedings, 3rd Int. Symp. Photo-interpretation ISP* (Commission VII), vol. I, pp. 281–317. Dresden.

Hudson, J. P. (1971*a*). Meadow orchards. *Agriculture*, **78**, 157–60.

Hudson, J. P. (1971*b*). Horticulture in 2000 A.D. In *Potential Crop Production* (eds P. F. Wareing and J. P. Cooper), pp. 187–201. Heinemann, London.

Jaynes, R. A. (1969). Long-term storage of chestnut seed and scion wood. *60th Ann. Rep. Northern Nut Growers Association*, pp. 38–42. Carbondale, Illinois.

Lauer, D. T., Benson, A. S. & Hay, C. M. (1971). Multiband photography – Forestry and agricultural applications. *Proceedings of the American Society of Photogrammetry Technical Sessions and Symposium on Computational Photogrammetry*, pp. 531–53. San Francisco, California.

Maggs, D. H. (1973). Genetic resources in pistachio. *Plant Genetic Resources Newsletter*, **29**, 7–15.

Matthews, J. D., Waller, A. J. & Potts, K. R. (1960). Propagation of Leyland Cypress from cuttings. *Quart. J. Forest.*, **54**, 127–40.

Pansiot, F. P. (1962). Exploration, récolte, introduction et échange de plantes en horticulture. In *Advances in Horticultural Science and their Applications* (ed. J. C. Garnaud), vol. 3, pp. 120–40; *Proc. 15th Int. hort. Cong.*, Nice, France, 1958.

Posnette, A. F. & Cropley, R. (1962). Further studies on a selection of 'Williams Bon Chretein' pear compatible with quince A rootstocks. *J. Hort. Sci.*, **37**, 291–4.

Rohde, W. G. & Olson, C. E. Jnr. (1972). Multispectral sensing of forest tree species. *Photogrammetric Engineering, Hanover*, **38**, 1209–15.

Smithberg, M. H. & Weiser, C. J. (1968). Patterns of variation among climatic races of red osier dogwood. *Ecology*, **49**, 495–505.

Snyder, W. E. & Hess, C. E. (1956). Low temperature storage of rooted cuttings of nursery crops. *Proc. Am. Soc. hort. Sci.*, **67**, 545–8.

Solovieva, M. A. (1966). Long term storage of fruit seeds. *Proc. 17th Int. hort. Cong.*, Maryland, **1**, p. 238.

Stott, K. G. & Campbell, A. I. (1971). Variation in self-compatability in Cox clones. *Report of Long Ashton Research Station for 1970* (1971), p. 23.

Sykes, J. T. (1972). A description of some quince cultivars from western Turkey. *Econ. Bot.*, **26**, 21–31.

Vaartaja, O. (1959). Evidence of photoperiodic ecotypes in trees. *Ecological Monographs, Durham, N.C.*, **29**, 91–111.

Whittle, T. (1970). *The Plant Hunters*. Chilton Book Co., Philadelphia, New York and London.

Whyte, R. O. (1958). *Plant exploration, collection and introduction*. FAO Agricultural Study no. **41**, FAO, Rome.

Wilcox, J. R. (1968). Sweetgum stratification requirements related to winter climate at seed source. *Forest Sci.*, **14**, 16–19.

Williams, R. R. (1966). Pollination studies in fruit trees. III. The effective pollination period of some apple and pear varieties. *Report of Long Ashton Research Station for 1965* (1966), pp. 136–8.

Yost, E. & Wenderoth, S. (1971). Multispectral colour for agriculture and forestry. *Photogrammetric Engineering, Hanover*, **37**, 590–606.

Zagaja, S. W. (1970). Temperate zone tree fruits. In *Genetic Resources in Plants – their Exploration and Conservation* (eds O. H. Frankel and E. Bennett), *IBP Handbook*, no. **11**, pp. 327–33. Blackwell, Oxford and Edinburgh.

Part IIC. National and regional exploration activities

---

10. Recent plant exploration in the USA

H. L. HYLAND

The history and general procedures of plant introduction and international exchange, as related to the United States agricultural research programs, have been reported on other occasions (see Hodge & Erlanson, 1955; Hyland, 1963; Creech *et al.*, 1971). Throughout the years since 1898, when permanent inventory records of introduced germplasm were initiated, there have been various changes in methods, priorities, recording procedures, quarantine requirements, evaluation, and maintenance of the germplasm involved.

Foreign exploration

Before World War II, field explorations were generally oriented toward collecting those species needed to support rather broad areas of research. Since that period research programs have become more sophisticated and limited in scope. This has resulted in explorations with more definitive purposes led by specialists, rather than by general collectors such as Meyer, Fairchild, Koelz, and others who collected almost everything of potential interest. This approach was most productive. The following examples indicate the diversity of explorations in recent years.

Major significance is attached to the three phases of bean collecting in Latin America from 1965 to 1968 by H. S. Gentry (USDA), with emphasis on broadening the base of available germplasm by screening for specific diseases. Halo blight was a major concern at that time among bean producers. Approximately 900 accessions were obtained, representing gene combinations not hitherto acquired. Fifteen accessions from a remote region of Mexico should prove especially valuable in clarifying the relationship between wild and cultivated *Phaseolus* (see Gentry, 1969).

Peanut breeders have become increasingly concerned over the recent loss of primitive endemic forms in South America. The United States has conducted three explorations since 1961, the last in 1968 by R. O. Hammons and W. R. Langford (USDA–Georgia), concentrating on limited areas of Argentina and Brazil. The excellent co-operation received from the Argentinian specialist Dr A. Krapovickas resulted in 370 accessions, representing 27 recognised species, four known previously only from herbarium specimens. A substantial part of this germplasm had not been available to US and world breeders previously. Especially important has been the success in clonally propagating many types normally sparse in fruiting.

In 1969 the pea growing industry of the western US became alarmed over the damage from a root-rot organism, *Fusarium solani* forma *pisi*. Very little resistance was present in the germplasm held at that time. The decision was made to collect wild forms and old village varieties from areas in southern Italy, Greece and western Turkey. The field work was done by H. S. Gentry. This phase was conducted in 1969, followed by a second phase in Ethiopia in 1970. While the total number of accessions was rather limited (200), the samples of *Pisum sativum* var. *abyssinicum* from remote areas of Ethiopia furnished the only lead in the search for resistance to this damaging root-rot (Gentry, 1971*a*, *b*).

Ornamental horticulture has greatly profited from plant-exploration activities. Some of the best records, giving descriptions, distribution, and evaluation data are found today with the older arboreta in the United States. The importance of continued field exploration is significantly illustrated by a series of expeditions under a co-operative USDA–Longwood Foundation agreement initiated in 1955. Many of these were conducted by John L. Creech (USDA) – the last two in Taiwan in 1967 and in the USSR in 1971.

The Taiwan exploration was undertaken in co-operation with the Botany Department, National Taiwan University, and the Joint Commission for Rural Reconstruction. Collecting areas in Taiwan included Mt Morrison (3997 m) and several lesser mountains. At the time of collection, the vegetation was at the peak of seed production, and a broad array of species was collected. A new road over Hohuan Pass opened up areas previously unavailable for collecting. Some species were noted to be rare, such as the epiphytic rhododendron, *R. hawakami*. Serious efforts are succeeding in re-establishing stands of the unique conifer, *Taiwania cryptomerioides*. Among the collections

140

were seeds of another rare rhododendron, *R. pseudochrysanthum*, gathered at high elevations on Mt Morrison.

The exploration in the USSR was undertaken with excellent co-operation provided by the N. I. Vavilov Institute of Plant Industry and several botanic gardens. This included the Leningrad region, the Crimea, and central and eastern Siberia, with emphasis on the regions near Novosibirsk, Irkutsk, and Yakutsk. Although the collections were principally of ornamental species, several collections of *Medicago falcata* from extremely cold areas were made, and a number of grasses were gathered. The co-operation of the Soviet scientists made this an especially rewarding exploration, with travel farther into Siberia than had been done by any recent foreign collector. Several exchanges of plant materials have subsequently occurred as a result of this exploration. These collections will be of particular interest to arboreta and plantsmen in the cold regions of the northern parts of the United States.

In 1970, H. F. Winters and J. J. Higgins (USDA) conducted another of the USDA–Longwood expeditions in the Territory of Papua and New Guinea. Most of the 840 collections were from wild stands, permitting a good sampling of the variability within species. Where possible, cultivated forms were collected when different from those presently known in the USA. Special emphasis was placed on *Rhododendron*, resulting in 137 accessions, 101 identified as belonging to 32 species. The remainder are still under classification. Another highly interesting group included the large-flowered and variegated-foliage forms of *Impatiens* that are presently contributing greatly to breeding and improvement programs in the United States. An interesting comment from the explorers' notes relates to their concern over the loss of useful germplasm even in the remote jungle areas of New Guinea. They state, 'Some of the desired species are becoming extremely difficult to find'.

Many international and national agencies have been involved in stressing the need for improvement in food crops adapted to developing countries. Among these crops, edible yams rank high in consideration. In 1971, F. W. Martin (USDA–Puerto Rico), with FAO financial support, undertook a survey and collecting trip to western African countries. He collected 380 accessions, representing seven species, which are under evaluation for foliar traits, tuber morphology, and cooking and processing attributes. With funds provided by USDA and the US Agency for International Development, Dr Martin has made considerable progress in selecting desirable types for further testing in tropical countries. Limited stocks are already under observation outside the USA. A major problem has been the prevalence of viruses in much of the germplasm. Research is under way to determine nutritional qualities. Tentative plans call for future exploration in south-east Asia.

Tropical grasses and legumes are not widely used in the continental United States, but there is need for compatible pasture combinations in the humid sectors of the south-eastern States, as well as for Hawaii and Puerto Rico. A. J. Oakes (USDA) has undertaken two explorations to South Africa to locate useful species. In 1964, emphasis was placed on *Digitaria* germplasm having winter-hardiness and disease resistance, especially pangola stunt virus, prevalent at that time in pangola grass. Dr Oakes collected 316 accessions of 24 species. Based on preliminary performance and diversity of the 1964 material, he undertook a second phase in 1971 to increase the genetic diversity, especially in *Digitaria*. Certain areas of Lesotho, Mozambique, South Africa, and Swaziland were covered. The results were approximately 950 accessions from 67 grass and 33 legume genera, and 311 additional digitarias, which provide an excellent germplasm base for future research. Another species, *Hemarthria altissima*, first collected in 1964, has indicated potential as a good warm-season grass under good management, and 30 more wild accessions were collected during the 1971 trip. Many of these African species are propagated clonally, which raises the eternal problem of how to maintain such germplasm over long periods.

Introduced wheat-grasses (*Aegilops, Agropyron*, etc.) and wild rye-grasses (*Elymus*) play a prominent role in the agriculture of the western United States. Much of the breeding and cytogenetic studies with these grasses was related to germplasm obtained in the 1930s. To broaden the base for future research, our most recent exploration was to Iran and north-eastern Turkey during the late summer of 1972. D. D. Dewey (USDA–Utah), a specialist in the cytogenetics of *Agropyron*, was accompanied by J. S. Schwendiman, Soil Conservation Service (USDA–Washington). Their collecting concentrated on genera of the tribe Hordeae: *Aegilops, Agropyron*, and *Hordeum*. Dryland grass and legume species were also collected. About 1700 accessions were obtained, representing 135 genera and 255 species. Excellent field data were recorded for *Aegilops, Agropyron*, and *Hordeum*. Certain hybrid complexes, such as *Agropyron cristatum-elongatiforme*, *A. elongatiforme-repens*, and *H. glaucum-leporinum* will require cytogenetic studies as an aid for taxonomic classification. All agropyrons have been planted under both greenhouse and field conditions for immediate observation. The remaining collections from this Iran–Turkey trip await processing, with field planting likely in early 1974.

In addition to the specific explorations referred to above, the Germplasm Resources Laboratory at Beltsville, Maryland, has had excellent co-operation from other individuals and agencies in documenting introduced plant and seed stocks. Since 1950, researchers on sabbatical or similar assignments abroad have shared their collections to permit the formation of permanent inventory records, and for storage at the National Seed Storage Laboratory, Fort Collins, Colorado. An example of this collaboration is the trip of the soybean geneticist R. L. Bernard (USDA–Illinois) to Japan and Korea in late 1972 to collect exotic germplasm. He obtained about 150 accessions of wild forms which will greatly expand the genetic diversity available to soybean breeders. Technical assistance programs, Public Law 480 projects, and international institutes are other sources of germplasm obtained for the use of US breeders or for preservation.

Explorations under consideration are for the collection of indigenous cultivars and wild relatives of tomato in Latin American countries, oilseeds in countries surrounding the Mediterranean, turf grasses in Europe, and a survey for available germplasm of small fruits, also in European countries. Priorities will be determined in accordance with funding and collaboration available.

Domestic exploration

The establishment of Regional Plant Introduction Stations, starting in 1947, provided opportunities for plant collecting within a specific group of States having a common research interest in native or naturalised species. These explorations have contributed to new crop releases, particularly in the category of range plants. The basic concept for initiating domestic exploration was to try to save endangered species whenever researchers had a specific need for collecting them, and at the same time to place portions of the seed accessions under long-term storage. While some success has resulted, failures have occurred with clonally propagated species.

Among the recent noteworthy explorations is the collection of *Vaccinium* species of the eastern and south-western States, brought together by G. J. Galletta, North Carolina State University. This project terminated in 1969, after bringing together in one location 220

6-2

143

accessions representing more than 20 species. The blueberry industry along the Atlantic coast requires improved varieties, with disease resistance as a significant objective. The presence of diploids, tetraploids, and hexaploids among the collections not only illustrates problems of taxonomic classification but also the vast potential for breeding improved types.

In 1970, Alaskan breeders made extensive collections of *Poa* and *Festuca* to determine ranges of cold-hardiness, seed production, and related factors. Earlier collections had demonstrated the value of indigenous types in contributing to the release of new varieties. The latest collections have been grown in other States and show desirable growth characteristics, disease resistance, and a fairly wide range of adaptation outside Alaska. Exploration is in progress for native warm-season grasses in the northern Great Plains States. Many ecotypes of *Andropogon gerardi, Panicum virgatum* and *Sorghastrum nutans* have already been lost because of cultivation of lands formerly used for range. Tall, leafy, agronomically desirable plants have been located in lowland roadside ditches, and these are further endangered by present-day road-making methods. Some 715 accessions are under observation, and several are in advanced stages of evaluation.

The pecan is one of the few native North American plants that have become commercial crops. Wild stands have disappeared at a very rapid rate within the past 30 years. A proposal has been approved for collecting native species of *Carya illinoensis* from the northern limits of the natural range throughout the North Central States. Emphasis will be placed on locating germplasm for improvement of present commercial varieties grown mostly in the southern USA, so that areas of production can be pushed northward. Here again, we shall have to face the problem of maintaining repositories of clonally propagated germplasm.

References

Creech, J. L., *et al.* (1971). *The National Program for Conservation of Crop Germ-plasm*. Joint publication of the Agricultural Research Service, USDA, and Cooperating State Agricultural Experiment Stations, 73 pp. USDA, Beltsville.

Gentry, H. S. (1969). Origin of the common bean, *Phaseolus vulgaris*. *Econ. Bot.*, **23**, 55–69.

Gentry, H. S. (1971*a*). *Pisum* resources – a preliminary survey. *Plant Genetic Resources Newsletter*, no. **25**, 3–13.

Gentry, H. S. (1971*b*). Pea picking in Ethiopia. *Plant Genetic Resources Newsletter*, no. **26**, 20–4.

Hodge, W. H. & Erlanson, C. O. (1955). Plant introduction as a federal service to agriculture. *Adv. Agron.*, **7**, 189–211.
Hyland, H. L. (1963). Plant introduction objectives and procedures. *Genetica Agraria*, **17**, 470–82.

Appendix

Germplasm collections in the USA

Two major steps have been taken by USDA to meet the problem of conserving valuable plant germplasm. The first was the establishment of four Regional Plant Introduction Stations (1947–52) and the Inter-Regional Potato Introduction Station (IR-1) in 1949. Each of these stations increases and maintains active stocks of those species needed for current research programs, with the additional responsibility of placing basic stocks under long-term storage.

The latter is accomplished through the second step implemented in 1958 by the Department in establishing the National Seed Storage Laboratory, Fort Collins, Colorado.

Before the establishment of the Regional Stations and NSSL, about 165000 accessions were recorded in our inventory. However, estimates have ranged from 5 to 10 per cent of that total as having been maintained by researchers, either as seed held under relatively poor storage conditions or in living repositories such as large arboreta. Since 1948 an additional 200000 accessions have been recorded, and the degree of survival has been estimated as from 60 to 75 per cent. Some indication of this conservation is represented by listing a few crop groups presently held by the facilities referred to above.

Regional Plant Introduction Stations

(North-Eastern) Geneva, New York		(North-Central) Ames, Iowa		(Southern) Experiment, Ga.		(Western) Pullman, Wash.	
Peas	1500	Tomato	3579	Peanuts	3816	Beans	6000
Perennial		Corn	2305	Sorghum	3423	Safflower	1317
clovers	550	Alfalfa	969	Peppers	1784	Lentils	500
Trefoils	341	Bromegrass	617	Canteloupe	1626	Lettuce	493
Onions	320	Cucumber	596	Cowpeas	1212	Cabbage	263
Timothy	225						

The Potato Introduction Station, in Wisconsin, now holds more than 2000 tuber-bearing solanums, representing 92 species. True seed of 70 per cent of these has been placed in the NSSL, Fort Collins.

The latest report from NSSL shows 82629 accessions in storage. These are broken down into the following crop categories. It should be noted that these totals account, also, for many of the groups recorded above with the Regional and Potato Introduction Stations.

Small grain	32806	Corn	1383
Vegetable	14480	Tobacco	1025
Oilseed	11310	Ornamental	514
Sorghum	9676	Chemurgic	339
Forage	5115	Sugar	117
Genetic	4102	Strategic	11
Cotton	1751		
			82629

11. Plant exploration in the USSR

D. D. BREZHNEV

The major task of the N. I. Vavilov Institute, which is the largest of its kind in the USSR and employs some 2000 persons, is the exploration and utilisation of world plant genetic resources for the development of new varieties of crop plants to meet modern agricultural requirements. This work is recognised by the government of the USSR as one of the greatest national importance.

The Institute, which comprises some 25 experimental stations in various parts of the USSR as well as the headquarters at Leningrad, introduces from 12000 to 16000 seed and vegetative samples annually from the Soviet Union and foreign countries. As an example, in 1972 a total of 16757 samples were introduced, of which 1746 were from the USSR and 15011 from abroad.

The world collection of plant resources at the Institute now contains over 200000 accessions of seeds, tubers, bulbs and living plants of wild species, primitive forms, cultivars, hybrids and breeding lines of wheat, rye, barley, maize, millet, rice, pulse crops, industrial crops, vegetables, cucurbits, fruits, nuts, subtropical crops, flowers and grapes.

The Institute collections belong to the plant scientists and breeders of the whole world. We place our collections with great pleasure and without any restrictions at the disposal of breeders of all countries. We already send from 6000 to 8000 samples to foreign countries annually, and are willing to consider the re-introduction of certain species to the regions where they originated and have subsequently been lost.

Recently, foreign scientists have shown increased interest in our production of new cultivars of wheat, sunflower and cotton. Endemic wild species have also attracted great attention.

The Institute exchanges extensively with scientific institutions in the USA, France, Sweden, Australia, Denmark, Holland, Bulgaria, Yugoslavia, Hungary and many other countries. About 50000–60000 seed samples are sent to research and breeding institutes in the USSR annually as well as thousands of tubers, bulbs and plants for use as initial stock in breeding work.

The main approach for the conservation of world germplasm is the exploration of plant material in the primary and secondary centres of origin of cultivated plants in the areas of their maximum diversity and in poorly investigated and poorly explored countries. This is why we have paid particular attention to the exploration of plant resources in African and Latin American countries.

In the last five years alone in Africa, including the north, west and central regions, seven exploring parties have travelled through 13 countries and have collected some 10000 samples of local varieties and forms of cultivated plants and their wild relatives for the Institute collections. In all these cases our exploring parties encountered all the difficulties mentioned by Harlan (Chapter 7).

Amongst the materials collected, the salt-tolerant, non-shattering forms of the local rice varieties Gambiaca, Mereke and Fossa are of great interest for breeding. Of equal interest are the *Helminthosporium*-resistant maize variety, Perta; the African sorghum species, *S. arundinaceum*, *S. gambicum*, *S. guineense* and *S. margaritiferum*; gossypol-free forms of cotton; very rare forms of perennial *Capsicum frutescens*; drought-resistant varieties of *Voandzeia* for binding sandy soils; local samples and wild forms of Cucurbitaceae; and wild species of soybean, *Glycine javanica* and *G. petitiana*.

Four exploring parties were recently sent to countries in South and Mesoamerica, including Peru, Chile, Bolivia, Ecuador and Mexico. Valuable samples of primitive cultivated and wild species of potato were collected, resistant to viruses A and Y, to wart and *Phytophthora*. Forms of the cultivated tetraploid *S. andigena* were found with high-starch and protein content, and with resistance to potato nematode and aggressive biotypes of wart. Large-grained early forms of maize, grown in high-mountain areas, local varieties and wild forms of cotton and a diversity of rust-resistant varieties of cereals have also been brought from South America.

Iran and Pakistan, being located in the primary centre of origin of soft wheats, equally with our Central-Asiatic republics, have been explored recently, as well as Australia and many European countries.

Exploration of plant material has been considerably intensified recently in different regions of the USSR. Thus, from 25 to 30 exploring parties are being sent annually to various areas. In 1972 alone, 26 exploration missions were organised to areas ranging from Byelorussia to the Far East, and from the polar regions of Krasnoyarsk territory to the humid subtropics of Georgia and the subtropics of Turkmenistan. Exploration of plant material has been carried out on the left bank of the Angara in the Ust-Ilim region, in the basin of the Aldan, in the foothills and the mountainous areas of the Crimea and the Caucasus, in the Alazan valley, in the foothills of the Altar and Tarbagatai mountains, in the Zaisan and Alankol depression, in the region of Gornyj Badakhstan and in many other places. Great diversity of local varieties and wild forms has been discovered in these regions.

Since all this work was carried out in our own country the difficulties of travel were overcome more successfully than on foreign expeditions.

Great attention has been paid by VIR scientists to the conservation of genetic resources in our own country, including local land races developed over many centuries by selection under primitive agricultural conditions, and endemic wild species which are of interest for breeding.

Considering the fact that, even at present, spontaneous hybridisation and mutations occur on a broad scale in Central Asia and the Caucasus particularly, these areas are being explored very thoroughly.

The whole diversity of liguleless and hooded wheats has been added to the Institute's collections from the Central-Asiatic republics, nine new varieties of liguleless wheats having been found. Of particular interest are early-maturing, cold-resistant, naked-grained barleys from the alpine districts of the Pamirs, as well as local melon varieties with crisp flesh from the Khoresm area and the Turkmen SSR.

Of great value also are sweet-kernelled apricot forms from Gornyj Badakhstan (varieties Mashiok, Ravshan Ali, Romatullo, Rovgannesh, Mamad-nur and Lavkadzhak), and extremely well-flavoured large-fruited grape varieties from south Tajikistan (Chilgui angur, Chilgui saphed, Teremoi, Asma maida and Angur saphed). Forms of almond with clearly defined trunk and some with shells transitional from hard to soft have been found for the first time in the Bakharden and Kara-Kala regions.

Unique endemic species of wheat (*Triticum macha, T. araraticum*) have been found in the Caucasus and the Transcaucasus. *Aegilops crassa* has been discovered in the Transcaucasus for the first time, and a new locality for *A. calumnaris* has been found. A wild unique species of pea – *Pisum formosum* – has been brought from Armenia, and wild species of beet – *B. corolliflora, B. trigyna, B. macrorhiza* and *B. intermedia* – from Azerbaijan.

Exploration in Kazakhstan has shown that particularly cold-, heat- and drought-resistant populations of cereals and legumes have evolved in this zone.

Siberia, with its great diversity of cultivated and wild plants, is of considerable interest to breeders. Wild fodder-grasses and small-fruit

149

crops are of extreme value there. Grasses with short vegetative period and resistance to long-term flooding, as well as local populations suitable for the improvement of meadows and pastures, have been revealed in the Tomsk region. Highly winter-resistant grasses have been brought from the Chita and Irkutsk regions. Of great interest also are wild forms of black and red currant, honeysuckle, sea buckthorn and other small-fruit crops with great diversity of biological and economic characters.

The scientists of the Institute have collected in the Far East a great diversity of large-fruited, seedless, winter-hardy Amur grape forms resistant to fungal diseases, as well as wild samples of *Actinidia*, magnolia vine, currants and Manchurian walnut.

The European zone of our country still preserves local variants of fruits and vegetables which have not yet been used for breeding work. A considerable number of wild fodder-grasses, of interest for breeding, grow in Karelia and the Baltic Sea republics.

In spite of the large number of samples added annually to the Institute's collections we do not consider them to be so complete as to stop exploring, both abroad and in our own country. We shall therefore do our best to establish contacts and organise systematic exchanges with all the world gene centres. Recently the importance of the so-called narrow-endemic microcentres has been revealed, and VIR scientists will renew efforts to obtain valuable materials from these centres.

In this report on collecting activities there is no space to describe the research activities of the Institute on seed viability, the storage of materials in air-tight boxes at 0 °C to 4 °C, the plans for constructing a large seed-bank at Kuban and other regions, the work of botanic gardens, and the establishment of reserves. Particularly interesting is the reserve now planned for the Caucasus in the zone of endemic wheat and fruit trees, and the reserve in the mountains of Kopet-Dag, where a centre of diversity of many species of fodder-grasses, apricot, pistachio and almond is situated. An information storage and retrieval system is also being established at the Institute to make information on the collections immediately available to the breeders, and a computer centre will shortly be set up.

This report gives only a brief summary of the enormous work now being undertaken by the Institute, in the exploration, documentation, conservation and utilisation of the plant genetic resources of the USSR and other parts of the world.

12. Recent and proposed exploration activities of the Izmir Centre, Turkey

H. A. SENCER

The exploration activities of the Izmir Centre started in 1964. During 1964 and 1965, a few cereal collections were made in the Aegean region, chiefly for the plant breeding programmes of the Centre. In 1966 and 1967, cereals, forage crops, tree-fruits and ornamental plants were collected in the Aegean and Mediterranean regions, as well as in eastern Turkey. These two years can be considered as a transitional period during which various techniques of collecting, sampling and recording were tried and experience accumulated.

Intensive and systematic collections were made from 1968 onwards of cereals (*Triticum* spp., *Hordeum* spp., *Secale* spp., *Avena* spp.), grain legumes (*Cicer* L., *Vicia* L., *Lens* Moench, *Phaseolus* L., *Vigna* Savi), forage crops (*Vicia* spp., *Trifolium* spp., *Medicago* spp., *Lolium* spp., *Phalaris* spp., *Dactylis* spp., etc.), vegetables (*Brassica* spp., *Cucumis* spp., *Raphanus* spp., *Daucus* spp., etc.), spices, and industrial crops. An intensive exploration programme was carried out during 1968 in the Marmara, Aegean and Mediterranean regions, from sea level to *c.* 600 m, to obtain the remaining local wheat cultivars which were being replaced rapidly by improved Mexican ones.

The exploration activities of the Centre were interrupted in 1969 because of the termination of the joint UN/SF–Turkish Government project during the collecting season.

In 1970 the exploration programme was resumed and collections were made in south-east and eastern Turkey. In 1971 Thrace and eastern Turkey were visited, whilst the 1972 programme included the Aegean region, Thrace, Marmara and south-eastern and eastern Turkey. In 1973 further collections of cereals, including *Zea L.* and *Oryza* L., grain legumes, vegetables and industrial crops were made in Thrace, the Black Sea region, south-east and eastern Turkey.

The total number of collections made in Turkey during the period of 1964–73 and original collections received from other institutions are shown in Table 12.1. The total numbers of major seed collections

151

Table 12.1. *Original collections of the Izmir Centre, 1964–73*

Origin	Type of collection	1964-5	1966	1967	1968	1969	1970	1971	1972	1973	Total
Turkey	C (V)	88	9	583	68	6	23	—	—	—	777
Turkey	C (V, S)	—	—	526	627	283	823	—	—	—	2259
Turkey	C (S)	446	2177	3064	73	56	3	1018	1924	1573	10334
											13370
Turkey	IC (V)	—	—	15	17	—	—	—	—	—	32
Turkey	IC (S)	672	60	202	134	53	—	5	89	400	1615
Iran	IC (S)	91	—	—	—	—	—	—	—	525	616
Ethiopia	IC (S)	54	—	—	—	15	—	—	—	—	69
Austria	IC (S)	154	—	—	—	—	—	—	—	—	154
Nepal	IC (S)	—	—	—	—	—	—	—	140	—	140
Other	IC (S)	15	—	—	—	—	—	—	—	—	15
											2641

C: original collections in Turkey made by the Centre. IC: original collections received from other institutions. S: seed collection. V: vegetative collection.

Table 12.2. *Major seed collections made in Turkey, 1966–73*

Group	1966	1967	1968	1969	1970	1971	1972	1973	Total
Cereals	113	263	405	—	545	361	229	711	2627
Grain legumes	55	131	138	55	82	438	126	430	1455
Forage legumes	954	1092	46	18	64	91	20	19	2304
Forage grasses	497	382	1	—	8	3	1	—	892
Vegetables–Spices	72	59	31	31	49	73	86	274	675
Industrial crops	5	10	3	7	12	9	1409	58	1513
	1696	1937	624	111	760	975	1871	1492	9466

Table 12.3. *Collections of major groups of crop plants made in Turkey from 1966 to 1973*

Wheat (diploid)	129	Rice	88
Wheat (tetraploid)	544	Maize	486
Wheat (hexaploid)	438	Beans (*Phaseolus*)	966
Barley (two-row)	216	Broadbeans	135
Barley (six-row)	171	Lentils	142
Oats	154	Chick-peas	217
Rye	267	Forage legumes	2146
		Vegetables	619

It should be noted that these are population samples and not selected lines. To illustrate the diversity within samples, attention is drawn to one example; 15 wheat (tetraploid and hexaploid) samples, originating from south-east Turkey, gave 2000 distinct lines.

made in Turkey during the period of 1966–73 are shown in Tables 12.2 and 12.3. Thus over 13 000 seed and vegetative collections were made in Turkey and over 2500 seed collections were received from other institutions. During the same period 9140 herbarium samples were collected. The Izmir Centre has provided assistance to some 19 overseas visiting collectors from seven different countries.

Future exploration work

Activities for the future are being planned on the basis of the following considerations:

(*a*) The collecting that has already been carried out for the various crops and regions.

(*b*) The threat or actual progress of genetic erosion.

(*c*) The scientific and economic importance of the material which it is proposed to collect.

(*d*) The urgency of plant breeding needs.

The crops are listed below in diminishing order of urgency for collecting:

(i) Cereals, grain legumes, forage crops, vegetables and spices. Further collections are needed on the central Anatolian plateau and at higher altitudes (600 m) on the Taurus, Pontic and Aegean mountains.

(ii) Industrial crops, i.e. oil and fibre crops (sesame, linseed, flax, hemp, safflower, sunflower), medicinal plants, perfume plants, and stimulants. These need to be collected from all over Turkey.

(iii) Exploration of local tree-fruit varieties and in-situ preservation

153

of wild species. There is a general need for collections of this material all over Turkey, though primarily on the Pontic mountains.

(iv) Exploration and/or in-situ preservation of native ornamental plants in all the mountain regions of Turkey.

Methods and techniques

Certain principles, based on theoretical considerations as well as field experience, are applied during the exploration work:

(1) *Collecting* of seed material is done in such a way as to ensure that the samples represent as large a proportion as possible of the genetic variation existing in the populations. The size of each sample is large enough for long-term conservation as well as for immediate utilization of the collections. The main ecological factors, i.e. altitude, topography, soil, plant community, etc., are taken into account in the selection of the populations to be sampled.

(2) *Herbarium samples* are made for all the collections, both to facilitate identification and to illustrate the phenotypic variation existing in the populations.

(3) *Standard documentation* of collections. The Centre has aimed at standard documentation of the collections from the very beginning. A standard form has been developed based on theoretical needs as well as on experience gained during exploration work. This provides a format which enables data on the name, origin, habitat, etc. of the collections to be put into computer storage with maximum efficiency.

(4) *Vegetation surveys*. The Centre conducted a number of vegetation surveys in Turkey, parallel to its plant collecting activities. Despite the outstanding contributions of taxonomists on the Turkish flora the distribution of several economically important plants is still poorly known. Increase in our knowledge along these lines is thus a step towards the proper conservation of species in natural reserves. This is especially important for the conservation of groups such as tree-fruits, ornamental plants and forage species.

Evaluation studies

The success of a genetic resources centre depends not only on the material which it holds in store but on the use to which this material is put.

Evaluation studies must generally be carried out in collaboration

154

with other laboratories, and the centre itself should hold fully documented records of the results.

The Izmir Centre has distributed material for evaluation and breeding studies to a wide range of countries in different parts of the world. Table 12.4 shows the extent to which material from the Turkish collections is being screened for resistance to diseases, for specific biochemical characters, and for adaptation to climatic and edaphic extremes. In a number of cases promising results have been obtained and several new varieties have been released.

Table 12.4. *Evaluation studies in progress on the Izmir Centre collections*

Material and recipient country	Year of despatch	Purpose of research
Triticum monococcum L. (wild, cult.)		
Netherlands (Wageningen)	1968	Search for sources of stripe rust (*Puccinia striiformis*) resistance. (Promising results obtained)
USA (Corvallis–Oregon)	1972	ditto
USA (California)	1972	Identification of amino acids and proteins
Sweden (Svalöv)	1973	Determination of protein and lysine content
Triticum turgidum L. emend. Bowden Gr. *durum*		
Sweden (Svalöv)	1966	Search for sources of mildew resistance
Canada (Winnipeg)	1970	Search for sources of disease resistance
Turkey (Izmir)	1971	Search for sources of *Puccinia graminis*, *Puccinia recondita* and *Puccinia striiformis* resistance. (Promising results obtained)
Israel	1974	Search for sources of *Septoria* resistance
Turkey (Ankara–WRTC)*	1974	Search for sources of winter hardiness
Turkey (Izmir)	1974	Search for sources of *Septoria* resistance
Australia	1974	Not defined
Triticum aestivum L. emend. Bowden		
Canada (Winnipeg)	1970	Search for sources of disease resistance
Turkey (Izmir)	1971	Search for sources of *Puccinia graminis*, *Puccinia recondita* and *Puccinia striiformis* resistance. (Promising results obtained)
Sweden (Svalöv)	1973	Determination of protein and lysine content. (Results available)
Israel	1974	Search for sources of *Septoria* resistance
Turkey (Izmir)	1974	Search for sources of *Septoria* resistance
Australia	1974	Not defined

* Wheat Research and Training Centre–Ankara.

155

Table 12.4 (*cont.*)

Material and recipient country	Year of despatch	Purpose of research
Hordeum vulgare L. emend. Bowden Gr. *vulgare, distichon*		
Canada (Winnipeg)	1970	Search for sources of disease resistance
Sweden (Svalöv)	1973	Search for sources of mildew resistance
Turkey (Izmir)	1974	Search for sources of *Ustilago, Helmintho-sporium, Erisiphe* and *Septoria* resistance
Turkey (Ankara–WRTC)	1974	Search for sources of winter hardiness
Hordeum spp. (wild)		
Canada (Winnipeg)	1970	Search for sources of disease resistance
Avena byzantina C. Koch, *A. sativa* L., *A.* spp. (wild)		
Canada (Winnipeg)	1970	Search for sources of stem-rust and crown rust resistance and determination of oil and protein content
Secale cereale L.		
Mexico (CIMMYT)	1970	To obtain *Triticale*
Germany (Braunschweig)	1972	Breeding of rye varieties
Turkey (Izmir)	1972	Taxonomic studies in genus *Secale*
Cicer arietinum L.		
Spain	1969	Cytogenetical and morphological studies
Netherlands	1971	Taxonomic studies
Turkey (Izmir)	1971	Taxonomic studies
Lebanon	1972	Variety improvement for arid lands
Iraq	1972	Breeding for high-yielding varieties
Australia	1973	Search for sources of protein
Israel	1973	Search for *Ascochita* blight resistance and large-seeded types
Turkey (Eskişehir)	1973	Breeding for high-yielding varieties
Lens culinaris Medik		
Spain	1969	Cytogenetical and morphological studies
Argentina	1972	Search for sources of resistance to *Fusarium oxysporium* Schl.
Iraq	1972	Breeding for high-yielding varieties
Lebanon	1972	Variety improvement for arid lands
Turkey (Eskişehir)	1973	Breeding for high-yielding varieties
Phaseolus vulgaris L.		
Iraq	1972	Breeding for high-yielding varieties
Turkey (Eskişehir)	1973	Breeding for high-yielding varieties
Vicia faba L.		
Sweden (Svalöv)	1969	Search for earliness
Germany	1971	Genetical studies
Lebanon	1972	Variety improvement for arid lands
Australia	1973	Search for sources of protein
Turkey (Izmir)	1973	Breeding for high-yielding varieties

Table 12.4 (*cont.*)

Material and recipient country	Year of despatch	Purpose of research
Citrullus vulgaris Forssk., *Cucumis sativus* L., *C. melo* L.		
Israel	1973	Search for sources of *Verticillium* and *Fusarium* resistance
Lactuca sativa L.		
Israel	1973	Search for sources of resistance to *Perenospora*, *Stemphylium* and LMV
Solanum melongena L.		
Israel	1973	Sources of *Verticillium* resistance
Capsicum annuum L.		
Israel	1973	Search for sources of resistance to CMV and breeding of hot and sweet cultivars
Vicia sativa L., *V. pannonica* Crantz, *V. villosa* Roth, *V. dasycarpa* Ten.		
Turkey (Izmir)	1968	Breeding of annual vetch varieties with high dry-matter in hay and suitable for rotation with cotton in Aegean region. (Four new varieties will be released during 1974, selected from the Centre's own collections)
Phalaris spp., *Lolium* spp., *Dactylis* spp., *Bromus* spp., *Arrhenatherum* spp., *Festuca* spp.		
Turkey (Izmir)	1968	Variety improvement for dry and irrigated land in Aegean region. (The Centre's own collections are being evaluated for characters such as winter-hardiness, drought resistance, and other morphological and biological attributes)

157

13. Exploration and survey in rice

T. T. CHANG

Exploration

Exploration for primitive cultivars in the Jeypore Tract of Orissa (India) during the late 1950s was the first organised effort to survey and to conserve diverse and indigenous rice germplasm (Govindaswamy, Krishnamurty & Sastry, 1966). From 1958 to 1964, the Rockefeller Foundation supported Japanese geneticists and taxonomists in exploring and collecting wild taxa and primitive forms of the genus *Oryza* in East and West Africa, South and South-East Asia, and Latin America (Kihara, 1959; Kihara & Nakao, 1960; Oka, 1964; Tateoka, 1965; Katayama, 1968). During the same decade Japanese researchers initiated explorations to Nepal, Sikkim, and South-East Asia (Nakao, 1957; Kihara & Nakao, 1960; Hamada, 1965).

Since 1962 the International Rice Research Institute (IRRI) has helped rice researchers collect wild and primitive types in West Africa and in tropical Asia (Tateoka & Pancho, 1963; Oka & Chang, 1964; Chang, 1970). J. R. Harlan made several trips to West Africa during the 1960s which culminated in a report that included a survey of rice germplasm in many West African nations (Harlan, 1973). In recent years Japanese scientists made extensive collection efforts in South and South-East Asia (Akihama & Watabe, 1970; Katayama, Watabe & Kuroda, 1972; Akihama & Toshimitsu, 1972; Katayama, Akihama & Weliwita, 1972). Between 1967 and 1971 Indian researchers made vigorous efforts in canvassing the indigenous germplasm of North-East India which resulted in more than 6000 rice samples (Sharma, Vellanki, Hakim & Singh, 1971). During 1971 and 1972, extensive efforts were made by the national research center concerned and IRRI in collecting minor and primitive cultivars as well as weed races from remote areas of Indonesia, Nepal, Philippines, Sri Lanka, and South Vietnam.

Germplasm of Africa

Fig. 13.1 shows the geographic distribution of the African rice (*Oryza glaberrima*) and its wild relatives (*O. barthii* and *O. longistaminata*). The less widely distributed *O. eichingeri* and *O. punctata* are also shown.

159

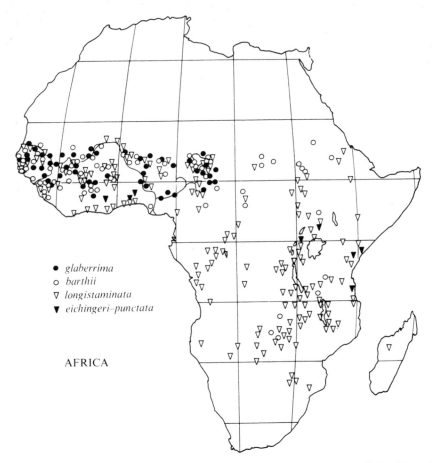

Fig. 13.1. Geographic distribution of *Oryza glaberrima*, its wild relatives *O. barthii* and *O. longistaminata*, and two other wild rice species in Africa.

The sites were based on the findings of Oka & Chang (1964), Tateoka (1965), and on unpublished work of J. R. Harlan (personal communication).

A recent FAO survey (Gullberg, 1971) indicated that about 280 *O. glaberrima* strains were maintained in several national collections of African countries. The IRRI collection includes about 375 strains of *O. glaberrima* and 220 strains of five wild African taxa. Considering the extensive distribution of *O. glaberrima* and *O. barthii*, there is an urgent need to make further collections of the African rices which are being rapidly replaced by the common rice of Asia, *O. sativa*. The countries concerned are Senegal, Gambia, Portuguese Guinea, Guinea,

Table 13.1. *Varieties of* Oryza sativa *maintained in national collections of tropical Asia, compared with the estimated total in each country and the accessions in the IRRI collection*

Country	National collection	IRRI collection	Estimated total
South Asia			
Afghanistan	—	30	—
Bangladesh	2600	805	4000 +
India	20000	4366	—
Nepal	910	961	1000 +
Pakistan	600	670	1800
Sri Lanka	2070	2080	2500
South-East Asia			
Burma	736	146	3500
Indonesia	4398	3906	6000 +
Khmer	820	798	1000
Laos	250	587	—
Malaysia	1250 +	504	—
Philippines	800	1216	1000 +
Thailand	2000	260	3500 +
Vietnam (South)	690	581	1000

Sierra Leone, Liberia, Ivory Coast, Upper Volta, Ghana, Mali, Togo, Dahomey, Nigeria, Niger, Cameroons, and Chad.

The African cultivars of *O. sativa* are relatively recent introductions, largely from tropical Asia. A substantial segment of the *O. sativa* cultivars (about 1340) has been assembled in the IRRI collection. Although the FAO survey (Gullberg, 1971) indicated that more than 4000 cultivars were kept in the various national collections, it is not clear as to how many accessions are viable.

Many *O. sativa* × *O. glaberrima* hybrids of different intergrades existing as mixtures on African farms may deserve the attention of rice researchers as potentially useful material. Several intermediate strains labelled as *O. sativa* were found in the IRRI collection.

Germplasm of South and South-East Asia

At IRRI, a great diversity in morpho-agronomic features has been found in accessions from Bangladesh, India, Indonesia, Thailand and Vietnam. Unfortunately, the IRRI collection of genetic stocks from Burma, temperate Nepal, and the mainland of China is insufficient to provide a balanced assessment of genetic diversity in these countries.

An earlier survey by IRRI of *O. sativa* cultivars found in tropical

161

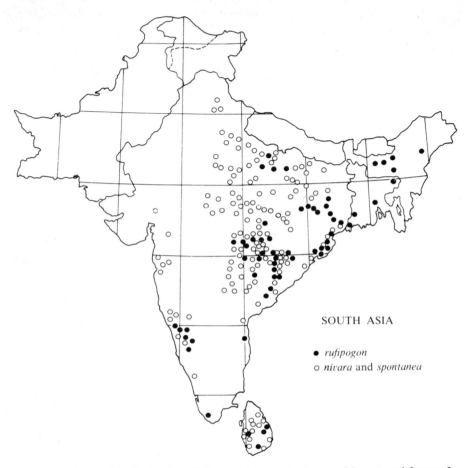

Fig. 13.2. Geographic distribution of *Oryza rufipogon, O. nivara* and '*spontanea*' forms of *O. sativa* in South Asia.

Asia (Chang, 1972) has been updated to show the distribution in South Asia and in South-East Asia (Table 13.1). By comparing the estimated total of indigenous varieties and the total number maintained by the national center concerned or by IRRI, it is apparent that further collections in Bangladesh, Burma, India, Indonesia, Laos, Malaysia, Thailand and Vietnam should be fruitful. A study by the participants in the International Rice Collection and Evaluation Project in 1970 and subsequent discussions with national rice research leaders, resulted in the recommendation of the following places as important collection sites: (1) all parts of Bangladesh for the *aus* and *aman* varieties; (2) the less accessible parts of Burma; (3) all parts of Sri Lanka (Ceylon);

162

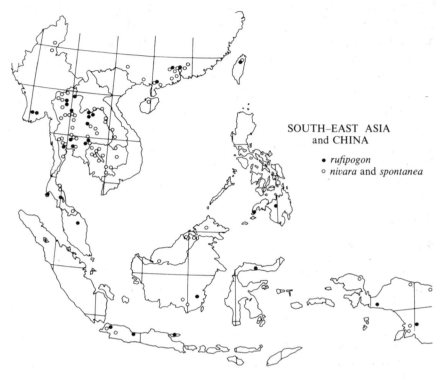

Fig. 13.3. Geographic distribution of *Oryza rufipogon*, *O. nivara* and '*spontanea*' forms of *O. sativa* in South-East Asia and China.

(4) areas in the sub-Himalayan regions, the central states, Dangs and Konkan of India for the minor varieties; (5) all parts of Indo-China; (6) the Kalimantan and Sumatra in Indonesia for the *tjereh* varieties and Bali for the *bulu* varieties; (7) East Malaysia; (8) Nepal, Sikkim, and Bhutan; (9) Pakistan for the minor varieties; (10) the hilly areas of the Philippines; (11) the hilly areas of Thailand; and (12) the various rice growing countries of Africa.

During 1971 and 1972 field collections were conducted in Indonesia, Khmer, Sri Lanka, Nepal, and South Vietnam. Field collections are planned for Bangladesh, Burma and East Malaysia. Genetic stock officers, breeders, agronomists, extension workers, college students, and missionary workers participate in the collecting activities.

The distribution of collection sites of the two principal weed races in tropical and subtropical Asia, *O. rufipogon* and *O. nivara*, and of their hybrids with *O. sativa* (= '*spontanea*' forms) is shown in Fig. 13.2 for South Asia and in Fig. 13.3 for South-East Asia and China. The infor-

163

mation was based on the findings of collectors cited under 'Exploration', other reports (Watt, 1891; Backer, 1946; Ting, 1949; Chatterjee, 1951; Sampath & Govindaswamy, 1958; Bor, 1960; Sharma & Shastry, 1965), and my observations. The sites undoubtedly represent only a small proportion of those areas where these taxa can be found in their natural habitats. The wild rice in the Taoyuen district of Taiwan disappeared from its restricted habitat during the 1960s. In other areas the wild taxa will soon be threatened by the construction of irrigation and drainage facilities, highways, housing, and industrial developments. It will require the concerted efforts of trained scientists to collect and maintain these weed races adequately.

References

Akihama, T. & Toshimitsu, K. (1972). Geographical distribution of morphological variation on wild rices in central and southern India. *Prelim. Rept. Tottori Univ. Sci. Survey*, 1971, pp. 48–59.

Akihama, T. & Watabe, T. (1970). Geographical distribution and ecotypic differentiation of wild rice in Thailand. *Tonan Ajia Kenkyu* (Southeast Asian Studies), **8**, 337–46.

Backer, C. A. (1946). The wild species of *Oryza* in the Malay Archipelago. *Blumea, Suppl.* **3**, 45–55.

Bor, N. L. (1960). *The Grasses of Burma, Ceylon, India and Pakistan* (excluding Bambuseae), 767 pp. Pergamon, London.

Chang, T. T. (1970). Rice. In *Genetic Resources in Plants – Their Exploration and Conservation* (eds O. H. Frankel and E. Bennett), *IBP Handbook*, no. **11**, pp. 267–72. Blackwell, Oxford and Edinburgh.

Chang, T. T. (1972). IRRI rice germplasm project and its relation to national varietal collections. *Plant Genetic Resources Newsletter*, no. **27**, 9–15.

Chatterjee, D. (1951). Note on the origin and distribution of wild and cultivated rices. *Indian J. Genet. Pl. Breed.*, **11**, 18–22.

Govindaswamy, S., Krishnamurty, A. & Sastry, N. S. (1966). The role of introgression in the varietal variability in rice in the Jeypore tract of Orissa. *Oryza*, **3**, 74–85.

Gullberg, U. (comp.) (1971). World list of germplasm collections: rice. *Plant Genetic Resources Newsletter*, no. **26**, 27–35.

Hamada, H. (1965). Rice in the Mekong Valleys. In *Indo-Chinese Studies: Synthetic Research of the Culture of Rice-cultivating Races in Southeast Asian Countries*, vol. I, pp. 517–86. The Japan. Soc. Ethnology, Tokyo.

Harlan, J. R. (1973). Genetic resources of some major field crops in Africa. *Survey of Crop Genetic Resources in their Centres of Diversity*, pp. 45–64. FAO–IBP, Rome.

Katayama, T. C. (1968). Scientific report of the rice collection trip to the Philippines, New Guinea, Borneo and Java. *Mem. Fac. Agr. Kagoshima Univ.*, **6** (2), 89–134.

Katayama, T. C., Akihama, T. & Weliwita, S. M. P. (1972). Distribution and some morphological characters of the wild rice in Ceylon. *Prelim. Rept. Tottori Univ. Sci. Survey*, 1971, pp. 60–4.

Katayama, T. C., Watabe, T. & Kuroda, T. (1972). Distributions and some morphological characters of the wild rice in the Ganga Plains (part I). *Prelim. Rept. Tottori Univ. Sci. Survey*, 1971, pp. 1–47.

Kihara, H. (1959). Considerations on the origin of cultivated rice. *Seiken Ziho*, **10**, 68–83.

Kihara, H. & Nakao, S. (1960). The rice plant in Sikkim. *Seiken Ziho*, **11**, 46–54.

Nakao, S. (1957). Transmission of cultivated plants through the Sino-Himalayan route. In *Peoples of Nepal Himalaya* (ed. H. Kihara), vol. III, pp. 397–420. Kyoto University, Kyoto.

Oka, H. I. (1964). Pattern of interspecific relationships and evolutionary dynamics in *Oryza*. In *Rice Genetics and Cytogenetics*, pp. 71–90. Elsevier, Amsterdam.

Oka, H. I. & Chang, W. T. (1964). *Observations of Wild and Cultivated Rice Species in Africa*, 73 pp. National Institute of Genetics, Misima, Japan (mimeographed).

Sampath, S. & Govindaswamy, S. (1958). Wild rices of Orissa – their relationship to cultivated varieties. *Rice News Teller*, **6** (3), 17–20.

Sharma, S. D. & Shastry, S. V. S. (1965). Taxonomic studies in genus *Oryza* L. III. *O. rufipogon* Griff. *sensu stricto* and *O. nivara* Sharma et Shastry *nom. nov. Indian J. Genet.*, **25**, 157–67.

Sharma, S. D., Vellanki, J. M. R., Hakim, K. L. & Singh, R. K. (1971). Primitive and current cultivars of rice in Assam – a rich source of valuable genes. *Curr. Sci.*, **40**, 126–8.

Tateoka, T. (1965). A taxonomic study of *Oryza eichingeri* and *O. punctata*. *Bot. Mag., Tokyo*, **78**, 156–63.

Tateoka, T. & Pancho, J. V. (1963). A cytotaxonomic study of *O. minuta* and *O. officinalis. Bot. Mag., Tokyo*, **76**, 366–73.

Ting, Y. (1949). Origination of the rice cultivation in China. *Agron. Bull.*, Sun Yatsen Univ., Ser. II, **7**, 1–18.

Watt, G. (1891). *Dictionary of the Economic Products of India*, vol. V, pp. 498–508. W. H. Allen, London.

14. Potato collecting expeditions in Chile, Bolivia and Peru, and the genetic erosion of indigenous cultivars

C. OCHOA

Although the varietal richness of the South American indigenous potatoes has been known for more than half a century (Cevallos Tovar, 1914; Leguas, 1897; Wight, 1916) it was only after the publication of the work of S. M. Bukasov in 1933 that the real importance of this material began to be understood. From that time onwards expeditions have been sent from various parts of the world to collect the wild and cultivated potatoes of the Americas for subsequent utilisation in plant breeding studies.

Unfortunately, this immense genetic reserve is not inexhaustible. On the contrary, it is in danger of partial or complete destruction in a short space of time, especially where the cultivated species are concerned. The danger of genetic erosion in potatoes has been seen for some time. Thus Brücher (1963a) stated that he had observed a great reduction of cultivated diploid potatoes in certain regions of Bolivia and Argentina. The same author (Brücher, 1969) indicated that he had found only tetraploid potatoes in the Venezuelan Andes of Mérida, Táchira and Trujillo, where it was previously said that varieties of the diploid cultivated *Solanum phureja* occurred. He ended by observing that the diploid 'criolla' potatoes of Venezuela were on the way to complete extinction.

Genetic erosion and exploration in Chilean potatoes

One of the areas most strongly affected by genetic erosion in primitive potatoes must surely be the Chiloé Archipelago in southern Chile. Vallega & de Santis (1938) mention that, of the forms collected by them in that year in Chiloé, particularly in Yutuy, all but one or two correspond to those found ten years before by the Russian botanist Juzepczuk. Some 200 samples were found on that trip in a relatively restricted area. On the other hand, Castronovo found not much more than half that

167

amount ten years later (Castronovo, 1949), whilst Brücher found even fewer in 1958 (Brücher, 1963*b*).

Zikin, a member of the latest Russian expedition to Chiloé, southern Peru and Bolivia in 1967, although unfortunately not giving a list of the material collected, relates that he was completely unsuccessful in his attempts to collect native potatoes in the region extending from central Chile to Puerto Montt. These potatoes had been almost entirely replaced by European varieties. He also states that the ancient cultivated forms of *S. tuberosum* from the island of Chiloé were practically unobtainable and were only to be found in abandoned kitchen gardens in a semi-wild state. Only in this way was he able to collect no more than a dozen tuber and herbarium samples in an abandoned garden south of Cucao (Zikin, 1968).

The last expedition to Chiloé was that made by the present writer in 1969, with financial aid from IICA (Instituto Interamericano de Ciencias Agricolas). On this expedition, in spite of the very bad weather conditions, the bad state or complete lack of roads in some areas, difficulties of obtaining vehicles, etc., it was nevertheless possible to travel extensively over Chiloé and certain neighbouring islands. As a result of this intensive work, which lasted for about a month, we collected some 35 to 40 samples of primitive potatoes, excluding duplicates. Furthermore, in the Chonos Archipelago we were lucky enough to find 'Darwin's Potato', almost in the place where Darwin himself collected it on 7 January 1835, at Low Bay on Guaytecas Island (see Darwin, 1845). We thus have in our gene bank the only living collection of this famous potato. Incidentally, we found it to be a tetraploid escaped form of *S. tuberosum*.

To sum up the situation in Chile, we can see clearly that the primitive potatoes of this country have been decimated by *Phytophthora* and largely replaced by European varieties. The few remaining forms are doomed to extinction in a very short time.

Genetic erosion in Peruvian and Bolivian potatoes

The primitive varieties of Peru and Bolivia are also unfortunately suffering from genetic erosion. Comparing the extensive lists of varieties published from 1933 to 1965 (Bukasov, 1933; Hawkes, 1944; Ochoa, 1955; 1958; 1964; 1965; Vargas, 1949; 1956) with those of the last six years one obtains a good idea of what is taking place. The causes are probably diverse, but certainly include such factors as fungal and bacterial attacks, extremes of drought and frost, natural disasters, introduction of improved varieties and conscious or unconscious selection practised by the farmers.

In general the losses are less in the south of Peru (including the Lake Titicaca basin) than in the centre and north. This is because of the more advanced agricultural practices and smaller Indian population of these latter areas. The Indian is by nature conservative and antagonistic to innovations, and will not adopt newly-bred varieties, especially if their culinary quality differs from that of the potatoes to which he is accustomed. Furthermore, the climatic conditions of the Lake Titicaca basin are unfavourable to late blight development and the varieties of that region are not therefore attacked by it.

One of the most dramatic cases yet observed in an important zone of primitive potato cultivation is seen in the district of Quero (province Paucartambo, department Cuzco) in south Peru. Here, in an extremely isolated Indian community at 3200 to 3400 m above sea level, the peasants follow the custom of discarding all diseased potatoes (called by them 'sojra-huajacc'). This means in effect that all *Phytophthora*-infected clones are destroyed. Unfortunately many of these belong to the extremely susceptible *S. stenotomum*, many of whose varieties in this district have in effect been completely eliminated. In northern Peru especially, the genetic erosion is greater than in any of the other Andean regions south of the equator. The principal causes for this erosion are *Phytophthora* and *Pseudomonas* infection, and replacement by improved varieties. Some 20 years ago we made our first exploration trip to northern Peru, collecting some 385 samples of cultivated potatoes and various wild species new to science. Fifteen years later we returned to a number of places we had visited previously, such as Colquimarca and Catac in Ancash department, where we had collected 25 indigenous varieties and on this occasion found not a single one. Similarly, in the high Andean regions ('jalcas') of Porcón and Quilcate in Cajamarca department we found nothing where previously we had collected 20 samples. In these and other departments the indigenous varieties have been replaced by the high-yielding, *Phytophthora*-tolerant cultivar Renacimiento. In this way we have not seen again, either in these or other localities, the old varieties Rambramina, Ila and Clavelina of *S. stenotomum* or Chilopa and Montañera of *S. tuberosum* subsp. *andigena*, which have now evidently been completely lost.

We shall finish by quoting one more example of genetic erosion. In the district of Leymebamba alone, in the little-known department of

Amazonas, we found, in 1952, more than double the number of varieties of potatoes than in the whole department of Piura. Unfortunately, even in this region, despite its isolation, genetic erosion takes place. Eighteen years later we found that the original varieties had been reduced to a quarter and a number of endemic biotypes had disappeared completely. Even though many herbarium samples have been preserved, the living material has disappeared for ever.

Genetic erosion in cultivated diploids, triploids and pentaploids

It seems certain that the cultivated diploid species are most affected by genetic erosion, followed closely by the triploids belonging to the species *S. chaucha*, especially in northern Peru.

On the other hand, the frost-resistant triploid *S. juzepczukii* and pentaploid *S. curtilobum* suffer the least genetic erosion. They are grown at high altitudes for the production of the dried potato product known as 'chuño', and their limited use and altitude at which they are grown probably account for their protection.

Genetic erosion in cultivated tetraploids

In the vast group of tetraploids belonging to *S. tuberosum* subsp. *andigena* we have seen most unfortunate and irretrievable losses of valuable cultivars, such as Naranja from north Peru, Tornasol and Amarilla from the centre, and Sajma from southern Peru. The form known as Amarilla came from Huasahuasi near Tarma at 3400 m, and in spite of its name, which generally indicates the diploid species *S. goniocalyx*, was a tetraploid with large round tubers and deep yellow flesh, though the skin was black. It was a well-liked variety, with good flavour and cooking quality. The variety Sajma from south Peru was first collected by us in 1948 at Umana in the important potato-growing zone of Paucartambo at 3600 m – in Cuzco department. Its chief distinguishing feature was the curious shape, similar to a clenched fist, very like a boxing glove. We have looked for it repeatedly during the last ten years on numerous occasions but without success. It appears to have become completely extinct.

Potato exploration by the author in Peru and Bolivia

We made very extensive potato collections from 1966 to 1972 in all the principal potato-growing regions of Peru as well as in the basin of

Table 14.1. *Species and samples collected in Chile, Bolivia and Peru from 1966 to 1972*

Species	2n	Samples			
		Chile	Bolivia	Peru	Total
S. *ajanhuiri*	24	—	10	5	15
S. *goniocalyx*	24	—	—	30	30
S. *phureja*	24	—	2	13	15
S. *stenotomum*	24	—	59	215	274
S. *chaucha*	36	—	—	37	37
S. *juzepczukii*	36	—	48	55	103
S. *tuberosum*					
Subsp. *tuberosum*	?	30	—	—	30
Subsp. *andigena*	48	—	220	1407	1627
S. *curtilobum*	60	—	11	35	46
Total		30	350	1797	2177

Lake Titicaca in north Bolivia. These collections amount to over 2000 samples, which have been classified and a chromosome count made in most cases.

The classification used has followed that of Hawkes (1963), except that we have retained *S. goniocalyx* as a species instead of placing it as a subspecies of *S. stenotomum*.

The species and samples collected are shown in Table 14.1.

We believe that our collections of cultivated potatoes made in Peru in the last six years are the most complete yet formed. Even so this work is not yet finished. The Andean potato-growing territories are so vast, with such an infinite number of localities that can only be reached with great difficulty, that much collecting work is still required to save the material from extinction.

Conclusions

The costly responsibility for safeguarding this unique material cannot be restricted to the developing Andean countries, with their reduced financial resources. Equally, these genetic resources do not belong uniquely to these countries, but rather to the whole of humanity. If the material is to be saved, rapid and immediate action is needed in the form of well-equipped expeditions to collect it and later to conserve, evaluate and distribute it to scientific institutes in all parts of the world.

These and others are in fact the purposes of the newly formed

Exploration: national and regional

International Potato Centre (CIP), recently established in Lima under the direction of Dr Richard Sawyer (Anon., 1972*a*, *b*; 1973). A number of these plans coincide with those set out by Hawkes (1970).

CIP has established a germplasm bank of potatoes which contains some 4000 accessions of wild and primitive cultivated species from the Peruvian and other South American gene centres. The material is being evaluated and multiplied so as to make it available to breeders in all parts of the world. Collections are constantly being added, so that it would be not altogether surprising if the number of accessions surpassed 12000 in five years' time. Thus this germplasm bank will save the potato gene pool in the Andes and elsewhere and will constitute the most complete and well-utilised collection of potato variability in the world.

References

Anon. (1972*a*). *The International Potato Center* (CIP), 38 pp. Lima, Peru.

Anon. (1972*b*). *Prospects for the Potato in the Developing World.* The International Potato Center, 275 pp. Lima, Peru.

Anon. (1973). *Workshop on Germ-plasm Exploration and Taxonomy of Potatoes*, 35 pp. Publication of the Centro Internacional de le Papa. Lima, Peru (mimeographed).

Brücher, H. (1963*a*). Das südlichste Vorkommen diploider, Kulturkartoffeln in Südamerika auf der Insel Chiloé. *Qualitas Plantarum*, **9**, 187–202.

Brücher, H. (1963*b*). Untersuchungen über die *Solanum* (*Tuberarium*). Cultivare der Insel Chiloé. *Z. PflZücht.*, **49**, 7–54.

Brücher, H. (1969). Venezuelas Primitiv-Kartoffeln. *Angew. Bot.*, **42**, 179–88.

Bukasov, S. M. (1933). The potatoes of South America and their breeding possibilities. *Bull. appl. Bot. Genet. Pl. Breed., Leningrad, Suppl.*, **58**, 192 pp.

Castronovo, A. (1949). Papas Chilotas. *Rev. Invest. agric., Buenos Aires*, **3**, 209–45.

Cevallos Tovar, W. (1914). Clasificación de le papa de Bolivia. *Bol. Dir. Estad., La Paz*, **10** (88), 1–8.

Darwin, C. (1845). *The Voyage of a Naturalist Round the World in H.M.S. 'Beagle'.* London.

Hawkes, J. G. (1944). Potato collecting expeditions in Mexico and South America. II. Systematic classification of the collections. *Bull. Imp. Bur. Pl.-Breed. Genet., Cambridge*, 142 pp.

Hawkes, J. G. (1963). A revision of the tuber-bearing Solanums. *Rec. Scott. Pl.-Breed. Stn.* (2nd edn), pp. 76–181.

Hawkes, J. G. (1970). Examples of crop exploration – potatoes. In *Genetic Resources in Plants – their Exploration and Conservation* (eds O. H. Frankel and E. Bennett), *IBP Handbook*, no. **11**, pp. 311–19. Blackwell, Oxford and Edinburgh.

Leguas, A. (1897). In *Estudios geograficos sobre Chiloé* (R. Maldonado), pp. 334–9.

Ochoa, C. (1955). Expedición colectora de papas al Norte del Peru. *Biota*, **1**, 47–64.

Ochoa, C. (1958). Expedición colectora de papas cultivadas a la Cuenca del Lago Titicaca. *Programa cooperativa de experimentación agropecuaria. Investigaciones en Papa*, no. **1**, 1–18. Min. Agric., Lima, Peru.

Ochoa, C. (1964). Recuentos cromosómicos y determinación sistemática de papas nativas cultivadas en el sur del Peru. *Anales Cientif.*, *Lima*, **2**, 1–41.

Ochoa, C. (1965). Determinación sistemática y recuentos cromosómicos de las papas indigenas cultivadas en el centro del Peru. *Anales Cientif.*, *Lima*, **3**, 103–63.

Vallega, J. & de Santis, L. (1938). Expedición a Chile en busca de semillas de plantas cultivadas aborígenes. *An. Prov. B. Aires*, *Argentina*, **6**, 183–212.

Vargas, C. (1949). *Las papas Sudperuanas*, vol. I, 144 pp. Publ. Univ. Ncl. Cuzco, Peru.

Vargas, C. (1956). *Las papas Sudperuanas*, vol. II, 66 pp. Publ. Univ. Ncl. Cuzco, Peru.

Wight, W. F. (1916). Origin, introduction and primitive culture of the potato. *Proc. 3rd ann. Meet. Potato Assn. Amer.*, **3**, 35–52.

Zikin, A. (1968). A trip to the potato country. *Potatoes and Vegetables*, **11**, 45–7.

15. Recent cocoa collecting expeditions

J. SORIA

The species *Theobroma cacao* L. originated in tropical America. According to Cheesman (1944) its chief centre of genetic diversity is in the Amazon basin between the headwaters of the Caquetá, Putomayo and Napo rivers. There is a secondary centre of genetic diversity in Mexico and Central America, where it was undoubtedly domesticated by the Mayas, thousands of years before the arrival of Columbus.

The beginning of cocoa cultivation for export dates from the seventeenth century when the Spaniards encouraged the plantation of superior varieties, which were apparently brought from Mexico and Central America to several countries in northern South America and Trinidad. According to Patiño (1963) cocoa cultivation was started in the valleys of the departments of Cauca, Caldas, Antioquia and Huila in Colombia, the extreme south of Lake Maracaibo, the northern coastal valleys of Aragua, the Chama Valley in Merida state and Sucre state in Venezuela, in the island of Trinidad, and on the western coast of Ecuador. In this last-mentioned region an Amazonian 'forastero' cocoa of high quality was grown. According to Pound (1938) and Soria (1966) the national Ecuadorean cocoa represents a sample of the wild cocoas which grow on the banks of the Napo and Pastaza rivers in the Ecuadorean eastern region.

Various collections were made in the 1930s by scientists from the Trinidad Imperial College of Tropical Agriculture in Mexico, Central America and the northern part of South America, collecting samples of cultivated varieties, and especially the superior native forms of Ecuador. Unfortunately, due to the low production, high susceptibility to fungus disease and a limited demand for good cocoa, most countries producing high quality cocoa have reduced or completely eliminated their production. Thus, for example, in Mexico, Guatemala and Nicaragua between 1900 and 1960 the Criollo varieties have become almost completely extinct without any attempts to save them. In the Teaching and Research Centre of the Interamerican Institute of Agricultural Sciences (IICA) at Turrialba, Costa Rica, the author has gathered together from these areas in the last 15 years the largest number of population samples in the world. However, this is still far from being a representative sample of the Criollo varieties which the Mayas and Aztecs cultivated, and new efforts are needed to prevent their complete disappearance.

Due to the appearance of the Witches' Broom fungal disease (*Marasmius perniciosus*), which attacked all varieties then cultivated, the Imperial College of Tropical Agriculture, Trinidad, sent Dr J. Pound on expeditions to the Amazon basin in 1937 and 1942–3 to collect cocoa resistant to this disease. Pound collected material along the whole length of the Amazon, from Belem at its mouth to near the headwaters of the river Marañón and its tributaries. He travelled in the surroundings of Iquitos, the river Nanay, and the mouths of the Marañón, Urcayali and Huallagas rivers; he ascended the River Napo (Ecuador) and thence continued to the headwaters of the Putumayo and the Caquetá (Colombia). In all these tributaries he found cocoa in a wild or semi-wild state. He made a large number of collections either as seeds or budwood for grafting, all of which he sent to Trinidad. This material has served as a basis for breeding programmes in nearly all parts of the world.

In 1952–4 the Imperial College of Tropical Agriculture at Trinidad sent Drs Becker, Bartley and Holliday on a further expedition to explore areas hitherto not satisfactorily covered by Pound in the Caquetá, Orteguaza and Vaupés rivers of Colombia.

In 1949, searching for material resistant to Witches' Broom disease, Srs Desrossiers and Buchwald, of the Experimental Station of Pichilingue, Ecuador, collected cocoa in the rivers Coca and Napo. Other expeditions organized by this Station visited the same and other areas in the region in 1958 (Doak, Ampuero and Buchwald) and in 1961 (Doak and Zumbrano), and brought back material to Pichilingue (see Soria, 1970).

In 1964 Soria collected cocoa seeds in the Iquitos region because resistance to *Ceratocystis fimbriata* had been detected in material from that area (Soria, 1970).

In 1965 CEPLAC (Comissão Executiva do Plano de Recuperacão Econômica da Lavoura Cacaueira) from Brazil and IICA from Costa Rica sent out an expedition to Brazil (Soria, Vello, Murça Pires and Medeiros) to explore the tributaries of the Madeiras and Purus rivers in the territories of Rondonia and Acre in south-east Brazil, and in the vicinity of Manaus and Obidos, Pará (Soria, 1970). Later, Vello & Rocha (1968) collected wild cocoa in the rivers Juruá and Moa in the State of Acre and the vicinity of Manaus. Vello & Ferreira da Silva (1968) collected cocoa in the river Jarí, which divides the State of Pará in a north–south direction, and in the Territory of Amapá in northern Brazil. All the collections are being multiplied in the experimental station of IPEAN (Instituto de Pesquisas Agropecuarias do Norte), Belem and the quarantine station of Salvador, Bahía.

Between 1968 and 1972 the Experimental Station of Pichilingue (INIAP) of Ecuador and the Imperial College of Tropical Agriculture at Trinidad carried out four expeditions to the headwaters of the rivers Putumayo, Napo, Coca, and Pastaza, the last of these being the principal centre of genetic variation of cocoa in its natural habitat. Similarly, in 1969, the author collected samples of wild cocoa in the headwaters of the rivers Orteguaza, Caquetá and Putumayo in Colombia.

The majority of collections made up to now have been entirely directed towards a search for resistance to the Witches' Broom disease and for the desirable agronomic characters of large fruits and seeds. Representative sampling of genetic variation of the species as a whole was not made, and our methods in this respect must be considerably revised for the future.

The efforts of recent years to collect material in the centre of diversity of cocoa is due to the discovery of large oilfields in this area. The exploration and exploitation of petroleum with the opening up of roads has brought about much settlement and deforestation. For this reason, if immediate and concerted efforts are not made to collect and conserve the indigenous cocoas of the region, we shall run the risk of losing this invaluable germplasm. It is in this region that Pound found all the gradations of form and size of fruits and seeds. The greatest concentration of incompatability alleles and possible sources of disease resistance have been found there, also. In the last few months various cocoa-producing countries in Africa, such as Ghana, Nigeria, Ivory Coast and Cameroun have expressed an interest in taking part in expeditions to this area.

The other areas of natural distribution of cocoa in the headwaters of the Amazonian tributaries are not so urgently in need of emergency collecting, but these areas also will need attention in the near future with the opening of new communications and the settlement of populations by various countries with access to the Amazon basin.

Genetic erosion in the secondary centre of diversity is perhaps as much or more serious than in the primary centre. There is no doubt that cocoa was domesticated in Mexico and Central America by the Mayas and developed into one of their most valuable crops. These cocoas form a distinct genetic group, even though included within the same species, and they exhibit a good range of variability. With the introduction of some of the Amazonian cocoas into certain parts of Mesoamerica, an enormous variability has been created as a result of hybridization and recombination.

In spite of the fact that the Criollo cocoas are considered as the finest in flavour and aroma, they yield poorly and are susceptible to diseases and pests. For this reason they have been gradually replaced by more profitable crops, such as bananas (Costa Rica, Honduras), cotton (Nicaragua) and other improved cocoa varieties with a strong influence from the Amazonian strains. It is thus very difficult to find in Mexico and Costa Rica even small plantations of pure Mexican and Central American Criollo cocoas. One occasionally finds isolated trees or groups of trees in certain areas, but because of their high susceptibility to *Ceratocystis fimbriata*, a disease which has now invaded the whole region, these fine quality varieties are in danger of disappearing completely. Unfortunately not a single institution, with the exception of Turrialba, has taken an interest in collecting this extremely valuable germplasm.

In the last ten years I have, whenever possible, been systematically collecting seeds and budwood of this group to add to the collections at Turrialba. Nevertheless the losses from *Ceratocystis* are great and make continual replacement of the material a matter of necessity. Although the use of superior cocoas is limited, a big international effort is needed to maintain this germplasm in *Ceratocystis*-free areas.

The chief germplasm collections of cocoa in the western hemisphere are maintained in the following institutions: (1) Marper Farm and the Imperial College of Tropical Agriculture, Trinidad; (2) Turrialba, Costa Rica; (3) Estación Experimental Tropical, Pichilingue, Ecuador; (4) CEPLAC, Itabuna, Brazil, and (5) Mayagüez, Puerto Rico.

Difficulties and limitations in collecting work

Among the principal difficulties in cocoa collecting are the following.

(1) The collecting sites are situated in very remote areas of the Amazonian forests, necessitating the use of many different types of transport; this lengthens the expeditions very considerably. Although it is not difficult to arrive at a settlement in the general collecting area by means of light aircraft, from there onwards one must travel on the rivers, at best by boats with outboard motors, or otherwise by canoe or on land by foot.

(2) In the collecting areas, the labour of making the actual collections is slow and difficult, since the points of occurrence of the cocoa plants are unknown. In most cases the Indians know the plants though they often confuse them with other species of *Theobroma*. For this reason one has to visit many localities before finding the species in question. However, in most cases the Indians readily agree to act as guides.

(3) Food and lodging are scarce in the regions to be visited, which makes it necessary to travel with camping equipment, since a convenient village is seldom to be found.

(4) Perhaps the greatest difficulty in obtaining a successful outcome to an expedition is the short viability of cocoa seeds and budwood, such that if immediate transport is not provided to carry the material to its ultimate destination it will die and the expedition will be valueless. In our experience this is the most critical factor, making it essential to improve methods to extend seed and budwood viability, to find better ways of collaboration with the local airways with a view to rapid transport of material, and finally, to take immediate action at the reception station to propagate or sow the material collected directly it is received.

References

Cheesman, E. E. (1944). Notes on the nomenclature, classification and possible relationships of cacao populations. *Trop. Agric., Trin.* **21** (8), 144–59.

Patiño, V. M. (1963). *Plantas cultivadas y animales domesticos en America Equinoccial* (first edn), vol. I. *Frutales*, pp. 268–334. Imprenta Departamental, Cali, Colombia.

Pound, F. J. (1938). *Cacao and witch broom disease* (Marasmius perniciosus) *of South America, with notes on the other species of* Theobroma; *report on a visit to Ecuador, the Amazon Valley and Colombia. April 1937– April 1938.* Port of Spain, Guille's Printerie. 58 pp. map.

Soria, J. (1966). Notas sobre las principales variedades de cacao cultivadas en America Tropical. *Turrialba, Costa Rica*, **16** (3), 261–6.

Soria, J. (1970). The latest cocoa expeditions to the Amazon basin. *Cacao, Turrialba, Costa Rica*, **15** (1), 5–15.

Vello, F. & Rocha, H. M. (1968). II. Expediçao a Amazonia Brasileira. CEPLAC, Centro de Pesquisas de Cacau. Rodovia Ilheus – Itabuna, Bahía. *Comunicação Técnica*, no. **4**, 20 pp.

Vello, F. & Ferreira de Silva, L. (1968). Relatorio de viagem a regiao amazonica. CEPLAC, Centro de Pesquisas de Cacau. Rodovia Ilheus – Itabuna, Bahía. *Comunicação Técnica*, no. **22**, p. 19.

PART III

Evaluation problems

16. The search for disease and insect resistance in rice germplasm

T. T. CHANG, S. H. OU, M. D. PATHAK, K. C. LING &
H. E. KAUFFMAN

The cultivated rices (*Oryza sativa* L. and *O. glaberrima* Steud.), their weed races, and the 18 wild species of the genus *Oryza* provide a rich pool of genes for rice improvement. In spite of the rapid adoption of the high-yielding semidwarf varieties in many countries, a substantial segment of the world's rice germplasm has been acquired by the International Rice Research Institute from the beginning of its operations (see Chang *et al.*, Chapter 35). IRRI scientists have evaluated the diverse gene pools in the germplasm bank and made allied studies of a fundamental nature (Chang, 1970). Among the IRRI screening programs for disease, insect and drought resistance, tolerance to adverse soil conditions, and improved yield, photosynthetic efficiency, and nutritive quality (Athwal, 1972), the identification of sources of resistance to the major diseases and insects ranks high in IRRI's contributions to rice improvement. The success of the IRRI evaluation programs has prompted several countries in tropical Asia to follow suit (All-India Coordinated Rice Improvement Project, 1969; Shastry, Freeman, Seshu, Israel & Roy, 1972; Fernando, 1972; Harahap, Siregar & Siwi, 1972; Pongprasert, Kovitvadhi, Leaumsang & Jackson, 1972; Seetharaman, Sharma & Shastry, 1972). This paper summarizes the progress of several screening programs on the principal rice diseases and insects from 1963 to 1972 at IRRI and affiliated national centers.

Blast disease

Blast disease caused by *Pyricularia oryzae* Cavara is one of the most widespread diseases of rice. It inflicts heavy damage on the rice plant by attacking young leaves (seedling blast), nodes of the culm (node blast), the base of the panicle (neck rot), and the panicle branches and the hulls.

Evaluation problems

Blast-resistant sources

Findings from more than 270 tests of the International Blast Nurseries have shown that the following varieties consistently produced resistant reactions of a 'vertical' type at most test sites: (1) Tetep, Nang Chet Luc and Trang Cut L. 11 from Vietnam; (2) Tadukan and Carreon from the Philippines; (3) Pah Leuad 111 from Thailand; (4) Huan-sen-go and Ta-poo-cho-z from China; (5) H-5 and M-302 from Ceylon; (6) D-25-4 and C-46-15 from Burma; (7) Ram Tulasi Sel. 1 from India; and (8) R-67 and Mamoriaka from African countries (Ou, Nuque & Ebron, 1970; Ou & Ebron, 1974). No varieties showing typical 'horizontal resistance' have been found.

Pathogenic variability

Isolates of *P. oryzae* vary in pathogenicity, cultural characteristics, and nutritional requirements. The fungus has a variety of genetic mechanisms that can continuously produce new variability: both uninucleate and multinucleate conidia, heterocaryotic cells, anastomosis of hyphae, parasexualism, and heterocytosomes. Critical cytological evidence related to genetic variability is lacking however (Ou, 1972a).

P. oryzae has many races, some of which appear in geographic groups as indicated by a regional pattern of varietal reactions. But either single conidial cultures, single-cell cultures or conidia from a single lesion can give rise to several races, both old and new (Ou & Ayad, 1968; Giatgong & Frederiksen, 1969; Ou, Nuque, Ebron & Awoderu, 1970).

Genetic control of resistance

Mendelian analysis of many crosses where single isolates of the fungus were used indicated that resistance in a variety is controlled by any one of the following mechanisms: (1) from one to four dominant genes that could be independent or cumulative in action, (2) one major dominant gene and a few minor genes, (3) linked dominant genes, (4) a recessive gene for resistance, (5) a dominant gene inhibiting a resistant gene, or (6) from five to nine effective factor pairs (Chang, 1964; Takahashi, 1965).

Studies in Japan have differentiated 13 major genes of the above nature, most of which confer the vertical type of resistance (Kiyosawa,

184

1972). For field resistance, Japanese workers have found that either one major gene or a major gene and a few minor genes can reduce the number of lesions (Kiyosawa, 1972; Toriyama, 1972).

Resistance controlled by the highly specific major genes tends to break down due to changes in the composition of prevalent races or mutations of the avirulence gene to virulence in the fungus (Kiyosawa, 1972). Therefore, varieties such as Tetep, Carreon, and Tadukan that have a broad spectrum of resistance to many isolates of the fungus not only offer a long span of usefulness but also tend to minimize the number of lesions on the leaf when inoculated with a virulent race (Ou, Nuque, Ebron & Awoderu, 1970). This type of stable resistance is being searched for by the use of the International Blast Nurseries (Ou, 1972*b*).

Tungro virus

Tungro is the most widespread virus disease of rice in tropical and sub-tropical Asia. Tungro and tungro-like diseases have been called leaf yellowing in India, *penyakit merah* in Malaysia, *mentek* in Indonesia, and yellow-orange leaf in Thailand. The tungro virus is primarily trans-mitted by the rice green leafhopper [*Nephotettix virescens* (Distant)]. Other species of leafhoppers, *Nephotettix nigropictus*, *N. parvus*, and *Recilia dorsalis*, are also capable of transmitting tungro, though at a much lower level of efficiency than *N. virescens* (IRRI, 1972*a*).

Sources of resistance

Among the 3000 IRRI accessions tested for tungro reaction by the mass screening technique (Ling, 1969; IRRI, 1973), a wide range of resistance was found. The most resistant variety was Pankhari 203 from India. Many other resistant accessions came from Bangladesh, China, India, Indonesia, Malaysia, Pakistan, Philippines, Sri Lanka, and Thailand (Ou, Rivera, Nararatnam & Goh, 1965; Rivera, Ou & Tantere, 1968; Ling, 1969; 1972).

By comparing the life span of insects on several varieties, IRRI researchers were able to separate the reaction of a variety to the virus from its reaction to the vector. Thus, Pankhari 203 is resistant to both tungro and its insect vector, while IR8 is moderately resistant to the green leafhopper but susceptible to tungro (Ling, 1968).

185

Interaction of virus strains and resistant sources

In the Philippines there are three strains of tungro virus: 'S', 'M', and 'T', of which 'S' is the most prevalent strain (Ou & Rivera, 1969; IRRI, 1972*a*). If a variety is resistant to the 'S' strain, it is also resistant to the other two strains. Indian workers have identified four substrains under two strains of tungro (Shastry, John & Seshu, 1972). Some of the Indian strains appear to be different from the Philippine strains, because test varieties show differential reactions at two sites (Shastry, John & Seshu, 1972; IRRI, 1972*a*).

Genetics of resistance to the tungro virus

Preliminary studies at IRRI on crosses involving Pankhari and susceptible varieties showed that the F_1 hybrids were generally resistant to the tungro virus. Some of the F_2 populations segregated as 9R:7S, showing the dominant nature of resistance (IRRI, 1967*a*). In the cross of Peta × I-geo-tze, F_3 lines showed a continuous distribution in the level of resistance and relatively few resistant semidwarf lines were obtained (T. T. Chang and K. C. Ling, unpublished). In the IR8 × Latisail cross, Indian workers found that the F_1 plants were resistant to tungro and the 20-day-old seedlings were scored as 9R:7S. But at 60 days following transplanting, the F_2 ratio became 15R:1S (Shastry, John & Seshu, 1972). Earlier observations at IRRI (IRRI, 1968) showed that among hybrid progenies some infected seedlings recovered from the disease and appeared healthy at a later stage, while some others became diseased as they grew older. This could indicate that seedling reaction and adult resistance are not controlled by the same gene or genes. In India the change in reaction is one-directional, from 'susceptible' to 'resistant', suggesting a recovery from infection.

Grassy stunt virus

Grassy stunt is found in India, Malaysia, Philippines, Sri Lanka, Thailand, and Vietnam. The virus is transmitted by the brown planthopper [*Nilaparvata lugens* (Stål)] and is of the persistent type. This disease is transmitted at a lower efficiency than the tungro and requires more viruliferous insects per seedling and a longer period of acquisition feeding (Ling, 1972).

Sources of resistance

After testing 10000 entries, including 2540 IRRI accessions, 100 wild forms, and more than 7000 IRRI breeding lines, the only source of resistance was found in a strain of *Oryza nivara* (Sharma and Shastry) from India (IRRI Accession No. 101508). It contained a few plants that are resistant to the grassy stunt virus (IRRI, 1970; Ling, Aguiero & Lee, 1970). The *O. nivara* strain is suceptible to the vector while it is resistant to the virus. Fortunately, this wild strain is relatively cross-fertile with the cultivars of *O. sativa* (Khush & Beachell, 1972).

Genetics of resistance

The resistance of the *O. nivara* strain to grassy stunt has been shown to be controlled by a single dominant gene (Khush, Torres & Aquino, 1971).

Bacterial leaf blight

Bacterial leaf blight (*Xanthomonas oryzae* Uyeda and Ishiyama) is widely distributed in all rice growing countries of Asia. The disease generally appears around heading, causing water-soaked stripes on leaf margins and particularly on the flag leaf. In severe cases the lesion may extend to the lower leaf sheaths. Seedlings can also be infected and killed by the pathogen – known as the *kresek* disease in Indonesia. Infection at certain stages of growth results in pale yellow leaves. The disease appears annually over such a wide area that its damage to the crop must be extensive. In severe cases, the yield losses range from 30 to 50 per cent.

Pathogenic variability

Strains of the pathogen differ considerably in virulence. Isolates from tropical countries are generally more virulent than those from Japan (Goto, 1965; Wakimoto, 1967; Buddenhagen & Reddy, 1972). While there is appreciable difference in virulence among strains within some countries (Ou, Nuque & Silva, 1971; Kauffman & Rao, 1972), the most virulent isolates came from Bangladesh, India, and Ceylon (Wakimoto, 1967; Buddenhagen & Reddy, 1972).

Sources of resistance

Among 9000 accessions screened at IRRI during 1965–9, resistance was first found in accessions from Bangladesh, China, Egypt, India, Indonesia, Japan, Philippines, Taiwan, and the USA (IRRI, 1968). Later, through international seed exchange and testing, TKM-6 and BJ-1 from India, Zenith from USA, and Semora Mangga and Sigadis from Indonesia were found to have high levels of resistance in several countries. The latest findings indicate that TKM-6, Sigadis, BJ-1, and IR22 have a broad spectrum of resistance over a wide geographic distribution in tropical Asia. During 1971–2, 3500 more accessions were screened by the leaf clipping method at IRRI (Kauffman, Reddy, Hsieh & Merca, 1973).

About 400 accessions from the Assam rice collection of India were screened by the All-India Coordinated Rice Improvement Project (1971).

Genetics of resistance

Studies in Japan indicate that resistance to the pathogen is controlled either by one dominant gene, two linked genes or two complementary genes (Toriyama, 1972). IRRI plant pathologists have found that resistance may be recessive or dominant in the F_1 hybrids, depending on the cross combination and the level of resistance in the resistant parent (IRRI, 1967*b*). Heu, Chang & Beachell (1968) found that the resistance of Sigadis to Philippine isolate 72 is controlled by a partially dominant gene and that it is loosely linked with the recessive semidwarfing gene in Taichung Native 1. The resistance of Lacrosse × Zenith-Nira selection to an Indian isolate is governed by a single recessive gene (All-India Coordinated Rice Improvement Project, 1969). Using the Philippine isolate B15-37, Murty & Khush (1972) found that the resistance in BJ-1 is controlled by an incompletely dominant gene, while the resistance of DZ-192 involves two recessive loci.

Sheath blight

Sheath blight caused by *Thanatephorus cucumeris* (Frank) Donk (= *Corticium sasakii* (Shirai) Matsumoto) infects the leaf sheath and blade. In severe cases it kills the plant. The disease is often associated with high temperatures, high humidity, and liberal nitrogen supply.

188

Ta-poo-cho-z from China and several varieties collected from Assam (India) and Bangladesh are the most resistant accessions among 1000 varieties tested, but the level of resistance is not as high as would be desired for breeding purposes, probably because the pathogen is a facultative parasite with broad host range and large inherent variability (Ou, 1972*a*).

Hashioka (1951) reported that resistance to sheath blight is inherited as a dominant character in Japanese varieties and in some crosses a 3R:1S ratio was found in the F_2 populations.

Bacterial leaf streak

Bacterial leaf streak (*Xanthomonas translucens* f. sp. *oryzicola*) is found only in tropical Asia. Disease incidence generally appears after typhoons and storms, resulting in linear lesions between veins and death of the infected leaf tissues.

Goto (1965) tested 102 varieties and found that most japonica varieties were relatively resistant while most indica varieties were susceptible. Among 700 varieties tested at IRRI, the most resistant accessions were DZ-60 from Bangladesh, three *spontanea* strains from Australia, and one strain of *O. rufipogon* from Malaysia. Other resistant accessions came from Bangladesh, China, India, Taiwan, and USA. Several strains of *O. glaberrima* and *O. rufipogon* were also resistant. Many strains exist in the pathogen but resistant varieties were resistant to all isolates tested (Ou, Franck & Merca, 1971).

Stem borers

The principal species of stem borers found in tropical Asia are the striped rice borer [*Chilo suppressalis* (Walker)], the yellow borer [*Tryporyza incertulas* (Walker)], the white borer [*T. innotata* (Walker)] and the pink borer [*Sesamia inferens* (Walker)]. Of these, the striped borer and the yellow borer have been more intensively investigated. Borer damage to young rice tillers is commonly called 'dead hearts'. Infestation of the internodes prior to heading results in 'white heads'.

Sources of resistance

Varieties resistant to the striped borer are: (1) DD48, DNJ-97, DV88, DZ41, and HBJ Boro 2 from Bangladesh; (2) Rusty Late, Su-Yai 20, and Szu-Miao from China; (3) Yabami Montakhab 55 from Egypt;

Evaluation problems

(4) TKM-6, CO-13, Patnai 6, and PTB-10 from India; (5) Ginmasari from Japan; and (6) Chianan 2 and Taitung 16 from Taiwan. The resistant varieties appear to differ in their reactions to dead heart damage or to white-head infestation, however (Pathak, Andres, Galacgac & Raros, 1971).

In field and screenhouse testings of several resistant varieties, the striped borer, the yellow borer, the pink borer, and the white borer caused consistently low dead-heart percentages on TKM-6, while the yellow, white, and pink borers caused few dead hearts on PTB-10, Taitung 16, Chianan 2, DV 139, Mudgo, IR8, MTU 19, and Su-Yai 20. Controlled feeding rates on young plants show that borers on TKM-6, Chianan 2, Taitung 16, DV 139, and Mudgo have the lowest mean values for larval weight, percentage of insect survival, and percentage of dead hearts (Pathak, 1972). TKM-6 has been extensively used in the IRRI breeding program and is a parent of IR20 which also has a moderate level of borer resistance.

Genetic control of borer resistance

Because borer resistance is a complex phenomenon, genetic studies require an understanding of the different components of resistance involved. Using borer infestation as a criterion of resistance, Koshiary· *et al.* (1957) showed that the field resistance of Giza 14 to stem borers was under multiple gene control but few genes appeared to be involved. Another report indicates that the field resistance of TKM-6 to stem borers, as measured by the incidence of white heads, was simply inherited (All-India Coordinated Rice Improvement Project, 1968). Studies at IRRI showed that the F_2 distribution with respect to larval weight in a resistant × susceptible cross could be ascribed to the effect of multiple genes. The distribution of such a cross according to dead-heart incidence was continuous in nature. Judging by the relative ease in recovering resistant progenies from resistant × susceptible crosses, borer resistance could be rather simply inherited and have relatively high heritability (Koshiary *et al.*, 1957; All-India Coordinated Rice Improvement Project, 1968; Athwal & Pathak, 1972).

Area of future research

Only moderate resistance to stem borers has been found. It consists primarily of antibiosis and to a certain extent the non-preference type of resistance. Larvae caged on resistant plants suffer high mortality

190

and have slower rates of growth than those on susceptible plants. Although this level of resistance can protect the crop from dead hearts, it often fails to provide adequate protection from white head. This is because white heads are often caused by the larvae with little feeding or, in other words, before varietal resistance begins to exert an adverse effect on the larvae. IRRI's research is therefore oriented towards incorporating the non-preference type of resistance that may minimize the number of eggs laid on the plants and towards combining resistance from various sources to increase the level of resistance in the plant.

Most of the studies have been restricted to the striped borer. Since several other species of borers, such as the yellow borer, the pink borer and the white borer are also important in certain areas, efforts will be made to develop lines resistant to all these species. The overall objective, however, is to develop varieties that are not only resistant to the common species of borers but to other important insect pests as well.

Leafhoppers and planthoppers

The principal species of this leaf-feeding and virus-transmitting group found in tropical Asia are the rice green leafhopper, *Nephotettix virescens* (Distant), and the brown planthopper, *Nilaparvata lugens* (Stål). The green leafhopper is a vector of the tungro virus and the brown planthopper transmits the grassy stunt virus. Both can cause death of rice plants when infestation becomes very heavy, a phenomenon commonly known as 'hopperburn' (Pathak, 1969*b*).

Sources of resistance

Many accessions are resistant to the green leafhopper: (1) DK 1, DM 77, DNJ 9, DNJ 97, and Jhingsail of Bangladesh, (2) Su-Yai 20, Bir-Tsan 3, and Lien-Tsan 50 of China, (3) Pankhari 203, ASD-7, PTB-18, PTB-21, CO-9, D-204-1, and Sukali of India, (4) Chao Phokha, Do Khao, and Pa Thong from Laos, (5) FB 123 and UCP 122 of the Philippines, (6) Sinnasuappu from Sri Lanka, and (7) Betong, Intan 2400, Mas, and Peta 2802 from Indonesia.

Accessions resistant to the brown planthopper are principally from India and Sri Lanka. Mudgo was the first variety shown to have brown planthopper resistance and it has been extensively used in several breeding programs. Accessions resistant to the white-back planthopper came mainly from China, India, and Sri Lanka. Seedling reactions to

191

the leafhoppers and planthoppers obtained in the screenhouse (Pathak, Cheng & Fortuno, 1969) agreed with the reactions of adult plants grown in the field (Pathak, 1972).

Nature of resistance to leafhoppers and planthoppers

In screening tests the insects showed a gustatory non-preference for certain varieties. The brown planthoppers caged on Mudgo starved to death rather than feed on the resistant plants.

Both green leafhoppers and brown planthoppers lost weight when caged on resistant varieties, the differences being greater in the brown planthopper. The brown planthoppers fed more frequently on the susceptible Taichung Native 1 and excreted more honey dew. Generally speaking, varietal resistance to the leafhopper or planthopper is indicated by the higher mortality of nymphs found among insects feeding on individual plants of the resistant variety (Pathak, 1969*a*; 1972).

Genetic control of resistance

The inheritance of resistance to the green leafhopper was studied by Athwal, Pathak, Bacalangco & Pura (1971) in crosses among Pankhari 203, ASD-7, IR8, and Taichung Native 1. Three independent dominant genes for resistance were found: *Glh-1* in Pankhari 203, *Glh-2* in ASD-7, and *Glh-3* in IR8. More recent studies involving other resistant sources showed that PTB-18 carries two genes for resistance, while DK-1 and Su-Yai 20 each have one gene for resistance. The gene for resistance in DK-1 does not appear to be allelic to *Glh-1* or *Glh-3* (IRRI, 1972*a*).

Genes for resistance to the brown planthopper were analyzed from crosses involving Mudgo, ASD-7, CO-22, MTU-15, MGL-2, PTB-18, and Taichung Native 1. The resistance gene in Mudgo, CO-22, MGL-2, and MTU-15 belongs to the *Bph-1* locus; that of ASD-7 and PTB-18 to *bph-2*. Recombination between *Bph-1* and *bph-2* was rare or non-existent. It appears that the two genes are closely linked or allelic at a complex locus (Athwal *et al.*, 1971; IRRI, 1972*a*).

The monogenic dominant (*Bph-1*) controlling the resistance in Mudgo was confirmed in Taiwan by Chen and Chang (1971). The resistance found in three IRRI breeding lines was postulated by C. R. Martinez (unpublished) to belong to either *Bph-1* or *bph-2*. The *Glh-2* and *bph-2* loci in ASD-7 are independent; the *Glh-1* locus in Pankhari 203 and *Bph-1* in Mudgo are non-allelic (Athwal & Pathak, 1972). Such genetic

192

information would help in combining resistance to both leafhoppers and planthoppers and in incorporating different resistance genes into improved varieties.

IRRI plant pathologists have noted that the average life span of brown planthoppers reared on Mudgo for ten generations improved from 4.2 days in the first generation to 16.0 days in the tenth generation (IRRI, 1970). The 16-day span is about equal to the life span of the insect on Taichung Native 1. The entomologists tested Mudgo plants against planthoppers which had been reared on Mudgo for 22 generations and found Mudgo nearly as susceptible as Taichung Native 1. Apparently a new biotype (biotype 2) developed from continuous rearing of the insect on resistant Mudgo. Varieties having the *Bph-1* gene for resistance such as CO-22, MTU-15, and MGL-2 were also susceptible to biotype 2, while those possessing the recessive *bph-2* gene remained resistant. The breakdown of a major source of resistance to a new biotype points to the need for using genetically diverse sources of resistance in breeding programs (Athwal & Pathak, 1972).

Rice whorl maggot

The rice whorl maggot [*Hydrellia philippina* (Ferino)] is widely distributed in South-East Asia. It feeds on the inner margin within the whorl of the youngest leaves while they are still unopened and scrapes off the tissues leaving only the two epidermal layers. The damaged leaves are shriveled and whitish.

Moderately resistant accessions have been identified among varieties collected from Assam (India), varieties originating from other parts of India, and varieties from China (Pathak, 1972).

Rice gall midge

The rice gall midge [*Pachydiplosis oryzae* (Wood-Mason)] is found in South China, India, Indonesia, Sri Lanka, Thailand, and Vietnam. The insect induces excessive tillering in young plants. The tillers turn into tubular galls ('silver shoot') which dry off without forming panicles. Infestation at a later stage causes branching of the culms at the higher nodes.

By testing thousands of varieties under field conditions, workers in India have identified Eswarakora, PTB-18, PTB-21, and several Assam rices as highly resistant (Israel, Rao & Prakasa Rao, 1963; Ventaka-

193

swamy, 1969; Shastry, Freeman, Seshu, Israel & Roy, 1972). W-1263, a hybrid strain selected from Eswarakora × MTU-15, combines gall midge resistance and resistance to other insects (Ventakaswamy, 1969). This strain has been extensively used in several national breeding programs and at IRRI.

Studies in India indicate that susceptibility is controlled by three complementary dominant genes, one of which is suppressed by a dominant inhibitor. Biotypes appeared to exist in the insect when the same set of resistant accessions was tested at different sites (Shastry, Seshu, Israel & Roy, 1972).

General discussion and conclusions

In reviewing the success obtained by IRRI scientists in evaluating and using rice germplasm, several factors contributing to the success may be discussed.

The germplasm being evaluated for resistance to a disease or an insect generally involves several thousand accessions, embracing different geographic regions and eco-genetic groups. Often the screening tests include wild forms and weed relatives of the cultivated rices. Moreover, for both *O. sativa* and *O. glaberrima* sufficient genetic diversity still exists today so that the screening programs include a full spectrum ranging from the primitive types to the modern elites. Testing large numbers of accessions gives better chances of finding resistant sources than testing a small number of entries. Outstanding sources of resistance were found in agronomically poor types such as Pankhari 203 and Mudgo and in the weed race *O. nivara*.

A resistant strain found in any testing program may be truly resistant, not infected by chance, or a product of genotype–pest–environment interactions under the prevailing test conditions. At IRRI any resistant accession found in the initial screening is generally retested, often repeatedly by more refined techniques, to ascertain the true nature of resistance.

Since rice germplasm has not been subject to intensive breeding for race-specific disease resistance as has taken place in wheats and potatoes, we can expect to identify more readily the horizontal type of disease resistance at these initial stages.

Genetic variability in the plant pathogen or insect pest is taken into consideration in all of the testing programs. For blast and bacterial leaf blight, rice accessions are inoculated with a composite of the prevalent isolates of the pathogen, or tested in succession with different isolates. Strains are also known to exist in the tungro virus and in the brown planthoppers. The IRRI pathologists and entomologists frequently collect samples from different provinces in the Philippines and compare the pathogenic isolate or the insect sample with the standard isolate or strain routinely used in the tests at Los Baños. In this manner, from samplings taken in the Philippines, varying numbers of distinct strains or biotypes were found to exist in the blast fungus, bacterial leaf blight pathogen, tungro virus, and the brown planthoppers. Similarly, workers in India have identified four substrains for the tungro virus (Shastry, John & Seshu, 1972).

Empirical search for pest resistance had been previously attempted in rice at several research centers and the results were often less than satisfactory. A lack of the necessary facilities could be one of the factors. When a clear understanding of the host–pest reaction in relation to environment factors, population density of the pest, and inherent variability of the pest was largely deficient, the researcher was unable to secure reliable and repeatable test results. At IRRI, while the main emphasis was the identification of resistance sources, research was also made on the nature of resistance and genetic variability in the pest so that more efficient screening methods could be formulated and the findings more clearly interpreted.

Several areas of research that greatly aided the screening program could be cited to illustrate the importance of fundamental research on the pathogen or insect. For the blast fungus, single conidial isolates were found continuously to produce pathogenically different progenies (Ou, Nuque, Ebron & Awoderu, 1970). This reveals a complex genetic mechanism within the fungus which may account for its enormous variability in terms of the number of isolates and the varying levels of virulence. Studies of International Blast Nurseries data, in conjunction with monitoring efforts on the prevalence and changes in the blast races, led to the search for a stable type of varietal resistance (Ou, 1972*b*). For the tungro virus disease, initial field screening results were rather variable. The identification of truly resistant stocks was made possible when the non-persistent nature of the virus was established (Ling, 1966). The life span of insects in controlled feeding tests separated a variety's reaction to the virus from its response to the insect vector (Ling, 1968; Ling, *et al.*, 1970). For the stem borers, entomologists were able to differentiate antibiosis from preference in the moths by comparing oviposition and the survival and pupation of larvae on test

195

varieties (Pathak, 1972). When the brown planthoppers were fed on Mudgo, the high mortality of nymphs and the slower increase in body weight ascertained the true nature of Mudgo's resistance to the insect (Pathak, 1969*a*; 1972).

The search for potentially promising genetic material is aided by the frequent contacts that the IRRI researchers maintain with their colleagues abroad through the attendance at international scientific conferences, visits to research centers, discussions with foreign scientists visiting at Los Baños, and free exchange of ideas and information at the annual IRRI international rice research conferences or at one of the six major symposia held at IRRI.

Outstanding sources of resistance to diseases and insects were found in those countries where the disease or insect is endemic and where varietal diversity is great. Examples are blast resistance found in varieties from the Philippines and Vietnam; tungro resistance in varieties from Bangladesh, India, and Indonesia; brown planthopper resistance in varieties from India and Sri Lanka. We are collecting more minor varieties from remote areas of those countries where an important disease or pest is known to prevail. Meanwhile, we screen intensively among accessions originating from such endemic areas for high levels of resistance.

The finding of many sources of resistance to a certain disease pathogen or insect pest does not necessarily indicate that the resistant accessions are genetically diverse in nature. By crossing a new resistant variety with a standard tester (resistant or susceptible type) or by intercrossing the resistant accessions, the allelic relationship between resistant cultivars can be readily determined. By such tests it has been established that among six sources, resistance to the green leafhopper is primarily controlled by three or more dominant genes, and resistance to the brown planthopper among six sources, by one dominant and one recessive locus (Athwal & Pathak, 1972). Such 'genetic cataloging' could effectively reduce the number of resistance sources that a rice breeder needs to use as parents and to maintain as breeding stocks. Multi-disciplinary collaboration is essential in such studies.

Interdisciplinary collaboration has established the sources of multiple resistance found in a single accession. Thus TKM-6 has resistance to the stem borers, bacterial leaf blight, and green leafhoppers. Sigadis has resistance to tungro and to bacterial leaf blight. Pankhari 203 is resistant to both the tungro virus and its vector. A truly virus-resistant variety requires resistance to both the virus and the vector.

196

While laboratory or screenhouse testing is invariably followed by field testing for verification, the resistant materials are also distributed to various collaborating national agencies for worldwide testing. At a symposium on rice breeding held at IRRI in 1971, a collaborative scheme for the exchange of resistant stocks for testing to blast, bacterial leaf blight, tungro virus disease, stem borers, the leafhoppers and planthoppers, and the gall midge was developed among rice researchers from many countries (IRRI, 1972*b*). Multi-site testing would not only identify genotypes of broad resistance but also foster international exchange of material and information. We envisage further expansion in the scope of such co-operative programs when useful findings are made known to researchers in other countries. Other important diseases and pests might also be included in the future. Multi-site and multi-disciplinary collaboration makes possible a dynamic program of evaluating and using the rice germplasm by recombining genetic resistance to the important diseases and insect pests.

References

All-India Coordinated Rice Improvement Project (1968). *Progress report, kharif 1968*, vol. I. Indian Council of Agricultural Research, New Delhi.

All-India Coordinated Rice Improvement Project (1969). *Progress report, kharif 1969*, vol. III. Indian Council of Agricultural Research, New Delhi.

All-India Coordinated Rice Improvement Project (1971). *Progress report of the All-India Coordinated Rice Improvement Project*, vol. III. Indian Council of Agricultural Research, New Delhi.

Athwal, D. S. (1972). IRRI's current research program. In *Rice, Science, and Man*, pp. 41–64. International Rice Research Institute, Los Baños, Philippines.

Athwal, D. S. & Pathak, M. D. (1972). Genetics of resistance to rice insects. In *Rice Breeding*, pp. 375–86. International Rice Research Institute, Los Baños, Philippines.

Athwal, D. S., Pathak, M. D., Bacalangco, E. H. & Pura, C. D. (1971). Genetics of resistance to brown planthoppers and green leafhoppers in *Oryza sativa* L. *Crop Sci.*, **11**, 747–50.

Buddenhagen, I. W. & Reddy, A. P. K. (1972). The host, the environment, *Xanthomonas oryzae*, and the researcher. In *Rice Breeding*, pp. 289–95. International Rice Research Institute, Los Baños, Philippines.

Chang, T. T. (1964). Present knowledge of rice genetics and cytogenetics. *Int. Rice Res. Inst. Tech. Bull.* **1**, 96 pp.

Chang, T. T. (1970). The description and preservation of the world's rice germplasm. *SABRAO Newslett.* **2** (1), 59–64.

Chen, L. C. & Chang, W. L. (1971). Inheritance of resistance to brown plant-hopper in rice variety, Mudgo (in Chinese, English summary). *J. Taiwan agr. Res.* **20** (1), 57–60.

Fernando, H. E. (1972). Biology and laboratory culture of the rice gall midge and studies on varietal resistance. In *Rice Breeding*, pp. 343–51. International Rice Research Institute, Los Baños, Philippines.

Giatgong, P. & Frederiksen, R. A. (1969). Pathogenicity, variability and cytology of monoconidial subcultures of *Pyricularia oryzae. Phytopathology*, **59**, 1152–7.

Goto, M. (1965). Resistance of rice varieties and species of wild rice to bacterial leaf blight and bacterial leaf streak disease. *Philippine Agric.*, **48**, 329–38.

Harahap, Z., Siregar, H. & Siwi, B. H. (1972). Breeding rice varieties for Indonesia. In *Rice Breeding*, pp. 141–6. International Rice Research Institute, Los Baños, Philippines.

Hashioka, Y. (1951). Inheritance of resistance to sheath blight in rice varieties (in Japanese). *Ann. Phytopath. Soc. Japan*, **15**, 98–9.

Heu, M. H., Chang, T. T. & Beachell, H. M. (1968). The inheritance of culm length, panicle length, duration to heading and bacterial leaf blight reaction in a rice cross: Sigadis × Taichung (Native) 1. *Jap. J. Breed.*, **18**, 7–11.

International Rice Research Institute (1967*a*). *Annual Report, 1966*, 302 pp. Los Baños, Philippines.

International Rice Research Institute (1967*b*). *Annual Report, 1967*, 308 pp. Los Baños, Philippines.

International Rice Research Institute (1968). *Annual Report, 1968*, 402 pp. Los Baños, Philippines.

International Rice Research Institute (1970). *Annual Report, 1969*, 266 pp. Los Baños, Philippines.

International Rice Research Institute (1972*a*). *Annual Report, 1971*, 238 pp. Los Baños, Philippines.

International Rice Research Institute (1972*b*). Discussions of international cooperation. In *Rice Breeding*, pp. 707–12. International Rice Research Institute, Los Baños, Philippines.

International Rice Research Institute (1973). *Annual Report, 1972*, 246 pp. Los Baños, Philippines.

Israel, P., Rao, Y. S. & Prakasa Rao, P. S. (1963). Reaction of wild rices and tetraploid strains of cultivated rices to incidence of gall fly. *Oryza*, **1**, 119–24.

Kauffman, H. E. & Rao, P. S. (1972). Resistance to bacterial leaf blight – India. In *Rice Breeding*, pp. 283–7. International Rice Research Institute, Los Baños, Philippines.

Kauffman, H. E., Reddy, A. P. K., Hsieh, S. P. Y. & Merca, S. D. (1973). An improved technique for evaluating resistance of rice varieties to *Xanthomonas oryzae. Pl. Dis. Reptr.*, **57**, 537–41.

Khush, G. S., Aquino, R. C. & Torres, E. (1971). Exploiting wild germ plasm of *Oryza* for improving cultivated rice. *Proc. Second Ann. Sci. Mtg. Crop Sci. Soc. Philippines*, pp. 311–20.

Khush, G. S. & Beachell, H. M. (1972). Breeding for disease and insect resistance at IRRI. In *Rice Breeding*, pp. 309–22. International Rice Research Institute, Los Baños, Philippines.

Kiyosawa, S. (1972). Genetics of blast resistance. In *Rice Breeding*, pp. 203–25. International Rice Research Institute, Los Baños, Philippines.

Koshiary, M. A., Pan, C. L., Hak, G. E., Zaid, I. S. A., Azizi, A., Hindi, C. & Masoud, M. (1957). A study on the resistance of rice to stem borer infestations. *Int. Rice Comm. Newslett.*, **6** (1), 23–5.

Ling, K. C. (1966). Non-persistence of the tungro virus of rice in its leafhopper vector, *Nephotettix impicticeps*. *Phytopathology*, **56**, 1252–6.

Ling, K. C. (1968). Mechanism of tungro-resistance in rice variety Pankhari 203. *Philippine Phytopath.*, **4**, 21–38.

Ling, K. C. (1969). Testing rice varieties for resistance to tungro disease. In *The Virus Diseases of the Rice Plant*, pp. 277–91. Johns Hopkins Press, Baltimore.

Ling, K. C. (1972). *Rice virus diseases*, 134 pp. International Rice Research Institute, Los Baños, Philippines.

Ling, K. C., Aguiero, V. M. & Lee, S. H. (1970). A mass screening method for testing resistance to grassy stunt disease of rice. *Pl. Dis. Reptr.*, **54**, 565–9.

Murty, V. V. S. & Khush, G. S. (1972). Studies on the inheritance of resistance to bacterial leaf blight in rice varieties. In *Rice Breeding*, pp. 301–5. International Rice Research Institute, Los Baños, Philippines.

Ou, S. H. (1972a). *Rice Diseases*, 368 pp. Commonw. Mycol. Institute, Kew, Surrey.

Ou, S. H. (1972b). Studies on stable resistance to rice blast disease. In *Rice Breeding*, pp. 227–37. International Rice Research Institute, Los Baños, Philippines.

Ou, S. H. & Ayad, M. R. (1968). Pathogenic races of *Pyricularia oryzae* originating from single lesions and monoconidial cultures. *Phytopathology*, **58**, 179–82.

Ou, S. H. & Ebron, T. T., Jr. (1974). The International Uniform Blast Nurseries, 1970–71 results. *Int. Rice Comm. Newslett.*, **23** (in press).

Ou, S. H., Franck, G. P. & Merca, S. D. (1971). Varietal resistance to bacterial leaf streak disease of rice. *Philippine Agric.*, **54**, 8–32.

Ou, S. H., Nuque, F. L. & Ebron, T. T., Jr. (1970). The International Uniform Blast Nurseries, 1968–1969 results. *Int. Rice Comm. Newslett.*, **19** (4), 1–13.

Ou, S. H., Nuque, F. L., Ebron, T. T., Jr. & Awoderu, V. A. (1970). Pathogenic races of *Pyricularia oryzae* derived from monoconidial cultures. *Pl. Dis. Reptr.*, **54**, 1045–9.

Ou, S. H., Nuque, F. L. & Silva, J. P. (1971). Varietal resistance to bacterial blight of rice. *Pl. Dis. Reptr.*, **55**, 17–21.

Ou, S. H. & Rivera, C. T. (1967). Virus diseases of rice in Southeast Asia. In *The Virus Diseases of the Rice Plant*, pp. 23–34. Johns Hopkins Press, Baltimore.

199

Ou, S. H., Rivera, C. T., Nararatnam, S. J. & Goh, K. G. (1965). Virus nature of 'penyakit merah' disease of rice in Malaysia. *Pl. Dis. Reptr.*, **49**, 778–82.

Pathak, M. D. (1969*a*). Stem borer and leafhopper–planthopper resistance in rice varieties. *Entomologia exp. appl.*, **12**, 789–800.

Pathak, M. D. (1969*b*). *Insect pests of rice.* 77 pp. International Rice Research Institute, Los Baños, Philippines.

Pathak, M. D. (1972). Resistance to insect pests in rice varieties. In *Rice Breeding*, pp. 325–41. International Rice Research Institute, Los Baños, Philippines.

Pathak, M. D., Andres, F., Galacgac, N. & Raros, R. (1971). Resistance of rice varieties to striped rice borers. *Int. Rice Res. Inst. Tech. Bull.*, **11**, 69 pp.

Pathak, M. D., Cheng, C. H. & Fortuno, M. E. (1969). Resistance to *Nephotettix impicticeps* and *Nilaparvata lugens* in varieties of rice. *Nature, Lond.*, **223**, 502–4.

Pongprasert, S., Kovitvadhi, K., Leaumsang, P. & Jackson, B. R. (1972). Progress in mass rearing, field testing, and breeding for resistance to the rice gall midge in Thailand. In *Rice Breeding*, pp. 367–71. International Rice Research Institute, Los Baños, Philippines.

Rivera, C. T., Ou, S. H. & Tantere, D. M. (1968). Tungro disease of rice in Indonesia. *Pl. Dis. Reptr.*, **52**, 122–4.

Seetharaman, R., Sharma, S. D. & Shastry, S. V. S. (1972). Germ plasm conservation and use in India. In *Rice Breeding*, pp. 187–200. International Rice Research Institute, Los Baños, Philippines.

Shastry, S. V. S., Freeman, W. H., Seshu, D. V., Israel, P. & Roy, J. K. (1972). Host-plant resistance to rice gall midge. In *Rice Breeding*, pp. 353–65. International Rice Research Institute, Los Baños, Philippines.

Shastry, S. V. S., John, V. T. & Seshu, D. V. (1972). Breeding for resistance to rice tungro virus in India. In *Rice Breeding*, pp. 239–52. International Rice Research Institute, Los Baños, Philippines.

Takahashi, Y. (1965). Genetics of resistance to rice blast disease. In *The Rice Blast Disease*, pp. 303–29. Johns Hopkins Press, Baltimore.

Toriyama, K. (1972). Breeding for resistance to major rice diseases in Japan. In *Rice Breeding*, pp. 253–81. International Rice Research Institute, Los Baños, Philippines.

Venkataswamy, T. (1969). A high yielding rice culture for resistance to gall midge and other insect pests. *Andhra Agric. J.*, **16**, 177–9.

Wakimoto, S. (1967). Strains of *Xanthomonas oryzae* in Asia and their virulence against rice varieties. In *Rice Diseases and Their Control by Growing Resistant Varieties and Other Measures*, pp. 19–24. Agriculture, Forestry and Fisheries Research Council, Tokyo.

17. Evaluation of sources of disease resistance

A. DINOOR

The current situation in disease-resistance breeding demands the continual study and mobilisation of new and different sources of resistance. Since most of the potential sources are probably to be found in wild species and primitive cultivars, we need sound and reliable means of collection and evaluation.

One of the main handicaps in the breeding for disease resistance is the basic philosophy behind evaluation methods. This applies not only to wild species and primitive cultivars but to advanced cultivars also.

Disease resistance, in contrast to many other characters being bred for, is not a stationary target. The pathogen populations constantly react to changes introduced into the host populations. Previously unimportant pathogens and isolates of known pathogens are constantly overcoming the various standards of resistance achieved by the breeder, and this is likely to continue into the future. Furthermore, resistance to pathogens or biotypes of pathogens known today may be useless against new ones in the future. Even the application of sophisticated methods of evaluation does not necessarily ensure that resistance will be preserved for the future since future changes in virulence cannot be predicted.

One way of satisfying future needs is to establish genetic reserve areas of wild species related to the cultigens in the centres of diversity of the host species (see Jain, Chapter 30). By this means the capability of the host populations to react to the continuous changes in the pathogen populations may be preserved.

Selection of resistant materials

Disease resistance must first be recognised and identified before selection can be undertaken. This can be done only by using the relevant pathogens. Since these may differ in different geographical regions, resistance needs cannot be satisfied by the development of a single, universally applicable programme.

Resistance to disease, as stated above, is not a fixed target. It is a

201

Table 17.1. *The phenotypes of host–pathogen interactions*

Resistance (R) is expressed whenever at least one pair of corresponding genes in host and pathogen are dominant. Susceptibility (S) is expressed only when in all pairs of corresponding genes at least one gene is recessive

	Host genotype			
Pathogen genotype	$R_1^- R_2^-$	$R_1^- r_2 r_2$	$r_1 r_1 R_2^-$	$r_1 r_1 r_2 r_2$
$P_1^- P_2^-$	R	R	R	S
$P_1^- p_2 p_2$	R	R	S	S
$p_1 p_1 P_2^-$	R	S	R	S
$p_1 p_1 p_2 p_2$	S	S	S	S

state activated by the pathogen, and the requirements for it are different both geographically and historically. Another basic principle in the expression of resistance is that it is phenotypic, and does not therefore necessarily reveal the genotype involved. The same expression may be the outcome of the interaction between many different host and pathogen genotypes. In many cases resistance is dominant and epistatic and therefore the reaction due to one gene might mask the presence of additional genes for resistance (Table 17.1). The identifiable resistance genes depend therefore on the pathogenic strains used in the identification of resistance.

Specific resistance against pathogens with stable races

Specific resistance is the expression of specific genes which are incompatible with some, but not necessarily all, of the avirulence genes in the pathogen. Therefore, when a variable population of the pathogen is being used in screening, the chances of identifying all of the resistance genes are slight. An individual host might have one or a few important genes for resistance that might be effective against many important biotypes of the pathogen. However, if only one isolate in the pathotype population possessed the compatible virulence gene, then these hosts would be phenotypically susceptible and therefore discarded. We have shown (Dinoor, 1967) that strains of *Avena sterilis* resistant to biotypes of the most virulent races of crown rust (races 264 and 276) are susceptible to biotypes of the very common race 202. Protection against biotypes of race 202 and others can easily be mobilised from Landhafer or Ukraine, both susceptible to race 264 and 276 (Table 17.2). The separate identification of resistance to biotypes of races 264 and 276 is therefore very important.

Table 17.2. *The reactions of 3 oat varieties and a selection of* Avena sterilis *to isolates of oat crown rust races common in Israel*

R = resistance, S = susceptibility. A synthetic could be produced by crosses between the oat varieties and A. sterilis. *When a bulk of the single cultures is used no resistance can be identified. The hypothetical synthetic would resist this bulk*

Race of pathogen	Host (species or variety)				
	Landhafer	Ukraine	Bond	*Avena sterilis*	Possible synthetic
202	R	R	S	S	R
203	R	S	S	S	R
263	S	R	S	S	R
264	S	S	S	R	R
276	S	S	S	R	R
277	S	R	R	S	R
286	S	S	R	S	R
A8–1	R	S	S	S	R
Bulk of the above	S	S	S	S	R

Genes conferring specific resistance should therefore be selected by the use of single pathogen biotypes. Relevant pathogenic biotypes might not be similar in different regions and therefore no universal screening programme should be adopted. Field testing of resistant material is often wasteful and inefficient, unless conducted in race-nurseries. A host genotype effective against all local pathogenic biotypes is not very likely to be found, but it can be constructed from a few other genotypes, each of which is resistant to a different fraction of the local biotypes (Table 17.2).

The identification of genes for resistance involves crosses and genetic analyses. The genetic analyses are based on classification into phenotypes based on the reactions of individual hosts to a population of pathogen biotypes. Such a procedure is likely to reveal only host resistance genes of a wide spectrum. A more thorough approach is to conduct tests separately with each of several important biotypes.

Wild plants, individuals of primitive cultivars, or individuals of an F_2 or further generations in a breeding programme each have a unique genotype. Separate tests conducted with single isolates will yield information for those isolates alone. More genetic information on resistance can be obtained by using a simultaneous inoculation technique, in which each individual plant is separately inoculated with several pathogen biotypes (see Plate 17.1).

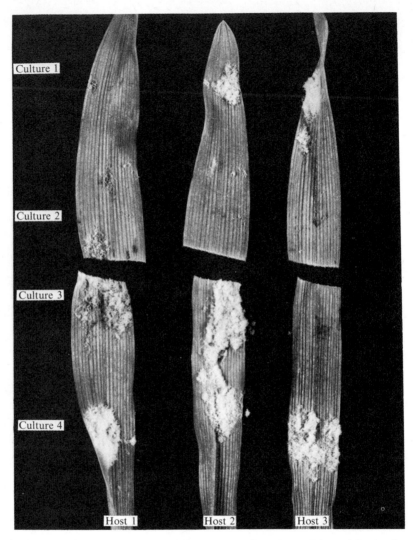

Plate 17.1. Simultaneous inoculation of four cultures of powdery mildew on the first leaves (prophylls) of three barley genotypes. Each prophyll was cut in two to avoid curling and was detached from the mother plant (Wolfe, 1972). The relationships between unique host genotypes and pathogen cultures can be clearly established by the use of this method.

When many plants need to be analysed genetically the amount of work involved is overwhelming. This may usefully be reduced by employing a method of selecting the most interesting representatives. For this purpose a computerised method of genetic classification of plants according to their resistance potentialities was developed

(Dinoor & Peleg, 1972). In this method, data on the reactions of many desirable plants to several relevant pathogen biotypes is fed into the computer. The computer output indicates how many loci are involved and the genetic situation at each locus, i.e. whether the host allele conditions resistance or susceptibility and whether the corresponding pathogen allele conditions virulence or avirulence.

Screening for resistance should, in our opinion, be undertaken in each region separately, according to local needs. The IBP has endorsed an International Survey of Virulence in several crops. The results of such surveys along with local experience could indicate the relevant pathogen biotypes to be used as testers of resistance in each region.

Specific resistance against a pathogen with unstable races

Pathogens with multinuclear spores and/or cells might pose additional problems. In such pathogens, heterokaryons may form and dissociate at times so that cultures derived from single spores or single hyphal-tips might not be genetically stable. The extent to which this phenomenon occurs is unknown. A few cases have been reported (see Chang *et al.*, Chapter 16) and they shed some light on the problem of ephemeral recombination among several possible genetic sets. Khair (1965) in his detailed studies on pathogenic variation in *Drechslera teres* has shown that cultures derived from single spores of a single-spore culture were pathogenically different. Even sister cultures derived from germ tubes of the same spore differed in their pathogenicity. Screening for resistance against such pathogens might therefore be erratic.

The first step in establishing a screening programme using pathogens with unstable races should therefore be the production of genetically pure cultures. Such cultures can be derived from mononucleate spores, from multinucleate spores produced by a conidiophore whose nuclei are of a mononuclear origin, or from spores produced through meiosis. The merits of genetically pure cultures could still be doubtful since in nature such cultures might readily recombine and overcome sources of resistance separately selected out against them. Genetically pure cultures can however serve in identifying single genes for resistance and in assembling the building blocks for the construction of planned combinations of resistances.

Non-specific resistance

This type of resistance is, by definition, effective against any biotype of the parasite concerned. As long as this definition holds, it means that evaluation and selection can be done in any place under favourable conditions, regardless of the biotypes used. Evaluation can be done by quantitative greenhouse tests (Umaerus, 1970; Johnson & Taylor, 1972) and/or by epidemiological field observations throughout the season (Van der Plank, 1968). This implies that screening for non-specific resistance can be centralised. The choice of cultures for screening, however, poses another critical problem. Any specific resistance is bound to obscure the tests for non-specific resistance since the former is epistatic to the latter. The necessity to avoid specific resistance forces some breeders and pathologists to the extreme of breeding and/or selection of more virulent biotypes that endanger the specific resistance being incorporated and used by others.

Since non-specific resistance results in the restriction of epidemics it implies that factors like climatic conditions and the amount of inoculum will affect the degree of resistance. These two factors are therefore critical during the tests for resistance. Further, when deciding upon levels of resistance at which to aim, local conditions such as climate, microclimate, length of the disease season, effectiveness of overseasoning and the like, should be taken into consideration. It is suggested that the levels of resistance accepted should be highest under the most favourable conditions for disease development and could be less under conditions less conducive to severe epidemics.

Collection and selection for resistance in wild species and primitive cultivars

During the planning of plant collecting expeditions the methods of selection for resistance should be borne in mind. Two alternative selection strategies can be adopted, as follows:

Selection for resistance in situ

The collection of selected material *in situ* is sometimes favoured (Niederhauser, Cervantes & Servin, 1954). However, selection *in situ*, whether for specific or non-specific resistance, cannot be recommended, for several reasons:

    1. Lack of information and lack of control of the biotypes of the

pathogen present in the area concerned. Biotypes of minor importance can obscure valuable specific genes for resistance (see p. 202).

2. Epiphytotics might be absent from the target area in the particular season when the material was sampled.

3. Promising plants need to be tagged, since selection should be made when reactions of hosts to pathogens are at their best expression, that is, before ripening. Hence visits to the area need to be repeated at least twice during the season where specific resistance is concerned (one for selection and tagging and one later on, for collection), and even more frequently where non-specific resistance is concerned. However, the access to some areas is sometimes difficult even for a single visit.

Random collection in situ *and directed selection elsewhere*

When collection and selection are separated each process can be made more efficient and more generally useful. Collection can then be carried out once in each location, at the optimal time. The value of selected material will not depend on the vagaries of the local disease situation and will not be biased by reactions to prevalent pathogens in the sampling area.

Collection can be planned and organised to facilitate sound sampling of the area. Appropriate samples can then be tested elsewhere for any desirable trait and according to local or regional needs and interests. Different selection pressures can be operated, and specific or non-specific resistance to any of a number of pathogens can be separately selected. Such a collection can serve many purposes not envisaged by the collector or impossible to accomplish at the time of collection. Subsequent selection from randomly collected material and the investigation of a particular character may be made by many more scientists than those who were present at the time of collection. Clearly also, a wider range of characters could be investigated in more places than could be achieved by a small team of collectors.

Random collection can be left to local professional people who are more familiar with the area, and for whom the area is more accessible at any time. The procedure of collection may thus be more readily adapted to professional needs rather than being dictated by problems involved in long-range foreign expeditions, limited by time and access.

Mapping the distribution of resistance

Carefully planned and organised random sampling provides a foundation for systematic screening for resistance. When any resistant material is encountered, its collecting point can then be traced and more intensive search be undertaken to clarify and amplify the finding.

Resistance of wild oats to crown rust in Israel was used as a model to illustrate this approach (Dinoor, 1967; 1970). A search for resistance *in situ* was usually unfruitful since oat plants in their natural habitats were susceptible wherever crown rust was prevalent. A grid of 5 km mesh was superimposed on the map of Israel and *Avena sterilis* and *A. barbata* were sampled from each square. Representative samples were then screened for resistance to biotypes of races 264 and 276. The results were set out in the form of distribution maps (Figs. 17.1–17.4) and were later verified by further sampling and testing (see also the distribution map of Ethiopian barleys resistant to barley yellow dwarf virus presented by Qualset in Chapter 5).

Distribution maps of resistance serve two purposes:

1. They provide accurate descriptions and locations of centres of diversity for disease resistance. Such centres can subsequently be more thoroughly investigated and exploited.

2. They pinpoint the most appropriate locations for the establishment of genetic reserves that will act as dynamic conservation centres of important wild host species and pathotype variability.

Suggestions for more efficient utilisation and preservation of resources

From the above discussion of the problems involved in the collection of plant material and its evaluation for disease resistance the following suggestions are put forward for the consideration of international or other organisations concerned with plant breeding and food production:

(*a*) Regional centres for the evaluation of disease resistance should be established. Specific expertise relevant to regional needs would make use of available collections, using sound methods of screening. It is suggested that plant material rather than pathogen cultures be exchanged between centres. There should be no directed production of super races for testing host resistance: when the need arises, plant material should be tested at the appropriate centre.

(*b*) Random collection of plant material should be organised in important regions of origin and diversity of crops. These collections, distributed later among screening centres, would be used for mapping the distribution of disease resistance. The distribution maps produced would serve as guidelines for future interests and needs.

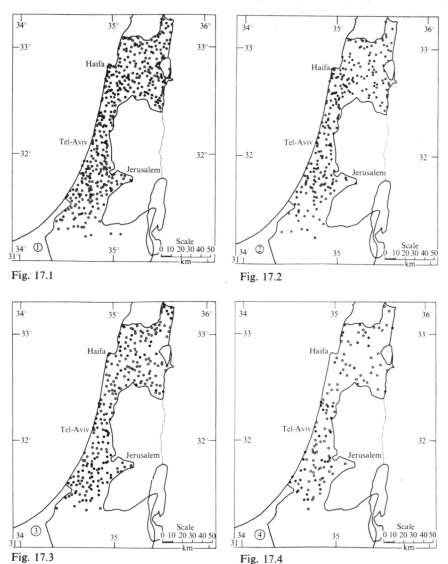

Figs. 17.1.–17.4. Distribution maps of wild oats resistant to oat crown rust races 264 and 276 (see Dinoor, 1970). 1, *A. sterilis*, seedling resistance. 2, *A. sterilis*, adult resistance. 3, *A. barbata*, seedling resistance. 4, *A. barbata*, adult resistance. ○, Sampling stations; ●, locations where resistance was found.

(c) Nature reserves (see Jain, Chapter 30) established anywhere and for any purpose should be linked also to the needs of plant genetic resources and improvement, by giving special consideration to the protection of wild plants of direct economic significance such as coffee,

209

or wild relatives of domesticated species such as *Aegilops*, *Hordeum*, etc. International organisations or others should confer with the governments or authorities concerned, to give guidance in the choice of areas and wild plants to be included. It would be worthwhile also to contact nature reserve authorites and to assemble available information on the existing flora. At a later stage, nature reserves should be sampled for the screening and mapping of disease resistance. Nature reserves could be used as dynamic reserves of evolving host and parasite populations.

(*d*) More intensive research should be undertaken on the genetics of host resistance and pathogen virulence and aggressiveness. The behaviour of host and pathogen genes in populations should also be carefully studied (Wolfe, personal communication). Genetical research into these problems should concentrate on the understanding of tactics for integrated disease control through resistance breeding and other means.

References

Dinoor, A. (1967). The role of cultivated and wild plants in the life cycle of *Puccinia coronata* Cda. var. *avenae* F. & L. and in the disease cycle of oat crown rust in Israel. Ph.D. Thesis, the Hebrew University of Jerusalem.

Dinoor, A. (1970). Sources of oat crown rust resistance in hexaploid and tetraploid wild oats in Israel. *Can. J. Bot.*, **45**, 153–61.

Dinoor, A. & Peleg, N. (1972). The identification of genes for resistance or virulence, without genetic analyses, by the aid of the 'gene-for-gene' hypothesis. *Proceedings of the European and Mediterranean Cereal Rusts Conference*, vol. II. Praha, Československo.

Johnson, R. & Taylor, A. J. (1972). Isolates of *Puccinia striiformis* collected in England from the wheat varieties Maris Beacon and Joss Cambier. *Nature, Lond.*, **238**, 105–6.

Khair, J. (1965). Investigations on physiologic specialization in *Drechslera teres* (Sacc.) Shoemaker (*Helminthosporium teres* Sacc.) the causal agent of barley net blotch. M.Sc. thesis, the Hebrew University of Jerusalem.

Niederhauser, J. S., Cervantes, J. & Servin, L. (1954). Late blight in Mexico and its implications. *Phytopathology*, **44**, 406–8.

Umaerus, V. (1970). Studies on field resistance to *Phytophthora infestans*. 5. Mechanisms of resistance and applications to potato breeding. *Z. PflZücht.*, **63**, 1–23.

Van der Plank, J. E. (1968). *Disease Resistance in Plants*. Academic Press, New York.

Wolfe, M. S. (1972). The forced evolution of cereal disease. *Outlook on Agriculture*, **7**, 27–31.

18. Evaluation of material for frost and drought resistance in wheat breeding

V. F. DOROFEEV

Three aspects in the scientific breeding of cereals for frost and drought resistance must be taken into consideration. These are the genetic bases, the physiological bases and the source material for breeding.

Frost resistance

The problem of frost-resistance breeding in wheat is particularly important in many countries of the northern hemisphere. However, there are several difficulties involved in the implementation of breeding programmes for winter hardiness. The plants can die from frost damage as such, or by ice damage, flooding, damping off, or root exposure through soil heaving. Thus resistance to these other factors must also be sought.

A further difficulty lies in the nature of the wheat plant itself, which has evolved under the relatively mild climatic conditions of South-West Asia. Nevertheless, in its spread from its centre of origin northwards, genes for frost resistance have appeared in it. Wheat breeders in many parts of the world, and especially in the USSR, have shown that cold resistance can be greatly improved by selection.

Theories of frost resistance

The physiological theory of frost resistance was established by N. A. Maximov and D. I. Tumanov and is based on the phenomenon of plant hardening. It was later supplemented after new data were acquired, and is generally accepted. According to this theory, death is due to the formation of ice in the plant tissues, with consequent cell damage, depending on the state of the plant and its innate resistance.

Three types of damage can be recognised. In the first type, ice is directly formed inside the cell, at first in the protoplasm and then in the vacuole. In the second type, ice is formed in the intercellular spaces

211

and as a result the cell dies, due to dehydration and deformation. The third type of damage is complex. Ice is first formed in the intercellular spaces, but during further lowering of the temperature supercooled water accumulates in the cell itself and later freezes there. In winter during severe frosts plants die through the third type of damage. In spring, when frosts recur from time to time, the first type of damage, with tissue necrosis, is usually observed.

Some workers support a biochemical theory of frost resistance. This states that under low temperatures cells are dehydrated, the water moving out into the intercellular spaces. The protein molecules are thus brought into close proximity and consequently the sulphydryl groups are oxidised, forming disulphide bonds, which leads to protein denaturation.

The physiological theory of frost resistance postulates three stages in the hardening or overwintering process. The first stage is connected with the dormancy period in which the plants complete their growth processes, and some aggregates formed from the water-soluble proteins and other substances appear in the protoplasm. At the second stage, the first real phase of plant hardening, certain protective substances accumulate in the liquid phase of the protoplasm. As a result, the freezing temperature of the protoplasm is lowered. At the third stage, the second real phase of plant hardening when ice is formed in the intercellular spaces and the cells are dehydrated, the protein aggregates are drawn together and the number of bonds are increased. This complex mechanism of hardening usually takes place under natural conditions in autumn, so that the plants are able to overwinter by stages.

This physiological theory of overwintering is used as a basis for working out effective methods for increasing frost resistance in winter cereal crops, and particularly in wheat. Agro-technical methods include the control of such factors as optional sowing dates, fertiliser composition, snow retention, drainage, and control of pathogenic soil micro-organisms.

Physiology of frost resistance

Current investigations on the physiological nature of frost resistance are directed towards the following areas: physiological processes of growth; dormancy condition; submicroscopic structures of cell protoplasts; physiological nature of plant death; peculiarities of water, carbohydrate, phosphate and protein metabolism; photosynthesis and respiration intensities; plastids; and the root systems of frost-resistant and frost-susceptible forms.

Much information has accumulated on the physiological peculiarities of frost-resistant plants. This may be summarised as follows.

1. The period of overwintering of frost-resistant cereal crops is characterised by the earlier cessation of growth, reduction of tissue flooding and increase of cell osmotic pressure. In this period the ratio of bound to unbound water is directly correlated with frost resistance.

2. The successful overwintering of winter cereal crops is determined to a great extent by the carbohydrate metabolism. Sugars play an active part in cell structure stabilisation, stimulate the activity of adenosine triphosphatases, and delay the denaturation of proteins which are necessary for subsequent spring growth. Frost-resistant winter wheat varieties accumulate a greater quantity of sugars and consume them more sparingly.

3. In winter the intensity of respiration in frost-resistant varieties is reduced at low temperatures. In warm winters with frequent thaws, their respiration process is stabilised.

4. Frost-resistant varieties are characterised by a more versatile nitrogen metabolism. They show greater variation in total and protein nitrogen, increased quantity of water-soluble proteins, changed electrophoretic protein spectra, and intensive accumulation of proline and asparagine. With the beginning of spring growth the quantities of these amino acids sharply decrease.

5. In frost-resistant varieties of winter cereal crops the activities of enzymes of both proteolysis and carbohydrate metabolism show increases at lowered temperatures. A specific change is observed in phosphorus metabolism, ensuring faster and more energy-saving reactions to the influence of low temperature.

6. Studies of the chromosomes of a number of varieties with differing frost resistance subjected to subzero temperatures reveal qualitative differences in chromatin characteristics, with those of the more resistant varieties showing increased levels of associated protein and lipid. The more highly condensed nature of the chromosomes of frost-resistant varieties at low temperatures is presumed to stabilise the DNA, and may restrict the biochemical activities of such plants.

Evaluation of frost resistance

More than ten characteristics of a frost-resistant wheat plant may be enumerated, and a corresponding number of methods may be listed for determining the degree of frost resistance. However, we cannot distinguish any single metabolic factor as being most responsible for the degree of resistance. The correlation of any physiological process with frost resistance only partly reflects the level of resistance in a plant. This level of resistance is due to the whole complex of metabolic processes, depending on the genotype and its interaction with the environment.

The complexity of the problem of frost resistance in cereal crops and its investigation for more than two centuries has evoked the development of numerous evaluation methods, both direct and indirect. At the N. I. Vavilov Institute a complex system for evaluating the basic breeding material is used, involving hardening in cold chambers as well as in seed boxes directly in the field. Laboratory methods are always used in combination with field evaluation of plants in regions with severe winters. The evaluation by stages of winter wheat for frost resistance is given in scheme 1 (see Table 18.1). Plates 18.1–18.4 show, respectively, the freezing chambers, boxes of seedlings for testing, bundles of plants for hardening, and boxes of plants for freezing under natural conditions.

Drought resistance

Drought resistance of cereal crops is determined by morphological characters (length and vigour of root system, characteristics of plant and rhythm of development, plant pubescence and leaf size), as well as by physiological properties (respiration intensity, elasticity of protoplasm, hydrophilic colloids of protoplasm, etc.).

As a result of drought a plant is submitted to a disturbance of a number of physiological functions, resulting in a retardation or cessation of the growth process. The nature of drought resistance differs widely in various plant species, as does the nature of the drought effects. The degree of resistance of a cereal variety is the result of the interactions of several morphological and physiological characters, enabling the plant to tolerate low soil moisture, high temperature and low humidity over a long period of time.

On the whole, drought resistance was not found in the plants obtained

Table 18.1

---

*Scheme 1: Stages in the evaluation of the winter grain crop
collection for frost resistance*

Wheat

Method I. *Seedling evaluation (see Plate 18.2)*

Stage
1. Seed soaking at +15 °C for 24 hours.
2. Germination in the dark at +15 °C for 48–72 hours, to coleoptile length of 2–3 mm.
3. Plant hardening without light. Phase 1 at +2 °C for 168 hours, phase 2 at −4 °C for 72 hours.
4. Freezing at −7, −10, −12, −15, −18 °C for 24 hours at each temperature.
5. Thawing at +2 °C for 24 hours.
6. Growing at +20 °C for 168 hours with 24-hour lighting.

   Stages 1–5: in cuvettes of thermostable vinyl plastic boxes, 70 × 35 × 8 cm in size.
   Stage 6: in 65 × 62 × 17 cm germinator.

Method II. *Evaluation of bundles of older seedlings (see Plate 18.3)*

Stage
1. Field sowing (August–September): Digging out of plants; bundle preparation (November); two bundles of one variety of 100 plants per two freezing temperatures.
2. Hardening of plants (first stage): Under artificial conditions in 15% sucrose solution at +2 °C (November–December).
3. Hardening of plants (second stage): Under natural conditions, bundle storage in boxes with soil (until January).
4. Preliminary freezing of control varieties in refrigerating chambers:
   (*a*) In November–December.
   (*b*) In December–January, at −10, −13, −16, −18, −21 °C.
5. Freezing of collection under two regimes:
   (*a*) In November–December.
   (*b*) In December–January, at −15, −18 °C.*
6. Thawing: At +2 °C for 24 hours.
7. Planting for growing: Growing for 20 days at +18 °C.
   Surviving plants counted in:
   (*a*) December–January.
   (*b*) February–March.

Method III. *Box freezing in field (see Plate 18.4)*

1. Box sowing: August–September.
2. Freezing: As in Method II.

---

* Freezing temperature is determined depending on hardiness of control varieties.

from Vavilov's Centres of Origin of cultivated plants. The majority of the 200 000 accessions of plant material in the Institute collections are from tropical and subtropical regions and have been shown to be chiefly moisture-loving species.

The main bread crops (wheat, barley, and others) are medium drought-resistant. Taking into consideration the various kinds of drought and

215

Plate 18.1. Refrigerating chambers for hardening and freezing of plants.

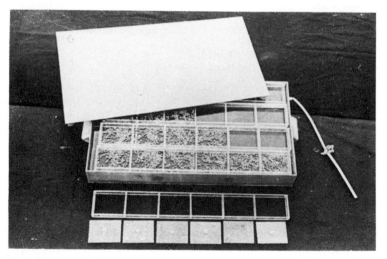

Plate 18.2. Complementary set of seedlings for primary frost resistance evaluation of winter cereal crops.

Plate 18.3. Bundle of older seedlings, prepared for hardening, using sucrose solution.

their characteristics as well as the different degrees of drought resistance, it is very difficult to breed a variety possessing universal resistance to all types.

To evaluate wheat, barley and other cereals more accurately for drought resistance, several methods are usually employed, based on different principles.

The most objective test for this complex character can be obtained directly under field conditions in the regions of well-expressed droughts. Schemes 2 and 3 (see Tables 18.2–18.4) and Plate 18.5 show the procedures for the evaluation of drought resistance in wheat varieties.

217

Plate 18.4. Seedling boxes with plants prepared for direct freezing in the field.

Table 18.2

Scheme 2: Evaluation of grain crop samples for drought resistance
(see Plate 18.5)

Wheat

Method I. *Primary mass evaluation; seed germination method in sucrose solution*

Species:	*T. aestivum*	germinated in sucrose solution of 16 atmospheres osmotic pressure (17.6 % concentration)		
	T. durum	germinated in sucrose solution of 10 atmospheres osmotic pressure (11.9 % concentration)		
Temperature:		+20–21 °C.		
Quantities:		50 seeds in each petri dish with 50 ml sucrose solution. 3–4 replicates		
Evaluation criterion:		germination on the fifth to sixth day		
Results:		High resistance	Medium resistance	Low resistance
	T. aestivum	70 %	20–70 %	20 %
	T. durum	80 %	30–80 %	30 %

General conclusions

During the past ten years the N. I. Vavilov Institute has tested some 10 000 spring wheat varieties and 9000 winter wheat varieties by laboratory methods combined with field evaluation under severe climatic conditions. Only 200 varieties with high frost-resistance and 150

Table 18.3

Scheme 2 (continued)

Method II. *Evaluation at a critical period of development*

Two tests are made in parallel.

(a) Determination of water-holding capacity of tissues

Cut leaf wilting method
The top leaf (first canopy level) is used
Temperature: +25–30 °C
Time: 1–5 hours
Replication: 5–10-fold
The water loss % is recorded.

(b) Determination of ability to withstand dehydration

Electrolytic method
Weighed portions: 0.5–1 g
Water quantity: 20–50 ml at +20–25 °C
Water loss no less than 45–50 %
Time of infusion: 2 hours
Replication: 3–4-fold

The yield of electrolytes from wilting leaves as recorded by a conductometer is compared
to the results of tests on fresh leaves.

Table 18.4

Scheme 3: Grain crop samples for heat resistance

Wheat

Method I. *Primary mass evaluation*

Imbibed seed heating method
    Temperature: +54 °C
    Time: 20 min
    Quantities: 50 seeds in petri dish
    Replication: 3–4-fold
    Evaluation criterion: germination on the seventh to eighth day
    Results: high resistance, 80 %; medium resistance, 50–80 %; low resistance, 50 %

Method II. *Determination of ability of tissues to withstand overheating at a
critical period of development*

Electrolytic method
    Weighed portion: 0.5–1 g
    Water quantity: 20–50 ml
    Temperature: +43 °C
    Control temperature: +20 °C
    Exposure time: 4 hours
    Replication: 3–4-fold
    Instrument: conductometer or rheochordic bridge
    Evaluation criterion: change in yield of electrolytes in the experiment as compared with
        control

219

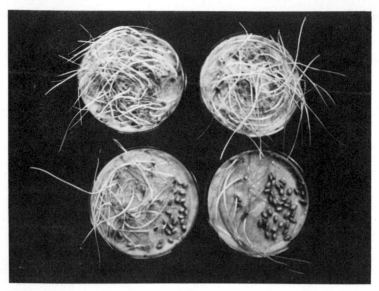

Plate 18.5. Treatment of seeds by high temperature at 54 °C (bottom row); controls (top row). Drought-resistant variety *Erythrospermum* 841 (left) and control variety Diamant (right).

varieties with good drought-resistance were identified. The greatest number of frost-resistant varieties is found in the USSR, Canada, and the USA. A high percentage of drought-resistant forms was found amongst accessions from Uzbekistan, Turkmenistan, Syria, Afghanistan, India, Chile, Mexico and Canada.

Drought resistance has been found in the following varieties of soft wheat (*T. aestivum*):
Erythrospermum 841, Saratovskaya 29, Saratovskaya 36, Saratovskaya 38, Tsezium 31, Tsezium 3, Onohoiskaya 4, Akmolinka 1 and Tselinogradka. Drought-resistant varieties in *T. durum* include the following: Lenkurum 33, Kharkovskaya 46, Melanopus 26, Bezenchukskaya 105 and Shark.

In conclusion it should be noted that frost and drought resistance are complex polygenic characters subject to considerable influence of modifier genes. The inheritance mechanisms have not been sufficiently studied, and there is little information about allelic interactions at multiple loci. Few data on genotype–environment interaction have been gathered either. The lack of such data adversely affects breeding programmes, and in consequence the breeding work is often much delayed. However, it is hoped that these problems will be solved by the efforts of scientists from many countries.

The collaboration of M. A. Barashkova and N. N. Kozhushko is gratefully acknowledged in the preparation of Tables 18.1–18.4.

Selected references

Winter-hardiness

Alden, J. & Hermann, R. K. (1971). Aspects of the cold-hardiness mechanism in plants. *Bot. Rev.*, **37** (1), 37–42.

Barashkova, E. A., Alekseyeva, E. N. & Vinogradova, V. V. (1971). *The estimation of winter wheat frost hardiness by the variation in electrolyte diffusion from tillering nodes.* Leningrad, VIR (Vsesoyuznyi Institut Rastenievodstva im. N. I. Vavilova).

Barashkova, E. A., Alekseyeva, E. N. & Vinogradova, V. V. (1973). Methods of evaluation of frost hardiness in plants used in the investigation of the World Collection at the All-Union Vavilov Institute of Plant Industry. *Methods of evaluating plant resistance to unfavourable environmental conditions.* VIR, Leningrad.

Dubinin, N. P. (1971). Genetic principles of plant selection. *Genetical principles of plant breeding.* Nauka, Moscow.

Golodriga, P. Y. & Osipova, A. V. (1972). A rapid method and instruments for diagnosing frost resistance in plants. *Fiziol. Biokhim. Kul'turnykh Rastenii*, **4** (6), 650–5.

Heber, U. & Santarius, K. A. (1971). Empfindlichkeit und Resistenz der Zelle gegen Frost. *Umschau*, **71**, (25).

Kolosha, O. I. (1969). The evaluation of winter crops for frost hardiness by the method of monolites. Collected articles: *Metody opredeleniya morozo- i zimostoikosti ozimykh kul'tur.* Moscow.

Law, C. N. & Jenkins, G. (1970). A genetic study of cold resistance in wheat. *Genet. Res.*, **15**, 197–208.

Medinets, V. D. (1972). The field evaluation method of varietal winter hardiness. *Selektsiya i semenovodstvo*, 20.

Pisarev, V. E. & Batrshina, A. A. (1971). On the solution of the problem of winter wheat in Siberia. *Selektsiya i semenovodstvo*, (1) 30–3.

Santarius, K. A. (1971). Ursachen der Frostschäden und Frostadaptation bei Pflanzen. *Ber. Dtsch. bot. Ges.* **84** (7/8), 425–36.

Tumanov, I. I. (1969). *Methods of estimation of plant winter hardiness.* Nauka, Moscow.

Udovenko, G. V., Oleinikova, T. V., Kozhusko, N. N., Barashkova, E. A., Vinogradova, V. V. & Alekseyeva, E. N. (1970). *Methods of diagnosing plant hardiness.* VIR, Leningrad.

Udovenko, G. V. & Tkachev, M. V. (1970). Resistance of winter rye varieties to soaking in relation to their physiological characters. *Trudy prikl. Bot., Gen. Selek.*, **43** (1), 79–87.

Evaluation problems

Vinogradova, V. V. (1968). A comparative study of methods of evaluating winter hardiness in winter wheats. *Doklady Vaskhnil*, **8**, 11–15.

Vlasyuk, P. A., Protsenko, D. F., Kolosha, O. I. & Ostaplyuk, E. D. (1968). The nature of winter hardiness in cereal crops. *Procedures and methods for increasing cold resistance in winter cereal crops*, pp. 146–59. Kolos, Moscow.

Zlobina, E. S. (1970). Cytophysiological diagnosis of depth of dormancy and frost hardiness in winter wheat. *Sel'skokhozyaistvennaya biologiya*, **5** (1), 21–5.

Drought resistance

Derera, N. F., Marshall, D. R. & Balaam, L. N. (1969). Genetic variability in root development in relation to drought tolerance in spring wheats. *Expt. Agric.*, **5** (4), 327–37.

Genkel, P. A. (1956). *Diagnosis of drought hardiness in cultivars and methods of its improvement* (Technical instructions). Institut Fiziologii Rastenii im. K. A. Timiryazeva. Moscow.

Genkel, P. A., Badanova, K. A. & Levina, V. V. (1970). On the new laboratory diagnostic method of heat and drought resistance for plant breeding. *Fiziologiya Rastenii*, **17** (2), 431–5.

Hurd, E. A. (1969). A method of breeding for yield of wheat in semi-arid climates. *Euphytica*, **18** (2), 217–26.

Kuz'min, V. P. (1970). Breeding spring wheats for drought resistance in North Kazakhstan. *Improvement of drought resistance in cereal crops*, pp. 6–17. Moscow.

Oleinikova, T. V. (1970). Physiological methods of evaluating wheat for drought resistance. *Improvement of drought resistance in cereal crops*, pp. 56–65. Moscow.

Oleinikova, T. V. & Kozhushko, N. N. (1970). Laboratory evaluation methods of drought resistance of some cereal crops. *Trudy prikl. Bot., Gen. Selek.*, **43**, 100–11.

Palfi, G. & Juhasz, J. (1971). The theoretical basis and practical application of a new method of selection for determining water deficiency in plants. *Pl. Soil*, **34** (2), 503–7.

Peterson, R. F. (1958). Twenty-five years progress in breeding new varieties of wheat for Canada. *Emp. J. exp. Agric.*, **26**, 104–22.

Uglov, P. D. (1970). On the methods of study of the spring wheat collection for its drought resistance. *Trudy prikl. Bot., Gen. Selek.*, **43** (1), 88–94.

Volkova, A. M. & Kozhushko, N. N. (1972). Investigation of wheat varieties and species by laboratory methods. *Bulleten' VIR*, **24**, 58–62.

Weisner, C. J. (1970). *Climate, Irrigation and Agriculture*. Sydney.

Williams, T. V., Snell, R. S. & Ellis, J. F. (1967). Methods of measuring drought tolerance in corn. *Crop Sci.*, **7** (3), 179–82.

19. The identification of high-quality protein variants and their use in crop plant improvement

D. E. ALEXANDER

Plant breeders are inevitably drawn to the work they do by biological reality and by economic necessity. It is not difficult to double the oil content of maize or to render its staminate flowers incapable of producing pollen or to breed hybrids capable of producing 150 quintals of grain per hectare. It is difficult, however, to do all these things within the economic support provided and maintain the kind of genetic diversity which is desirable. There is no reason to believe that independent programs, each restricted to unrelated gene pools, would not produce equally satisfactory varieties and in addition spread the risk of catastrophe brought about by special races of pests. However, the extra funding to support such 'duplicate' effort would be too great.

Good plant breeding programs always have 'almost as good' alternatives, varieties that are only slightly related to the 'best' variety but for one reason or another are never introduced to practical use. Maize breeders have literally dozens of excellent inbreds, all tested in thousands of trials, that are almost as good as the six elite inbreds that appeared in 70 per cent of US corn-belt hybrids in 1972. Should the elite combinations fail, many others would be able to take their place, but with some penalty in performance, of course.

Thus a great deal of built-in safety exists in the sophisticated plant breeding systems and this positive aspect should be kept in mind.

In the following sections I want to give some illustrative examples of maize breeding programs in which the protein quantity and quality have been of paramount importance. Some examples from other crops will be given also.

Potential for increasing protein content in maize by selection

Divergent selection for protein content has been carried out in Burr's White, an open-pollinated variety, at the University of Illinois since 1896 except for a three year period between 1942 through 1944. Pertinent data on the experiment are shown in Table 19.1.

Table 19.1. *Progress in selection for high and low protein content in Burr White, 1896–1969*

Generation	Illinois High Protein strain (% protein)	Illinois Low Protein strain (% protein)
0	10.9	10.9
5	13.7	9.6
20	16.1	7.3
50	19.3	4.9
70	26.6	4.4

This classical experiment vividly demonstrates the effectiveness of selection, as well as the inherent wide range that exists in the species for protein content. J. W. Dudley (personal communication) reports that the 1970 mean of Illinois High Protein was twelve standard deviations beyond the mean of the original population. He also predicts that had 6000 ears of the original population been analyzed, the most extreme ear would have been expected to be no greater than 3.8 standard deviations above the mean.

Progress continues in the Illinois High Protein strain in an approximately linear fashion. Dudley & Lambert (1969) found that significant genetic variation existed within the sixty-fifth generation population. Further evidence that significant heritable variation existed within the forty-eighth generation has been provided through the reverse selection experiment.

Unfortunately, the Illinois High Protein strain owes its high protein content to heightened level of prolamines (zein), a group of proteins low in lysine and tryptophan, and the strain has not been used in practical corn breeding programs in any substantial way.

Yield is quite low in the Illinois High Protein strain. In 1970–71, it produced 20 q/ha, about one-third that of the average held for the state of Illinois for those years. Two factors contribute to this unpromising performance: the coefficient of inbreeding has been estimated by Dudley (personal communication) to be 0.82–0.83; and there appears to be an inverse relationship between protein content and yield.

In the 1930s and early 1940s, Woodworth and several of his contemporaries outcrossed the elite inbreds of the day to the Illinois High Protein strain and in segregating backcross and F_2 populations they selected for protein content and for combining ability for agronomic performance. Hybrids with protein content as high as 18 per cent were encountered but never reached commercial significance because of their generally inferior performance, and perhaps more importantly, because inexpensive high-quality protein supplements were plentiful in the US.

These pioneering efforts are now of considerable interest. Corn gluten meal is selling for $250 a short ton today in the US, yet a year ago it sold for a third of that price. Attitudes are changing about the value of the so-called low-quality proteins, particularly in human diets, and it is not at all unlikely that renewed interest in the breeding of high-protein maize will occur, quite independently of that focused on the creation of high-quality protein varieties.

High-quality protein maize

Selection for protein quantity in maize had been under way for more than 50 years before serious attention was paid to the improvement of quality. Feeding experiments involving non-ruminant animals and the Illinois High Protein strain had been disappointing. The amino acid requirement of several animals was known, and analytical work had shown that the inferiority of corn endosperm protein was due to the low level of two amino acids, lysine and tryptophan. In a series of penetrating studies, G. F. Sprague and his student, K. J. Frey, were able to show that tryptophan content was under genetic control (Frey, Brimhall & Sprague, 1949). Their efforts to breed high-quality protein maize were abandoned, however, because of the difficulties of analyzing a sufficiently large number of samples by a bioassay system.

It is worth pointing out that Hopkin's protein selection program depended on a nitrogen analysis system developed by Kjeldahl a few years before.

In November 1963, Nelson, Mertz & Bates (see Mertz *et al.*, 1964) using the relatively new automatic amino acid analyzer in a routine analytical run, found an obscure endosperm mutant, opaque-2, which was high in both lysine and tryptophan. In a matter of a year, a second high-quality mutant, floury-2, was discovered, and rat-feeding trials had demonstrated the high biological quality of both these mutants (Nelson *et al.*, 1965). After seed stocks had been increased, feeding experiments with swine and with young and adult humans demonstrated the high nutritional value of the opaque-2 strain. Most spectacular among these experiments was one conducted by Pradilla and Harpstead in Columbia. A kwashiorkor child, destined to die in perhaps a month, was placed on a diet of opaque-2 maize and vegetables. In a

225

matter of weeks the child had recovered and was gaining weight comparably to kwashiorkor children fed on traditional recovery diets containing milk and meat.

It is an interesting and revealing sidelight to note that opaque-2 was among the first seven mutants selected by Nelson for analysis. The thought process which led to selection of the group is a classical example of reasonableness and scientific judgement, but which did not 'hold water' upon later examination. Nelson reasoned that an attack on the problem of low quality must be centered on the endosperm, since 80 per cent of the protein was carried there. Since starch is embedded in a protein matrix, he assumed that a major change in the composition of the matrix might bring about changes in its optical properties. Why not examine endosperm mutants? Of course, maize endosperms similar to opaque-2 or floury-2 are not appreciably different in protein quality from ordinary dent or flint types. Furthermore, modifiers have been found which transform the soft endosperm of opaque-2 corn to a hard, vitreous sort, indistinguishable from ordinary low quality varieties.

An early performance trial (Lambert, Alexander & Dudley, 1969) showed that opaque-2 hybrids were 15 per cent lower in yield than their normal counterparts. Steady improvement in performance has been encountered and some opaque-2 hybrids are now possibly yielding not less than 5 per cent below ordinary types. A growing acceptance of opaque-2 hybrids is now occurring in the USA, since the lower yield is offset by a reduction in expensive protein feed supplements.

Protein–amino acid screening methods

The Nelson–Mertz discovery would possibly not have been made if the automatic amino acid analyzer had not been developed. The availability of an analyzer stimulated thought on the important problem of protein quality and, of course, made it possible to carry out the work. However, automatic amino acid analyzers are slow, cumbersome devices, ill-adapted to the restless and demanding requirements of plant breeders. More rapid and inexpensive methods were promptly adapted to their needs. Most successful among these has probably been the TNBS scheme, which is dependent on binding with epsilon groups of lysine, arginine and histidine after acid hydrolysis of the proteins. Correlation between the automatic analyzer and the TNBS is 0.7–0.8 in experienced hands.

Approximately 3000 samples are analyzed for lysine by a modified

TNBS scheme in our laboratory in 5–6 winter months with approximately one man year of time expended in the effort. The method is accurate enough for practical plant breeding, the coefficient of variation of individual samples from three replicate trials, for example, being approximately 7 per cent.

Bioassay for lysine has also been used successfully but with somewhat greater error. The assay is based on growth rate of the lysine-requiring bacterium, *Leuconostoc mesenteroides*, in solutions of extracted protein. The turbidity of test tube cultures is measured by photospectroscopy and the values translated to lysine concentration by means of curves constructed from growth rates encountered on media containing known concentrations of lysine. The coefficients of variation are high (13 per cent) and the system is not widely used.

Gas–liquid chromatography is also used to assay for lysine. Sample preparation is a substantial part of the cost with this method, and its usefulness as a plant breeding tool is dependent on it.

Some preliminary work has been carried out using wide-line nuclear magnetic resonance spectroscopy. ^{15}N resonance was believed to be sufficiently different from that of ^{14}N in the epsilon group of lysine that quantitative estimates might be made. This scheme proved to be impractical, but with continuing and inevitable development of physical-chemical technology, it seems highly probable that rapid and accurate quantitative analysis of almost any compound, even in living tissue, will ultimately be possible.

An inexpensive and perhaps satisfactory rough screening method for protein quality involves quantitative analysis of prolamines. Prolamines, such as zein in maize, are low in lysine and tryptophan; hence the lower the proportion of prolamine in a sample, the higher will be the quality of the protein as a whole.

Munck (personal communication) has pointed out that joint application of the Udy system of protein quantification and the Kjeldahl system could be used to screen for lysine content. Samples are run by the Udy method, one based on dye binding with basic amino acids, and by the standard Kjeldahl scheme. The ratio of Udy and Kjeldahl values can be used to estimate lysine and other basic amino acids in the sample. Hiproly barley was discovered through application of this principle.

Maize protein quality improvement in 'non-mutant' genetic backgrounds

As pointed out earlier, variation in amino acid composition is known to exist in standard dent corns. Nordstrom found that maize with 7–8 per cent oil promoted satisfactory growth in swine with reduced levels of protein supplement intake (Nordstrom, Behrends, Meade & Thompson, 1972). The lysine content of advanced cycles of Alexho Synthetic, a high-oil strain, was higher than that of the original low-oil population. It is assumed that the advantage is associated with larger embryo size of the high-oil population.

Zuber (in Choe, Zuber, Helm & Hildebrand, 1973) selected for favorable lysine–protein ratio from two heterogeneous dent corn populations. The mean L/P values for second cycle populations approached those of the opaque-2 strains. These unexpected results need to be re-evaluated after the cycles have been grown together in the same season. Some substantiation is provided in Zuber's report (personal communication) that o_2/o_2 recoveries of certain high-lysine Logan selections have exceedingly high-lysine dry-matter ratios. The high values are due in part to the unexpectedly high protein level of 13–15 per cent, as compared to the 9–10 per cent level of the original cycle.

Wolf and co-workers (see Wolf *et al.*, 1972) found that Corico (also known as Piricinco), a South American lowland race, contained two to six layers of aleurone cells as compared with the single layer in ordinary dent corn. Because the lysine level of aleurone protein is higher than that of the remaining endosperm proteins, the overall quality of multiple aleurone types is considerably improved. The inheritance of multi-layered aleurone is polygenically controlled.

I agree with Munck's statement (personal communication): 'It . . . seems appropriate also to utilize minor genes affecting the balance between lysine-rich and lysine-poor protein subfractions of cereal endosperm (in plant breeding programs).' Evidence provided by the Zuber selection experience, the suggestion that protein quality is superior in high-oil dents, and the multi-layer aleurone finding, all strongly lead one to believe that appropriately designed breeding programs will ultimately lead to maize protein superior to any we now have.

Protein variants in barley and oats

Hageberg & Karlson (1969) reported that a certain Ethiopian barley appeared to have superior quality endosperm proteins. Subsequent genetic analyses led to the conclusion that a recessive allele (*lys*) conditioned the 'high-lysine' trait. Breeding programs are now under way to provide backcross *lys/lys* recoveries in elite varieties. It is likely that improvement in barley proteins through the *lys* allele will be less spectacular than that in maize through the o_2 allele, because standard barley endosperm proteins are of reasonably good nutritional quality.

The maize and barley findings have led to a further consideration of protein in oats, a cereal with excellent protein quality. Hybrids of *Avena sativa* and *A. sterilis* have been somewhat disappointing, although segregants with as much as 22 per cent groat protein of excellent quality are under agronomic evaluation. Surveys with *Avena sativa* have uncovered varieties of superior agronomic value with protein content approaching many of the *sativa × sterilis* derivatives. Dal, a new Wisconsin variety, and Otee from Illinois, have 20–22 per cent protein, a value appreciably above the 18 per cent level encountered in popular varieties.

Long-chain sugars and other undesirable metabolites in the soybean

Although these are outside the prescribed scope of this chapter, I wish to mention briefly two recent pieces of research that are of interest.

Many legume seeds contain appreciable levels of the oligosaccharides, raffinose and stachyose. They have been implicated as the causal agents for flatulence in humans, presumably because α-galactosidase is not present in the digestive tract. They pass intact into the large intestine where they are anaerobically fermented, and gas is produced. Hymowitz and co-workers (see Hymowitz, Collins, Panczner & Walker, 1972) surveyed several domestic and exotic soybean varieties. Raffinose was found to range from 0.1–0.9 g/100 g seed, and stachyose from 1.4–4.1. Similar ranges were observed in common beans and in the cowpea. Interestingly enough, most peanut cultivars were found to be low in oligosaccharides, although a few varieties were potential gas producers. The prospects seem promising, therefore, for breeding varieties low in these oligosaccharides and hence removing the 'flatulence factor' from various leguminous food plants.

The other example concerns the work reported by Mertz (personal

229

communication) who found that opaque-2 endosperm proteins in maize were devoid of certain human allergins (2, 3, 5) and carried 10 per cent as much of the strongest one (8) as did standard dents. The trypsin inhibitor level, unfortunately, was four times greater than in dent endosperm. Whether the amount is of biological consequence, or whether the higher level is a characteristic of opaque-2 is still unknown.

References

Choe, B. H., Zuber, M. S., Helm, J. L. & Hildebrand, E. S. (1973). Comparison of genetic systems for lysine synthesis in normal and opaque-2 corn. *Am. Soc. Agron. Abstr.*, p. 19.

Dudley, J. W. & Lambert, R. J. (1969). Genetic variability after 65 generations of selection in Illinois High Oil, Low Oil, High Protein and Low Protein strains of *Zea mays* L. *Crop Sci.*, **9**, 179–81.

Frey, K. J., Brimhall, B. & Sprague, G. F. (1949). The effects of selection upon protein quality in the corn kernel. *Agron. J.*, **41** (9), 399–403.

Hageberg, A. & Karlson, K. E. (1969). Breeding for high protein content and quality in barley. New approaches to breeding for improved plant protein. *Int. Atomic Energy Comm. Publ.*, pp. 17–21.

Hymowitz, T., Collins, F. I., Panczner, J. & Walker, W. M. (1972). Relationship between the content of oil, protein and sugar in soybean seed. *Agron. J.*, **64**, 613–16.

Lambert, R. J., Alexander, D. E. & Dudley, J. W. (1969). Relative performance of normal and modified protein (opaque-2) maize hybrids. *Crop Sci.*, **9** (2), 242–3.

Mertz, E. T., Bates, L. S. & Nelson, O. E. (1964). Mutant gene that changes protein composition and increases lysine content of maize endosperm. *Science*, **148**, 1741–2.

Nelson, O. E., Mertz, E. T. & Bates, L. S. (1965). Second mutant gene affecting the amino acid pattern of maize endosperm proteins. *Science*, **150**, 1469–70.

Nordstrom, J. W., Behrends, B. R., Meade, R. J. & Thompson, E. H. (1972). Effects of feeding high oil corns to growing finishing swine. *J. Anim. Sci.*, **35** (2), 357–61.

Wolf, M. J., Cutler, H. C., Zuber, M. S. & Khoo, V. (1972). Maize with multilayer aleurone of high protein content. *Crop Sci.*, **12** (4), 440–2.

20. Screening for oils and fats in plants

G. RÖBBELEN

Deposition of oils and fats in plant tissues is universal and not limited to specific taxonomic groups. Oil crops belong to such distantly related families as Leguminosae, Euphorbiaceae, Compositae and Palmae. Furthermore, in the production of oilseeds, no ecological area takes precedence over others. The production of dry matter in plant populations is, of course, correlated to temperature, sunshine and other climatic conditions, but fat, which compared to other reserve substances is the most highly condensed form of energy, is no more frequently stored in seeds of tropical plants than in those from temperate regions (see Table 20.1).

Economic value of oils and fats

Man has long used vegetable oils and fats for various purposes. In the world market vegetable fats and oils currently amount to about 29 million tons of a total world production of 45 million tons (see Table 20.2). Since 1938 the world fat production has more than doubled, the major increase occurring with edibles. Thus, in spite of the rapidly increasing world population the annual *per capita* supply improved by about 2 per cent each year. The world average of the annual consumption of edible fats in the order of 8.8 kg per capita, however, obscures the actual deficiency of countries like India (with a mean of 5 kg) through regions of surplus consumption (about 25 kg) as in Western Europe. During the same period the relative share of vegetable oil from the total edible fats and oils increased from 55.6 per cent before World War II to about 64 per cent in 1972. This trend, which still continues, originates from the rising demands for polyenoic fatty acids (especially vitamin F) in the food of populations with higher incomes and more sophisticated tastes.

On the other hand, the industrial use of fats and oils has long remained relatively stable at an average of 15 per cent of the total world production. But within this sector the demand shifted considerably through important developments in modern oleochemistry. While

231

Table 20.1. *Distribution of some main oil crops on a taxonomic and climatic basis (after Godin & Spensley, 1971)*

Species	Family	1	2	3	4	5	6	7	8
Soybean	Leguminosae	+	.	.	.	+	.	+	.
Sunflower	Compositae	.	.	.	.	+	(+)	+	+
Groundnut	Leguminosae	.	+	+	.	+	+	+	.
Cotton	Malvaceae	.	.	+	+I	+	+	.	.
Rapeseed	Cruciferae	.	.	(+)	.	+	.	+	.
Olive	Oleaceae	.	.	.	(+)	.	+	.	.
Sesame	Pedaliaceae	.	.	+	(+)	+	+	+	.
Maize	Gramineae	(+)	.	(+)	+	+	+I	+	.
Safflower	Compositae	.	.	+	.	.	+	.	+
Coconut	Palmae	+	+	(+)	.	.	.	.	.
Oil palm	Palmae	+	.	.	.	.	.	.	.
Babassu palm	Palmae	+	.	+	.	.	.	.	.
Linseed	Linaceae	.	.	(+)	.	+	.	+	.
Castor oil plant	Euphorbiaceae	.	.	+	.	(+)	.	+	.
Tung oil tree	Euphorbiaceae	.	.	.	.	+	.	+	.
Oiticica	Rosaceae	.	.	.	+	.	.	.	.
Niger seed	Compositae	.	.	+	.	.	.	.	.
Hemp	Cannabaceae	.	.	.	.	.	.	+	.
Perilla	Labiatae	.	.	.	.	.	.	+	.
Poppy	Papaveraceae	.	.	.	.	.	.	+	+

( ) Limited cultivation possible; I, irrigation necessary. 1, tropical rain forest; 2, tropical monsoon; 3, tropical savanna; 4, dry tropical, i.e. steppe and desert; 5, humid subtropical; 6, dry subtropical, i.e. Mediterranean; 7, humid temperate; 8, dry temperate.

Table 20.2. *World production of oils and fats (1000 metric tons in oil equivalent)*

	1938	1958	1968	1972
Liquid oils*	7787	11750	18975	22315
Palm-type oils*	2659	3295	4115	4955
Industrial-type oils*	1559	1550	1810	1780
Animal fats	8752	11125	14225	14890
Marine oils	830	840	1180	1305
World total	21587	28560	40305	45245

* Compare Table 20.4.

soap lost its dominating position within this area, new uses were created for fats and fatty acids in the production of resins and plastics, paints and lubricants, rubbers and coatings, cosmetics and pharmaceuticals. Except for a limited amount of oils used exclusively for industrial purposes (cf. Table 20.2) much of the necessary raw material is derived from the unpalatable or spoiled portions of edible plant fats or as a

Table 20.3. *Net exports of the principal vegetable oils and oilseeds from primary producing countries (1000 metric tons in oil equivalent). (After Commonwealth Secretariat, Anonymous, 1970)*

	1960	1964	1968
Canada	164	162	217
Ceylon	74	155	77
West Malaysia	134	139	354
Nigeria	600	669	520
Argentina	285	230	244
Brazil	41	110	122
China	364	141	255
Congo	236	168	210
Indonesia	276	267	275
Philippines	732	769	684
Rumania	43	42	135
Senegal	226	245	258
Soviet Union	111	202	852
Spain	137	105	43
Sudan	84	135	94
United States	1426	1952	1946
Total	5835	6603	7450
Of which from:			
Developed countries	1808	2366	2393
Developing countries	3445	3782	3640
Centrally planned economies	578	455	1417

by-product of their processing (e.g. de-acidification of salad oils). Another important amount is absorbed by industry as the result of certain developments in the market. Thus the industrial requirements for vegetable oils and fats are obviously in most cases neither increasing nor particularly demanding in quality. Screening efforts and plant improvement programmes may here be required only in specific sectors of minor importance.

Another feature with a bearing upon adjustment of oil-crop cultivation becomes apparent when considering the export figures, though these reflect only a little more than one quarter of the total production. Since the beginning of this century the exports from India, Ceylon, South-East Asia and the Pacific area have been proportionally decreasing steadily. On the other hand, not only did the USA considerably increase its oilseed production but so did Canada and the Soviet Union, leaving the developing countries far behind (see Table 20.3). The Provisonal Indicative World Plan for Agricultural Development of FAO assumes that the world consumption of oilseeds will double from 1962 to 1985, but that oilseed production by the developed

Table 20.4. *World production of the main seed oils in 1000 metric tons (after Commonwealth Secretariat, Anonymous, 1970)*

Oils	1959–60	1964–65	1968–69
Liquid			
Soybean	27301	31820	42350
Cottonseed	18821	20636	21277
Groundnut	12398	16215	15000
Sunflower	5188	8345	9600
Rapeseed	3521	4506	5350
Sesame	1455	1716	1600
Olive oil	1225	968	1485
Palm type			
Copra	3189	3252	3100
Palm kernels	888	859	700
Palm oil	699	708	900
Industrial type			
Linseed	3026	3341	3030
Castor seed	513	695	830
Tung oil	106	94	105

countries will increase only up to 40 per cent during the same period. However, I see no indications as yet that the export balance of the developing countries will improve in this way, for two reasons: (1) Several main oil crops in the tropical areas are perennials like the various palm trees, for which the establishment of new productive plantations (e.g. in Malaysia) takes a long time and is very expensive. Annual crops, like soybeans or sunflower, respond much more quickly to improvement through breeding, agronomic measures and economic demands (see Table 20.4). This is especially true since a long steady decline has been in progress in the price-levels for most oils and fats, with the exception of olive oil. (2) In their residue after oil extraction, oilseeds contain from 20 to 40 per cent protein of high nutritive value. The soybean boom in the USA, for example, was stimulated by the rapidly increasing protein demand for live-stock feeds. As this development is characteristic for most of the 'rich' countries the competition against the tropical oil seeds should not be expected to slow down in the foreseeable future.

Improvement of production through breeding

These few hints on the world situation of oil crops must suffice to frame the main question of this paper: what can genetic variability contribute to improve the production of oil crops? This question aims in two directions, a quantitative and a qualitative one.

Quantitative variation

Regarding the quantitative aspect, the enormous variation in oil content and oil yield between different species of oilseeds is well known (see Table 20.5). These differences are entirely independent of the type of organ or tissue in which the fat deposits occur, be it the embryo (cotyledons, as in cruciferous and leguminous species), the endosperm (as in the coconut or poppy seed) or the carpel (mesocarp), which may contain as little as 23 per cent oil in the olive or as much as 56 per cent in the fruit of the oil palm. The higher the total oil content of the fruit or seed the more economical becomes the processing, unless the demand for high-grade protein meal justifies additional expenditure. Therefore, one of the breeder's main concerns has long been the selection of varieties with an ever higher oil content in the harvested product. For example, remarkable success has recently been achieved with sunflower seed in Russia, not only by reducing the proportion of shell in the achene to below 35 per cent, but also by increasing the oil content in the embryo itself from 25–30 per cent up to 40 per cent and more. The 'chemical composition of the seed of oleiferous plants' may also vary 'in dependence on geographical factors', as Ivanov has already stated in 1926. In addition, the productivity of an oil crop depends on its average yield per hectare; olive trees may yield below 10 kg or up to 300 kg of fruit per year. But without appropriate experimental data the impact of the genetic variance cannot safely be estimated. There is, however, little doubt that successful breeding can shift the order of productivity even between crop species. Yet except for oil content, the decisive characteristics to be sought in the genetic resources are not typical of the capacities of the plant for synthesis of fat or oil. For example, in soybean an increase of oil yield depends on the insensitivity of the yield to the photoperiod prevailing during the time prior to flowering; in the olive plant it depends on the capability of the young trees to bear the first fruit six and not, as normally, 15 years after planting, whilst in the monoecious oil palm, the sex ratio determines the number of fruit bunches.

Qualitative variation

Thus the special screening for oils and fats in crop plants is mainly a problem of quality. In Table 20.6 the example of the main margarine ingredients indicates that the quality requirements are obviously, in

Table 20.5. *Oil extraction rates for oilseeds (after Commonwealth Secretariat, Anonymous, 1970)*

Babassu kernels	63	Olives	15
Castor seed	45	Palm kernels	47
Copra	64	Perilla seed	37
Cottonseed	16	Poppy seed	41
Groundnuts, shelled	45	Rapeseed	35
Groundnuts, unshelled	32	Safflower seed	30
Hempseed	24	Sesame seed	45
Kapok seed	18	Shea nuts	45
Linseed	34	Sunflower seed	35
Mustard seed	23	Soybeans	17
Niger seed	35	Tung nuts	16
Oiticica seed	45	Others	30

Table 20.6. *Margarine ingredients in five selected countries, 1963 (percent of total ingredients)*

	United States	United Kingdom	Netherlands	Sweden	France
Soybean oil	75	10	13	2	*
Groundnut oil	—	7	—	—	4
Coconut oil	—	6	11	36	} 45
Palm kernel oil	1	1	10	—	
Palm oil	—	9	15	—	15
Cottonseed oil	7	3	—	4	*
Rapeseed oil	—	—	3	23	*
Sunflower seed oil	—	—	3	—	*
Corn (maize) oil	8	—	—	—	*
Other vegetable oils	2	6	2	5	10
Marine oils	—	38	39	27	23
Lard	6	19	} 4	} 3	3
Tallow	1	1			—
Total	100	100	100	100	100

* Included in 'Other vegetable oils'.

general, met by a number of seed oils and that the raw material is highly interchangeable. The physical and nutritional properties of the various seed oils are determined by the proportion and nature of the fatty acids which they contain in their triglycerides. These differ chiefly in the number of carbon atoms (e.g. 18 in oleic acid) forming the chain of the long molecule, and in the number of double bonds between the carbon atoms (e.g. 1 in oleic acid, which is therefore specified as 18:1 in Table 20.7). The higher the number of carbon atoms and the lower the number of double bonds, the higher is the melting point; palmitic

Table 20.7. *Composition of vegetable oils and fats (per cent of total fatty acids; t = traces, less than 1 per cent). (Composed from Godin & Spensley, 1971)*

Oils	Satd. under C_{12}	Lauric 12:0	Myrist. 14:0	Palmit. 16:0	Stearic 18:0	Oleic 18:1	Linoleic 18:2	Linolenic 18:3	Others
Liquid									
Soybean	—	t	t	7–14	2–6	23–34	52–60	2–6	
Sunflower	—	—	—	3–7	1–3	22–28	58–68	—	
Groundnut	—	—	t	6–12	2–4	42–72	13–28	t–11	⎰ 5–15 Eicosenoic
Cottonseed	—	—	t	20–25	2–7	18–30	40–55	3–10	⎱ 40–55 Erucic
Rapeseed	—	—	t	t–5	t–4	14–29	9–25	—	
Olive	—	—	t	7–20	1–3	65–86	4–15	—	
Sesame	—	—	—	7–9	4–5	37–50	37–47	—	
Maize	—	—	t	8–12	2–5	19–49	34–62	—	
Safflower	—	—	—	5–8	1–3	11–15	74–79	—	
Palm type									
Coconut	10–19	44–52	13–19	7–10	1–3	5–8	1–3	—	
Palm kernel	6–11	46–52	14–17	6–9	1–3	13–19	0–2	—	
Palm oil	—	—	t–3	32–47	1–9	40–53	2–11	—	
Babassu kernel	8–12	40–52	14–19	6–9	1–2	13–20	0–1	—	
Industrial type									
Linseed	—	—	4–7	3–8	3–8	13–36	10–25	30–60	
Castor bean	—	—	—	1–2	1–2	t–9	4–5	—	82–95 Ricinoleic
Tung	←——	3.5–8	—	—	↑	4–16	8–16	—	77–81 Eleostearic
Other vegetable									
Oiticica	←	9–12	—	↑	↑	4–12	0–5	—	70–82 Licanic
Niger	←	9–17	—	↑	↑	31–39	51–55	—	
Hempseed	←	4–10	—	↑	↑	14–16	45–65	15–30	
Perilla seed	←	6–8	—	↑	↑	10–23	16–35	41–70	
Chinese vegetable tallow									
Mesocarp	—	0.3	4.2	62.3	5.9	27.4	—	—	
Seed	1.5	—	0.9	2.8	1.0	9.4	53.5	30.1	

and stearic acid are semisolid at 15 °C, whereas oleic and linoleic acids are liquid.

A comparison of the fatty acid composition reveals characteristic differences between various oilseeds (see Table 20.7). Palm-seed oils, for example, have a high proportion of medium-long saturated fatty acids, like lauric acid, while cotton seed or the fruit flesh of the Chinese vegetable tallow (*Sapium sebiferum*) is relatively rich in palmitic acid. Liquid seed-oils show higher percentages of polyenoic fatty acids, e.g. linolenic acid in linseed and rapeseed or linoleic acid in safflower and sunflower. In addition to these 'major fatty acids' there are 'minors' like eicosenoic or erucic acid in rapeseed which occur less frequently. A third class of so-called 'unusual fatty acids' occurs sporadically throughout the plant kingdom. In rare cases large amounts are accumulated, e.g. ricinoleic acid in the castor bean or licanic acid in the rosaceous oiticica. But most of them occur only in traces in the seed oils. Up to now more than 300 of such compounds have become known. Some have been found in bacteria and fungi, but most of them are typical for specific taxonomic entities in the dicotyledons: alkin fatty acids in Oleaceae, Santalaceae; hydroxy fatty acids in Compositae; epoxy fatty acids in Euphorbiaceae; cycloprenoic fatty acids in Malvaceae, Sterculiaceae. The molecular variation frequently starts from an end product of the biogenetic chain, many unusual fatty acids being derived from linolenic acid which is also particularly reactive because of its trienoic nature. In this connection it should be mentioned that in the early sixties the USDA sponsored an extensive screening for unusual fatty acids, which was headlined as 'Search for New Industrial Oils' (cf. Miller, Earle, Wolff & Barclay, 1967).

Evolutionary changes in the fatty acids

The seed oils of higher plants fall into groups conforming with their taxonomic relationship, and this is also true throughout the plant kingdom. In a very thorough study Wagner & Pohl (1966) demonstrated that this grouping has no relation to habitat, e.g. fresh-water and marine forms of the same group showing the same fatty acid composition. But an evolutionary arrangement (Fig. 20.1) reveals the most interesting feature, that the number of saturated fatty acids possible within the plant kingdom was almost complete in the Bacteriophyta, while the ability to synthesize unsaturated fatty acids evolved much more slowly. Thus bacteria form only monoenoic acids (up to oleic acid, here given

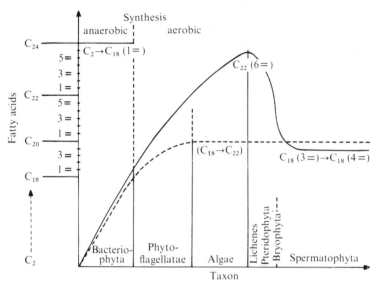

Fig. 20.1. Evolution of fatty acid biogenesis in the plant kingdom (after Wagner & Pohl, 1966). Solid line, unsaturated fatty acids; dashed line, saturated fatty acids.

as C_{18} (1=)) by means of an anaerobic process. In higher systematic groups additional aerobic systems of fatty acid synthesis originated. In anaerobically grown yeast still only oleic acid occurs; but *Lipomyces*, *Penicillium*, *Astasia* or *Polytoma* additionally develop considerable amounts of linoleic acid as well as some α-linolenic acid. It is, however, only in the green cells of *Euglena* and *Chlamydomonas* that γ-linolenic acid appears for the first time. Apparently the synthesis of this polyenoic fatty acid is bound to chlorophyll, oxygen and light. This may be the reason why animal organisms are unable to form even linoleic acid (vitamin F), though *Euglena* is said to have an additional biosynthetic pathway for polyenoic acids, which is light-independent. As soon as the aerobic synthesis of polyenoic acid originated, the evolution of the biogenesis of fatty acid quickly proceeded to its maximum in the green algae with C_{22} (6=). With the transition of plants from water to land these abilities were reduced, apparently in consequence of ecological needs. The synthesis of the C_{20} and C_{22} acids was abandoned and also the highly unsaturated types disappeared (e.g. all of the C_{16} polyenoics), leaving in existence the α-linolenic acid as the highest unsaturated form of the Spermatophyta.

This evolutionary arc prompts the question as to what the physiological role of these compounds is. We have so far only discussed the

239

structurally simplest members, the triglycerides, in which all three hydroxyl groups of the alcohol are esterified with fatty acids. But there are also mono- and diglycerides, with only one or two positions of the glycerol bound in this way, while the other is esterified to galactose derivatives as in the glycosylglycerides, or choline as in the lecithin molecule (which was the first plant phospholipid to be completely characterised). In addition, a multitude of similar constituents, of diollipids, wax-, carotenoid- and terpenoid-esters or highly condensed cutins are in existence in plant tissues.

The function of lipids within the cell

This wide variety of lipid structures takes care of a vast number of specific functions within the cell. All known biological membranes are constructed of lipids and proteins in proportions which vary from $1:3$ in bacterial membranes to $4:1$ in the myelin sheath of the animal nerve cells. These striking differences must be related to the specific properties of the interfaces, e.g. permeability; therefore the composition of a specific type of membrane is relatively constant. The thylakoids of the chloroplasts in tobacco leaves, for example, contain 83 per cent of the total cellular amount of monogalactosyl diglycerides, 88 per cent of the digalactosyl diglycerides, 76 per cent of the sulpholipids, and 74 per cent of the phosphatidyl glycerol. As these lipids are specially rich in linolenic acid, thylakoids generally contain more than 70 per cent of their total fatty acids in the form of linolenic acid.

For mitochondria, convincing evidence is available for the essentiality of membrane lipids (especially phospholipids) in electron transport systems. Mitochondria are known to be capable of cyclical contraction or swelling, and it is believed that these conformational changes are fundamental to the energy transfer within the organelle. But not only in this case does the flexibility of the membrane appear to be an important functional characteristic. Their constitution is therefore adjusted to any given situation by a varying ratio of saturated to unsaturated fatty acids in the lipids, favouring rigidity through a higher percentage of saturated, and extensibility through an increase of unsaturated fatty acids. In order to maintain constant physicochemical properties of the cellular membranes this ratio can change with the environmental conditions. It is well known that the composition of seed oils may likewise react in the same way to different climatic conditions during ripening (see Table 20.8).

Table 20.8. *Composition of linseed oils (variety Bison) affected by climatic conditions of ripening (data from Painter & Nesbitt, 1943 cited by Hilditch & Williams, 1964)*

Grown in	Iodine value	Component fatty acids			
		Satd.	18:1	18:2	18:3
Nebraska	155.4	12	32	21	35
Minnesota	162.8	11	30	18	40
N. Dakota	164.7	10	34	12	44
S. Dakota	171.5	10	27	19	44
Montana	177.0	10	25	18	47
Oregon	182.4	9	22	20	49
Saskatchewan	187.0	9	22	15	54
Nova Scotia	196.0	9	16	16	59

The function of seed oils as storage products

The tabulation of lipid functions could easily be continued as they are also effective in enzyme reactions (lipid cofactors), in the process of healing wounds giving rise to the wound hormone (traumatic acid), in ethylene production (fruits) or in others. In contrast, the function of seed oils appears to be much less specific. It is nothing more than to provide a source of carbon-skeleton precursors as well as a source of energy (ATP) for assembling these precursors into the carbohydrates, proteins and fats needed for the differentiation of cells in a germinating seed. These functional requirements are largely independent of the specific fatty acid composition of the triglycerides which are the dominating constituents (95 per cent) of all seed oils. There has therefore been no severe selection pressure against those unusual fatty acids during evolution (cf. p. 238) which sporadically appeared in isolated taxa probably in consequence of mutations. Similarly it should be possible to shift the fatty acid composition of seed oils on a considerable scale, if such changes do not affect other cellular lipid compounds via common pools of precursors or biosynthetic pathways. There is ample evidence that different tissues of even the same fruit can build distinctly different oils, as is the case in the mesocarp and the kernel of the oil palm or the Chinese vegetable tallow (cf. Table 20.7). But the chances of plant improvement are definitely limited by the degree of functional interdependence between the various sites of cell lipid metabolism.

Breeding for the improvement of rapeseed oil quality

The plant breeding programme which we have run in Göttingen since 1966 in order to improve the quality of rapeseed oil serves well to summarise the foregoing deductions.

Criteria

The first prerequisite for any efficient screening programme is a clear-cut formulation of the desired criteria (Röbbelen & Leitzke, 1970). In rapeseed oil, the food industry complains of three major defects as indicated in Table 20.9. Higher amounts of erucic acid fed to experimental animals resulted in the deposition of fatty acids in the heart and skeletal muscle fibres as well as other noxious effects, suggesting the need for selecting varieties free from this compound. Linoleic acid, on the other hand, compensates for a surplus of cholesterin, retaining the permeability of cellular membranes and counteracting atherosclerotic disorders such as heart disease; it is also a precursor of the prostaglandins. In linolenic acid the three double bonds create definite chemical instability, especially rapid autoxidation; the breakdown not only results in the bad 'green' taste of the product, but also makes it suspect of causing cancer (cf. Fukuzumi, 1972).

Screening methods

The second necessity for the plant breeder is the availability of effective and efficient screening methods. Since the elaboration of the versatile gas–liquid chromatographic (GLC) technique, any fatty acid can be determined precisely in relatively small samples of plant material (cf. Thies, 1971a). But for a broad screening project not only accurate, but also rapid, safe and cheap serial tests are desirable. An extensive effort has, therefore, been made to develop such techniques of pre-selection for each of the more interesting fatty acids in addition to the GLC method for the exact determinations which must follow.

For erucic acid such a quick test is based on paper chromatography, the principle of which is the separation of free fatty acids obtained by alkaline hydrolysis on a paper impregnated with paraffin oil, developed in acetic acid and stained with copper acetate and rubeanic acid (Thies, 1971a). This procedure allows one person to screen more than

Table 20.9. *Fatty acid composition of rapeseed oil* (*per cent of total fatty acids*)

Fatty acid		Content in present-day varieties (%)	Desirable content (%)
Erucic	22:1*	45	0
Linoleic	18:2	15	>25
Linolenic	18:3	11	0
Saturated	n:0	5–15	5–15
Oleic	18:1	rest	rest

* Number of carbon atoms: number of double bonds (per chain)

500 samples per day for erucic-acid-free types without expensive equipment (in comparison with 60 GLC analyses per apparatus per day).

For the ratio of linoleic to linolenic acid another semiquantitative screening technique was elaborated in our laboratory by Rakow & Thies (1972) using photometric measurements at 233 and 268 nm after isomerisation of the double bonds in the polyenoic acid methyl-esters by potassium tertiary butyrate and ethylene glycoldimethyl-ether. More recently a spot test on a new reagent paper was developed for estimating the staining reaction with linolenic acid visually or more exactly densitometrically. In this case the procedure is particularly quick (1000 samples per day), reducing the preparation of the material to some simple manipulations in polystyrene microtitre trays (Thies, 1972).

Selection programmes

An advantage in selection procedure for the fatty acids is the possibility of determining the genetic constitution of the plant already in the seed stage, as the fatty acid composition of the embryo is not dependent on the mother plant, but fully represents its own genotype, though with a lower heritability of the higher unsaturated fatty acids (cf. p. 240, Table 20.8). Moreover, most of the analytical techniques are sensitive enough to estimate the contents of the respective fatty acids within one (the outer) cotyledon. Thus the remaining part of the embryo (inner cotyledon, radicle and plumule) can be grown to a mature plant and immediately be used for further breeding work (Downey & Harvey, 1963).

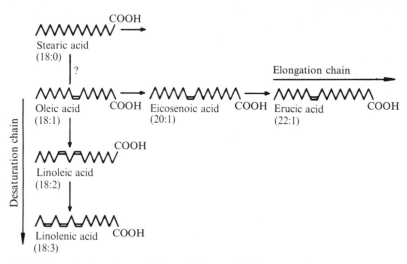

Fig. 20.2. Biosynthetic pathway of the main fatty acids in rapeseed.

With such methods at hand the first improvement required of rapeseed oil was relatively easy to accomplish. In 1963 Downey had selected a mutant with seed oil low in erucic acid having a genetic block in the elongation chain of the fatty acid synthesis (Fig. 20.2). Obviously such types are relatively frequent and not apparently eliminated by natural selection. As already stated (p. 238), the erucic acid is an evolutionary relic in the angiosperms and no longer plays any essential role in cell metabolism.

The desired changes within the desaturation chain (Fig. 20.2), however, proved to be much more difficult to achieve. The urgency of this problem was even greater in the new 'zero erucic' varieties as the increased substrate flow also added to the amount of linolenic acid in the rapeseed oil. As yet nothing is known about the mechanism of the desaturation process, especially whether one or two different enzyme systems control the two steps in question. In the latter case, every mutant which lacks one of the polyenoic C_{18}-acids should also be devoid of the other. In a large screening experiment no 'zero linolenic' type was found among more than 500 cruciferous species and in an equal number of *Brassica* varieties. This was surprising, as the desired forms with no linolenic and a high relative amount of linoleic acid are widespread in other plant families such as Compositae (sunflower, safflower), Papaveraceae (poppy), Malvaceae (cotton) or Gramineae (maize) (cf. Table 20.7; Thies, 1968). Thies (1971*b*), however, pointed out that

Table 20.10. *Polyenoic fatty acids (percent) in cruciferous plants (from Thies, 1970)*

	18:2	18:3	18:3/18:2
Malcolmia littorea	13	66	5:1
Alyssum montanum	14	60	4:1
Hutchinsia auerswaldii	13	39	3:1
Camelina microcarpa	25	45	2:1
Cakile maritima	15	17	1:1
Barbarea vulgaris	25	13	1:2
Lunaria rediviva	21	7	1:3
Sisymbrium loeselii	35	9	1:4
Conringia orientalis	29	2	1:15

Table 20.11. *Chemically induced linolenic acid mutants in rapeseed*

Mutant	Phenotypic value (9 = best)	18:2	18:3/18:2
8783-M 57	7.6	21.4	0.24
8767-M 6	4.9	23.0	0.16
8773-M 11	7.0	32.0	0.22
5007-M 1	Winter-type	22.7	0.25
Oro	8	16.8	0.48

the linolenic acid appears exclusively whenever the seed oil is formed within a green embryo, which is the case in rapeseed, soybean, etc. This finding strongly supports the idea already mentioned (p. 239) that linolenic acid is not only an essential constituent of the green chloroplast membranes, but that photosynthetic potentials are also needed for the further desaturation at least of the higher polyenoic fatty acids. If this is so, the chance of ever reaching the original breeding objective is nil. However, quantitative changes appear to be possible (see Table 20.10). Indeed, by mutagenic treatment genotypes were created in which regulatory factors obviously changed the light reactivity of the biogenetic system (Rakow, 1972) drastically improving the polyenoic acid contents in the rapeseed oil (Table 20.11).

In conclusion, it can be said that natural and induced genetic resources may not always be able to provide the plant breeder with immediately useful genotypes. However, the higher the available genetic variability, the better the chance of satisfying the breeding objective. Detailed comparative studies of the physiological and biosynthetic mechanisms, which are the basis of the qualitative traits in question, may save much unnecessary screening effort. Furthermore, such investigations, based on the greatest possible diversity of genotypes, may in many instances provide decisive clues to our understanding of the underlying biological processes.

Summary

The production of plant oils and fats, which has mainly been concentrated in tropical and subtropical regions in the past, is presently undergoing drastic changes. Simultaneously, the various requirements of the world fats market in respect of quantity and quality of plant oils fluctuate according to the changing industrial and nutritive utilisation. Plant breeding programmes have been efficient in increasing the oil contents in plant seeds and fruits, but selection for quality was successful only in some cases where definite demands were formulated by the processor. In several instances genetic variability appears to be exploitable, but in others physiological reasons limit variation. This is indicated by the major, not overlapping groups of oil crops with specific metabolic characteristics, the basic principles of which are not yet fully understood. Successful screening, however, largely depends on a clear understanding of the inherent physiological mechanisms linked to the development of quick, cheap and simple analytical techniques.

Much valuable information from Dr W. Thies on the scientific aspects of this chapter and useful advice from E. von Hippel in preparing the English text are gratefully acknowledged.

References

Anonymous (1970). *Vegetable Oils and Oilseeds. A review.* Prepared at the Commonwealth Secretariat, London.

Downey, R. K. (1963). Oil quality in rapeseed. *Can. Food Industr.*, 1–4 June.

Downey, R. K. & Harvey, B. L. (1963). Methods of breeding for oil quality in rape. *Can. J. Pl. Sci.*, **43**, 271–5.

Fukuzumi, K. (1972). On the relation between lipoperoxide, atherosclerosis, and cancer. Abstr. of Papers No. 59, *11th World Congr. Int. Soc. Fat Res.* Göteborg, Sweden, 18–22 June.

Godin, V. J. & Spensley, P. C. (1971). Oils and Oilseeds. No. 1. *Crop and Product Digests.* The Tropical Products Institute, London.

Hilditch, T. P. & Williams, P. N. (1964). *The Chemical Constitution of Natural Fats*, 4th edn. Chapman and Hall, London.

Ivanov, N. N. (1926). Variation in the chemical composition of the seeds of oleiferous plants in dependence on geographical factors. *Bull. appl. Bot. Pl. Breed.*, **16** (3), 1–88.

246

Miller, R. W., Earle, F. R., Wolff, I. A. & Barclay, A. S. (1967). Search for new seed oils. xv. *Lipids*, **3**, 43–5.

Rakow, G. (1972). Selektion auf Linol- und Linolensäuregehalt in Rapssamen nach mutagener Behandlung. Diss. Landw. Fak., Univ. Göttingen, 67 pp.

Rakow, G. & Thies, W. (1972). Schnelle und einfache Analysen der Fettsäurezusammensetzung in einzelnen Rapskotyledonen. ii. Photometrie der Polyenfettsäuren. *Z. PflZücht.*, **67**, 257–66.

Röbbelen, G. & Leitzke, B. (1970). Weshalb Rüböl nicht vollwertig und Züchtung auf Ölqualität beim Winterraps notwendig ist. *Saatgutwirtschaft, SAFA*, **22**, 544–6, 588–90.

Thies, W. (1968). Die Biogenese von Linol- und Linolensäure in den Samen höherer Pflanzen, insbesondere Raps und Rübsen, als Problem der Pflanzenzüchtung. *Angew. Bot.*, **42**, 140–54.

Thies, W. (1970). Chloroplast development and biogenesis of linolenic acid in ripening cotyledons of rapeseed. *Proc. Int. Conf. Rapeseed*, St Adele, Que., Canada, pp. 348–56.

Thies, W. (1971*a*). Schnelle und einfache Analysen der Fettsäurezusammensetzung in einzelnen Rapskotyledonen. i. Gaschromatographische und papierchromatographische Methoden. *Z. PflZücht.*, **65**, 181–202.

Thies, W. (1971*b*). Der Einfluss der Chloroplasten auf die Bildung von ungesättigten Fettsäuren in reifenden Rapssamen. *Fette, Seifen, Anstrichmittel*, **73**, 710–15.

Thies, W. (1972). Improvement in oil and meal quality of rapeseed by plant breeding: new methods for selection of strains free from erucic acid, linolenic acid and glucosinolates. *Abstr. of Papers*, no. **28**, *11th World Cong. Int. Soc. Fat Res.*, Göteborg, Sweden, 18–22 June.

Wagner, H. & Pohl, P. (1966). Eine These: Fettsäurebiosynthese und Evolution bei pflanzlichen und tierischen Organismen. *Phytochemistry*, **5**, 903–20.

21. Secondary metabolites and crop plants

R. HEGNAUER

Introduction: secondary metabolites and man

Proteins, sugars, starches, fats and vitamins are nutritionally the most important plant constituents. They are generally taken into consideration in plant breeding programmes. On the other hand many agriculturists do not pay much attention to the so-called secondary metabolites, because their functions are not well understood. The contribution of secondary plant constituents to the quality of our most important crop plants cannot yet be adequately estimated.

The distinction between primary and secondary metabolism in plants was introduced by Czapek (1913–21). Later, several authors tried to define more accurately what are generally known as natural products or secondary metabolites (Geissman & Crout, 1969; Luckner, 1969; 1971; Paech, 1950; Zenk, 1967).

Primary metabolites are sugars, amino acids, fatty acids, the acids of the citric acid cycle, pyrimidines and purines and their fundamental derivatives and polymers, such as polysaccharides, proteins, nucleic acids and fats. These products are essential to life and occur in every organism. On the other hand most phenolic compounds, quinones, essential oils, resins, saponins and alkaloids are secondary metabolites of plants. They occur in easily detectable amounts only in certain taxa, and in most instances their functions are not known precisely. Furthermore, high concentrations of many natural products are toxic to cells and organisms. Therefore plants that accumulate such products remove them from their metabolically active centres. Secondary metabolites are deposited in vacuoles, in lysigenous or schizogenous cavities or ducts or are excreted by glandular hairs or through the cuticles of leaves and young stems (Schnepf, 1969). In other instances natural products are deposited in dead tissues, such as heart-woods or the outer regions of barks.

However, the distinction between primary and secondary metabolism is by no means a sharp one. Many plant constituents essential to life, such as vitamins, several coenzymes, tetrapyrroles, sterols and carotinoids are synthesized along pathways of secondary metabolism

249

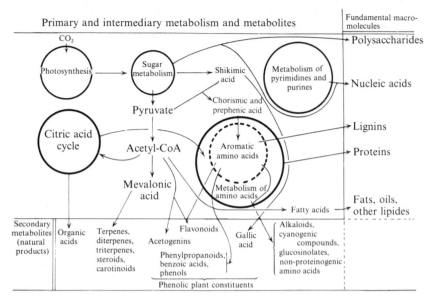

Primary and intermediary metabolism and metabolites | Fundamental macro-molecules

Fig. 21.1. The position of secondary metabolites (natural products) in plant metabolism (modified from Zenk, 1967).

(Bernfeld, 1967; Geissman & Crout, 1969; Luckner, 1969) (see Fig. 21). It would perhaps be logical to accept a proposal of Moritz & Frohne (1967) who distinguish central, intermediate, and peripheral pathways of plant metabolism. In this scheme, secondary metabolites could be defined as resulting from peripheric pathways. They are all metabolically more or less inert and are deposited in some way in the plant body or are excreted.

Notwithstanding its lack of exactness the distinction between primary and secondary plant constituents is a very practical one. It is understood by everyone and does not cause confusion if one realizes that every man-made classification is not perfectly good in every respect. This is especially true for all types of classification used in biology. In fact, classifying all plants according to uniform rules implies a certain degree of artificiality and subjectivity which causes much disagreement between taxonomists. Nevertheless each sound taxonomic revision is extremely useful to all scientists working with the groups of plants concerned.

Secondary metabolites are the most important constituents of many plants used by man for purposes other than nutrition and fibre production.

Polyphenolic compounds of distinct structure and molecular weight, the so-called tannins, make plants and plant parts suitable for leather production if they occur in large amounts (Dekker, 1913; Gnamm, 1949; Haslam, 1966).

Many natural products of extremely varying structure are highly toxic to fish and insects. Plants and plant extracts containing them, such as pyrethrum, derris, and acetum sabadillae, are used to destroy or repel insects (Frear, 1948*a*, *b*; Jacobson & Crosby, 1971) and as fish poisons (Greshoff, 1893–1913).

Colouring matters were and still are derived from many plants. They are all typical secondary metabolites and belong to very different classes of compounds, such as flavonoids, anthraquinones, phenyl-propanoids, carotinoids, and betalains (Goodwin, 1965).

Throughout history man has used spices to give odour and flavour to food and to make it more palatable. Many of our spices possess anti-septic properties besides odour and flavour. The former depend upon phenolic substances, such as rosmarinic acid in spices derived from Labiatae or upon eugenol in cloves. Eugenol is at the same time the main flavouring constituent of cloves, in which it is present in amazingly large amounts (up to 20 per cent). In general the volatile phenylpro-panoids and the mono- and sesquiterpenes, the so-called essential oils, contribute most to the odour and flavour of our spices. Pepper, chilis and ginger may serve as examples of pungent spices. In pepper and ginger the pungent constituents are accompanied by large amounts of essential oil whereas chilis are almost wholly pungent. The large number of cultivars of *Capsicum* shows clearly that secondary metabolism is under genetic control. The non-pungent cultivars selected by man for use as vegetables or as colouring and weakly flavouring spices are variants with no or very little capsaicins. Pickersgill (1971) showed clearly that the wild relatives of modern cultivated Capsicums are most important for future selection and breeding and for understanding the history of present cultivars. The same is true for all crop plants, as was stressed by Anderson (1967).

As old as spices are cosmetics. Fats, waxes, resins, essential oils and colouring matters, all derived from plants, were, and in many instances still are, the principal raw materials of cosmetic preparations (Guenther, 1948–52; Naves & Mazuyer, 1939).

The influence that plants and natural products have had on the development of modern medicine can hardly be overestimated (Kreig, 1964; Leuenberger, 1972; Swain, 1972). Medicine was the mother of many sciences. This is particularly true for several branches of botany

251

and for phytochemistry. In present times most therapeutic agents are prepared in the laboratories of pharmaceutical industries. It should not be forgotten, however, that the development of many drugs was guided by a knowledge of the constituents of medicinal plants. The alkaloids cocaine and quinine are good examples of this. The former helped chemists in their search for powerful local anaesthetics lacking its addictive and toxic properties. Quinine, the famous alkaloid of *Cinchona* bark, served as a model for the preparation of several chemo-therapeutic drugs. Similar interactions between plants, plant constituents and medicine still exist. This is exemplified by the present intensive screening of plants for psychotropic substances (Efron, 1967) and anti-tumor drugs (Hartwell, 1967–71; Kupchan, 1972).

It will never be possible to forecast the potential value of a taxon to different branches of science or to future plant utilization. The story of *Colchicum autumnale* and related taxa may serve to illustrate this point. *Colchicum* seeds and corms were used long ago for the treatment of gout but became more or less obsolete owing to their high toxicity. Interest in the plant revived after the discovery of the antimitotic and cytotoxic properties of its main alkaloids. Colchicine became an important tool in plant breeding and in the study of microevolution because it facilitated polyploidization in plants. At the same time it has made a by no means negligible contribution to cancer research. Among plant alkaloids colchicine and minor alkaloids of *Colchicum autumnale* have a rather peculiar structure. They all contain two seven-membered rings, one of which has a tropolone structure. Biogenetic investigations advanced rapidly when the structures of several alkaloids occurring in related taxa, especially *Colchicum speciosum* and species of *Bulbocodium* (monotypic), *Kreysigia* (monotypic), *Schelhammera* (three species) and *Androcymbium* (35 species) were elucidated. These alkaloids guided chemists in proposing a reasonable hypothesis for the biogenesis of colchicine. Later, this hypothesis was verified by experiments. Investigations on colchicine, a plant constituent of great scientific and practical value, would have been severely retarded if the mono- and oligotypic genera mentioned above, as well as *Androcymbium melanthioides*, had no longer been available for research.

Secondary metabolites and plant systematics

Secondary metabolites and pathways of secondary metabolism are characters of taxonomic interest (Hegnauer, 1962–73). This is illustrated by *Colchicum* and the allied taxa which were just discussed. All taxa mentioned belong to the Colchicaceae of Huber (1969), whereas Buxbaum (1937) and many other authors place *Kreysigia* and *Schelhammera* in another main group of liliaceous plants. Without any doubt colchicine and biogenetically related alkaloids represent an important character of the Colchicaceae. The attention of botanists and chemists has already been drawn by A. P. de Candolle (1816) to the fact that plant forms and plant constituents are very often intimately connected. Nevertheless, it is only in recent times that plant systematists became interested in plant chemistry. Today many taxon-specific patterns of constituents are known. Moreover, combined with biogenetical information about the compounds present in a taxon, indications concerning its phylogenetic affinities may be gained, as was just shown for *Colchicum* and related genera.

In this connection Salicaceae may be mentioned too. They produce a large number of phenolic glucosides and accumulate condensed tannins, but no galli- and ellagitannins. Their metabolic characters point to flacourtiaceous ancestors rather than to amentiferous ones (Hegnauer, 1973).

Chemical differentiation within the limits of a genus is well illustrated by *Ulmus*. Seed oils containing much capric acid, and flavonols, and large amounts of carbonate of lime seem to occur in all species. The constituents of barks and woods are of interest in several respects. They were investigated in connection with Dutch elm disease and with difficulties in the precise identification of commercial elm woods. Hitherto, barks have been investigated less intensively than woods; they contain triterpenic esters (lupeol cerotinate) and glycosides of catechin ([+]-catechin-5-xyloside). These constituents seem to be attractants for European elm beetles (*Scolytus multistriatus*), the vectors of the pathogenic fungus, *Ceratocystis ulmi*. Today we are much better informed about wood constituents. These seem to be correlated, at least to some extent, with those sections generally accepted by taxonomists, i.e. *Blepharocarpus, Madocarpus, Trichoptelea, Microptelea* and *Chaetoptelea*. 7-Hydroxycalamenene and 7-hydroxycadelene are widespread throughout the genus *Ulmus*. The same holds for the closely related sesquiterpenic quinone mansonone-C. Other wood constituents have a more restricted distribution. Species of *Madocarpus* contain cadalenal and derivatives of cadalenal, whereas lignans like thomasinic acid and lyoniresinol occur in species belonging to the sections *Trichoptelea, Microptelea* and *Chaetoptelea* (Hegnauer, 1973).

In all probability several of these phenolic wood constituents are involved in some way in susceptibility or resistance to Dutch elm disease.

One could expect indeed that the phylogenetic classification of plants should be man's best guide for plant utilization if plant form and plant metabolism evolved together. Generations of agriculturists and phytochemists can confirm that each biologically adequate classification contains much more information than was used by the older taxonomists. I should like to give just one example: *Buxus sempervirens* is a medicinal plant whose use became obsolete long ago. In the past century a toxic alkaloid named buxin was described. Some years ago interest in the alkaloids of this plant arose again. Buxin proved to be a complex mixture of mono- and diamino steroids and nortriterpenoids, some of which have cytotoxic and (or) tumor-inhibiting properties (Kupchan, 1972). Chemists began to investigate other species of *Buxus*, and members of the genera *Pachysandra* and *Sarcococca*, which were classified together with *Buxus* in Buxaceae by botanists. All proved to contain the same type of alkaloids as *Buxus sempervirens*. In their search for new sources of buxin-like alkaloids chemists were admirably guided by the botanical classification. One exception has to be made, however, for *Simmondsia californica*, doubtfully included in Buxaceae by many plant taxonomists.

Genetically controlled variation of secondary metabolites

Highly characteristic patterns of secondary metabolites occur in many taxa which are the result of genetic differentiation during evolution. For genetic investigations good hybrid fertility is needed, and plant breeding is possible only within the limits of the coenospecies. Conclusive experimental evidence for genetic control of secondary metabolites and secondary metabolism is still rather scanty (Hess, 1968) and is totally lacking in many instances.

There is, however, plenty of indirect evidence. Infraspecific chemical variation was demonstrated many times by investigations of individual plants and by comparing various local populations (topotypes, ecotypes). Chemodemes seem to be common in land plants (Tétényi, 1970) and many classes of compounds are involved in this type of variation. Genetically controlled chemical variation within species seems to be much more common for secondary metabolites than for primary metabolites. The secondary metabolites are not essential to life but are frequently involved in the adaptation of plants to particular habitats and may therefore be subjected to very different kinds of selection pressure.

As examples of chemical variation at the infraspecific level may be mentioned the essential oil-bearing plants (Mc Kern, 1965; Stahl, 1971; Tétényi, 1970), *Solanum dulcamara* (steroidal alkaloids) (Hegnauer, 1973), Composites containing sesquiterpenic lactones (Mabry, 1973), the withanolide-producing solanaceous species *Withania somnifera* (Hegnauer, 1973) and *Duboisia myoporoides*, a solanaceous species producing mainly atropin-type alkaloids (Hegnauer, 1973). It is true, of course, that some of the variation may be due to ontogenetic changes and phenotypic plasticity, which can only be distinguished from genotypic variability by experimentation.

There is, however, much empirical and experimental evidence for the conclusion that chemical variation within species and between related species is often governed by gene action.

A large number of cultivars has been bred mainly by selection. In this connection sweet lupins (von Sengbusch, 1953), the coca plant (*Erythroxylum coca*) of the Indians of South America (Hegnauer, 1966), low-nicotine tobacco (Hegnauer, 1973), low-coumarin sweet clover (Hess, 1968), quinine-rich *Cinchona* trees (Hegnauer, 1973), morphine-rich strains of *Papaver somniferum* (Hegnauer, 1969), cucumbers without cucurbitacins (Andeweg & de Bruyn, 1959), mountain ash trees with fruits lacking the bitter parasorboside (Hegnauer, 1973) and cotton plants low in gossypol content (Hegnauer, 1969) may be mentioned. Chemical investigation of interspecific hybrids points in the same direction. Species of *Malus* contain phloridzin and related dihydrochalcones, and species of *Pyrus* are characterized by arbutin and esters of arbutin (Hegnauer, 1973). The hybrid between apple and pear synthesizes both types of constituents, which proves independent genetic control of the two biosynthetic pathways (Hess, 1968).

Flower pigments were thoroughly investigated by geneticists (Crane & Lawrence, 1952; Harborne, 1967). Examples concerning another class of natural products can be found in cyanogenic plants. Sweet almond trees with no or very little amygdalin in the seeds produce bitter almonds when pollinated by pollen of bitter almond trees. Much work was performed with *Trifolium repens* and *Lotus corniculatus*. Both species are chemically polymorphic containing acyanogenic, weakly cyanogenic and strongly cyanogenic plants. In both species, genes controlling the production of the glucosides linamarin and lotaustralin were demonstrated. An independent gene which controls synthesis or activation of linamarase, the enzyme which releases prussic acid from these glucosides, is also present, and cyanogenic plants contain the dominant alleles of these genes. The number of cyanogenic plants in local populations varies from zero to 100 per cent. It has been possible to correlate allele frequencies with ecological factors. Cyanogenic plants are favoured by selection because they are rejected by several herbivorous animals. In cold regions, however, autotoxicity of prussic acid, which is readily released after frost damage in plants containing glucosides and linamarase, becomes important, and under such conditions acyanogenic plants are favoured (Jones, 1972*a*, *b*; 1973).

Besides mutation and selection, introgression through hybridization may be the cause of chemical polymorphism in plant species. Chemical races were produced experimentally in *Linaria vulgaris* by the introduction of genes for cyanogenesis from *L. repens* (Dillemann, 1953) and in *Datura stramonium* and *D. ferox* by activation or suppression of alleles which induce the transformation of hyoscyamine to hyoscine (Romeike, 1961; 1966).

Summarizing, we can conclude that intraspecific variation of patterns of secondary metabolites is a common feature in plants. This variation is due not only to ontogenetic changes of metabolism and to phenotypic plasticity but is frequently under genetic control.

Biological significance of secondary metabolites

In the foregoing discussion allusion was made to the biological functions of natural products. It seems worth treating this aspect a little more thoroughly since secondary metabolites fulfil many tasks. Their role in pollination ecology is rather well understood, and it is known that flower odours and colours attract and guide pollinators.

The belief that natural products are defence substances of plants predominated about a hundred years ago (Stahl, 1888). It is, however, very difficult to provide definite proof for such a hypothesis. Only in recent times has convincing experimental support for the involvement of secondary metabolites in the defence strategy of plants become available. Natural products often contribute significantly to the fitness of plants in various environments (Harborne, 1972; Levin, 1971; Sondheimer & Simeone, 1970; Steelink & Runeckles, 1970; Whittaker & Feeny, 1971; de Wilde & Schoonhoven, 1969). There is evidence that phloridzin in *Malus* and arbutin in *Pyrus* are concerned with resistance to microbial diseases. Many experiments have been performed with tulips. Thus, members of the genus *Tulipa* are not attacked by *Botrytis cinerea*, a pathogenic fungus using several host plants, because they produce large amounts of tuliposides. After cell damage the latter are converted to α-methylene butyrolactones which have strong antibiotic properties. *Botrytis tulipae*, which lives only in tulips, has overcome this defence mechanism. It inhibits the formation of lactones in tulip tissues and even uses the γ-hydroxy acids replacing the lactones, as growth stimulators (Schönbeck & Schröder, 1972; Schröder, 1972*a, b*). Tuliposides, the characteristic phytonzides of *Tulipa*, also occur in *Erythronium* and *Alstroemeria*; their distribution seems to agree very well with the ideas of Huber (1969) and Buxbaum (1937) about relationships in liliiflorous plants (Slob, 1973). α-Methylene-γ-butyrolactone derived from tuliposide A can cause severe allergenic reactions in man. It would be worthwhile to pay attention in plant breeding to plant constituents affecting health.

Phytoalexins (see Stoessl, 1970; Ingham, 1972) are phytonzides of a very particular nature. They are present in plants in easily detectable amounts only after infections. Their chemical structure seems to depend upon the taxa in which they are formed. For example, the Leguminosae–Lotoideae produce pterocarpanoids (Hijwegen, 1973). The phytoalexins of European orchidaceous plants, hircinol, loroglossol and orchinol, are dihydrophenanthrene derivatives (Fish, Flick & Arditti, 1973), and in solanaceous plants the sesquiterpenoid rishitinol-capsidiol group of phytoalexins seems to predominate (Hegnauer, 1973). Phytoalexins may be secondary metabolites, of which the ability to accumulate was lost during evolution. Their production, however, is still inducible by damage. If this is true, phytoalexin production could be looked on as a kind of metabolic atavism.

The ranunculins of Ranunculaceae are very similar to tuliposides in chemical structure and biological action. They readily generate the painfully acrid, vesicant and antibiotically active protoanemonin. The structures of the genuine forms of ranunculins which are present in intact cells have not yet been wholly elucidated. On the other hand protoanemonin is a chemically well-known compound. Its significance as a defence substance against herbivorous animals and phytopathogenic micro-organisms seems to be beyond any doubt. If we consider the family Ranunculaceae as a whole, several lines of evolution with respect to defensive secondary metabolites become apparent. The following classes of such constituents are known at present: Ranunculins, alkaloids of the aporphine and berberine groups, cyanogenic glucosides, saponins, cardenolides, bufadienolides, acrid and highly toxic diter-penic alkaloids, and thymoquinone (Hegnauer, 1973).

Strong feeding-deterrents to insects were demonstrated in teak wood (anthraquinones) (Hegnauer, 1973), in several meliaceous plants (aza-dirachtin-type hexanortriterpenoids) (Morgan & Thornton, 1973) and in many other plants (several classes of compounds) (Fraenkel, 1969; Jones, 1972*b*; Rehr, Bell, Janzen & Feeny, 1973; Rothschild, 1972; Wada & Munakata, 1971; Whittaker, 1970).

In most instances, however, we still have no information about the biological meaning of secondary metabolites. This may be due to the fact that events facilitating an ecological evaluation have not yet happened; that such circumstances may be helpful is illustrated by the work of Markham (1971) on Salicaceae which are not indigenous in New Zealand. However, species of *Salix* and *Populus* are frequently planted to prevent soil erosion. He observed that some taxa were heavily damaged by herbivorous opossums whilst others were avoided. A good correlation between bitter taste, the content of salicin and esters of salicin in the leaves, and the feeding behaviour of opossums was found. Manifestly a high content of salicin and salicin derivatives lends protection against grazing by opossums. Evidently the new conditions only induced an ecological evaluation of the natural products concerned. It seems highly probable that the phenolic constituents of *Salix* and *Populus* have a similar ecological significance in the natural areas of these genera where opossums do not occur. At the same time Salicaceae can serve as an example of a kind of pre-adaptation of a taxon to conditions not existing in biotypes within its own area. All taxa tend to enlarge their area and success or failure of area extension may possibly often depend to some extent upon patterns of secondary metabolites.

Remarks concerning conservation of gene pools

If we look at plants with their secondary metabolites in mind, it is abso-lutely impossible to indicate which of the taxa that are becoming rare through the action of man should be given highest priority for conserva-tion. Our information about the total numbers of secondary metabolites in plants is still so scanty that every prediction about the future scienti-fic or practical value of a taxon must be considered as being highly arbitrary. Moreover, we are not yet sufficiently aware of the manifold functions in biocoenoses of the plant constituents known to us at present.

I must confine myself, therefore, to some general remarks on this subject. I should like to show that every taxon which is becoming rare urgently needs conservation. Some monotypic taxa and highly specialized biocoenoses may serve to show what would have happened if man had already exterminated them.

Ginkgo biloba is the only remnant of a very old gymnospermous group of plants. This tree is an extraordinarily rich source of secondary metabolites, some of which have a wider distribution in Gymnospermae (shikimic acid, pinitol, lignans, biflavones, ginnone, ginnol, sesquiterpenes), but others seem to be unique in gymnosperms or even restricted to *Ginkgo*. Alkyl phenols like bilobol and ginkgolic acid, bilobalide and ginkgolides, and sesquiterpenic and diterpenic polylactones, belong to the latter group. The extraordinary resistance of *Ginkgo* to animal feeding and to diseases is believed to depend on ginkgolides (Major, 1967; Nakanishi, 1967).

Sciadopitys is another monotypic genus of Gymnospermae, now restricted to Japan. As *Sciadopitys verticillata* is seldom cultivated, this tree is much less known in Europe and in the United States. *Sciadopitys* too, contains a number of constituents frequently occurring in Gymnospermae. At the same time several unique diterpenes such as verticillol, sciadin, dimethyl sciadinonate, methyl sciadopate and sciadone have been isolated recently from the leaves and stems. These oxygen-rich compounds may serve several purposes (Kaneko, Hayashi & Ishikawa, 1964; Miyasaka, 1964).

Lilaea scilloides is a small plant living on wet places in the coastal range of California and in some regions of Middle and Southern America. According to some taxonomists it is the only representative of the family Lilaeaceae; others include it in Juncaginaceae (= Scheuchzeriaceae s.l.). With its strangely polymorphic flowers and fruits *Lilaea* is one of the marvels of evolution. At present only triglochinin, a biogenetically and taxonomically most interesting cyanogenic glucoside, occurring also in *Triglochin* and *Scheuchzeria*, is known from this species (Ettlinger & Eyjolfsson, 1972; Hegnauer & Ruijgrok, 1971). Most probably, however, *Lilaea* has much more to offer to science and possibly even to agriculture. It would be worthwhile to learn more about morphogenetic processes and their regulation in *Lilaea*. For instance, what are the causes of the production of two clearly different flowers (a subsessile basal type and a morphologically strongly deviating type in stalked spikes) and fruits (derived from the two flower types)?

259

Evaluation problems

Phryma leptostachya is a polytypic species growing in Northern America and Eastern Asia. It is the sole representative of Phrymaceae, included by some taxonomists in Verbenaceae. The roots are used in Japan as an insecticide and very recently leptostachyol acetate and phrymarolin-I and -II could be isolated from it. The former has insecticidal properties, and the phrymarolins, like sesamolin from sesame oil, stimulate the action of some true insecticides (Taniguchi & Oshima, 1972a, b, c).

It is self-evident that the destruction of a plant species implies extinction of all organisms which depend completely upon it. Interactions and interrelations in nature are highly complex. The consequences of total extermination cannot be foreseen with any degree of precision.

Relations such as exist between *Yucca* and the *Yucca* moth, figs and fig wasps, and species of *Drosophila* and giant cacti, serve to illustrate this point because they are rather well understood.

Drosophila pachea is monophagous. It lives on the cactus *Lophocereus schottii* only. The females deposit their eggs in stem wounds which produce a mucilaginous exudate after bacterial infection. The larvae feed on the bacteria, yeasts, plant debris and exudates of the rotting wounds. American scientists showed that *Drosophila pachea* is wholly dependent upon this cactus because for normal development it needs a host-specific sterol, schottenol ($=$ stigmast-7-en-3β-ol). At the same time *Drosophila pachea* is not affected by the senita alkaloids pilocereine, piloceridine and lophocereine, which are toxic to many other species of *Drosophila*. These observations explain the strange ecological specialization of an insect; it uses a niche in which there is no competition from another species and upon which it is wholly dependent on account of its specific sterol requirement. Similar explanations probably apply to the oligophagous *Drosophila nigrospiracula* which uses *Pachycereus pringlei*, *P. pectenaboriginum* and *Carnegia gigantea* as breeding sites. These giant cacti contain other types of sterols and alkaloids (see Kircher & Heed, 1970).

Giant cacti not only sustain species of *Drosophila* but also many other insects and a number of birds and mammals. Destruction of these plants implies a more or less total loss of whole biocoenoses. The same applies to every plant species. In most instances we simply do not know how many other organisms depend partly or totally upon a given species; nor can we forecast its future scientific or practical importance.

Finally I should like to draw attention to the ecologically highly specialized parasitic *Rafflesia* species of the rain forests of tropical

Asia. They are a marvel of evolution which is far from being understood and their loss would really be unworthy of humanity.

Our generation struggles with many problems, a major one being the extremely rapid growth of human population. Adequate and stringent measures for birth control must be considered to be much more humane and much more sound and expedient than an extreme exploitation of the flora and vegetation of our earth. Nobody can foresee the ultimate consequences of the immense destructions and extirpations which would result from a utilization of plants directed only by materialism, by short-term gains without thought for the future, and by the necessity of sustaining far too large a population.

References

(With a few exceptions only reference books and articles are mentioned; these can be used as guides to original papers.)

Anderson, E. (1967). *Plants, Man and Life*. University of California Press, Berkeley.

Andeweg, J. M. & Bruyn, J. W. de (1959). Breeding of non-bitter cucumbers. *Euphytica*, **8**, 13–20.

Bernfeld, P. (ed.) (1967). *Biogenesis of Natural Compounds*, 2nd edn. Pergamon Press, Oxford.

Buxbaum, F. (1937). Entwicklungslinien der Lilioideae I, II and III. *Botan. Archiv*, **38**, 213–93, 305–98.

Candolle, A. P. de (1816). *Essai sur les propriétés médicinales des plantes, comparées avec leurs formes extérieures et leur classification naturelle*, 2nd edn. Chez Crochard, Paris.

Crane, M. B. & Lawrence, W. J. C. (1952). *The Genetics of Garden Plants*, 4th edn. Macmillan and Co. Ltd, London.

Czapek, F. (1913–21). *Biochemie der Pflanzen*, vols I–III, 2nd edn. G. Fischer, Jena.

Dekker, J. (1913). *Die Gerbstoffe*. Borntraeger-Verlag, Berlin.

Dillemann, G. (1953). Recherches biochimiques sur la transmission des hétérosides cyanogénétiques par hybridisation interspécifique dans le genre Linaria. Thèse (Sci. Nat.) Univ. Paris.

Efron, D. H. (ed.-in-chief) (1967). *Ethnopharmacologic Search for Psychoactive Drugs*. Washington, US Dept of Health, Education and Welfare, Public Health Publ. No. 1645.

Ettlinger, M. & Eyjolfsson, R. (1972). Revision of the structure of the cyanogenic glucoside triglochinin. *J. C. S. Chem. Commun.*, 1972, p. 572.

Fish, M. H., Flick, B. H. & Arditti, J. (1973). Structure and antifungal activity of hircinol, loroglossol and orchinol. *Phytochemistry*, **12**, 437–41.

Fraenkel, G. (1969). Evaluation of our thoughts on secondary plant substances. In *Insect and Host Plant* (eds J. de Wilde and L. M. Schoonhoven), pp. 473–86. North-Holland, Amsterdam.

Evaluation problems

Frear, D. E. H. (1948*a*). *Chemistry of Insecticides, Fungicides and Herbicides,* 2nd edn. D. Van Nostrand Co., New York.

Frear, D. E. H. (1948*b*). *A Catalogue of Insecticides and Fungicides,* vol. II, *Chemical Fungicides and Plant Insecticides.* Chronica Botanica Co., Waltham, Mass.

Geissman, T. A. & Crout, D. H. G. (1969). *Organic Chemistry of Secondary Plant Metabolism.* Freeman, Cooper and Co., San Francisco.

Gnamm, H. (1949). *Gerbstoffe und Gerbmittel,* 3rd edn. Wissenschaftl. Verlagsgesellsch., Stuttgart.

Goodwin, T. W. (ed.) (1965). *Chemistry and Biochemistry of Plant Pigments.* Academic Press, London.

Greshoff, M. (1893–1913). *Beschrijving der giftige en bedwelmende planten bij de visvangst in gebruik (Monografia de plantis venenatis et sopienti-bus, quae ad pisces capiendos adhiberi solent).* Batavia, part I, Meded. uit's Lands plantentuin no. **10**, 201 pp.; part II, *ibid.* no. **21**, 251 pp.; part III, Meded. Dept. van Landbouw no. **17**, 370 pp.

Guenther, E. (1948–52). *The Essential Oils,* vols I–VI. D. Van Nostrand Co., New York.

Harborne, J. B. (1967). *Comparative Biochemistry of the Flavonoids,* chapter 8, pp. 250–79. Academic Press, London.

Harborne, J. B. (ed.) (1972). *Phytochemical Ecology.* Academic Press, London.

Hartwell, J. L. (1967–71). Plants used against cancer, a survey. *Lloydia,* **30**, 379–436; **31**, 71–170; **32**, 79–107, 153–205, 247–96; **33**, 97–194, 288–392; **34**, 103–60, 204–55, 310–61, 386–438.

Haslam, E. (1966). *Chemistry of Vegetable Tannins.* Academic Press, London.

Hegnauer, R. (1962). *Chemotaxonomie der Pflanzen,* vol. I; *Thallophyten, Bryophyten, Pteridophyten, Gymnospermen.* Birkhäuser-Verlag, Basel.

Hegnauer, R. (1963). *Chemotaxonomie der Pflanzen,* vol. II; *Monokotyledonen.* Birkhäuser-Verlag, Basel.

Hegnauer, R. (1964). *Chemotaxonomie der Pflanzen,* vol. III; *Dikotyledonen, Acanthaceae–Cyrillaceae.* Birkhäuser-Verlag, Basel.

Hegnauer, R. (1966). *Chemotaxonomie der Pflanzen,* vol. IV; *Dikotyledonen, Daphniphyllaceae–Lythraceae.* Birkhäuser-Verlag, Basel.

Hegnauer, R. (1969). *Chemotaxonomie der Pflanzen,* vol. V; *Dikotyledonen, Magnoliaceae–Quiinaceae.* Birkhäuser-Verlag, Basel.

Hegnauer, R. (1971). Pflanzenstoffe und Pflanzensystematik. *Naturwissenschaften,* **58**, 585–98.

Hegnauer, R. (1973). *Chemotaxonomie der Pflanzen,* vol. VI; *Dikotyledonen, Rafflesiaceae–Zygophyllaceae.* Birkhäuser-Verlag, Basel.

Hegnauer, R. & Ruijgrok, H. W. L. (1971). *Lilaea scilloides* und *Juncus bulbosus,* zwei neue cyanogene Pflanzen. *Phytochemistry,* **10**, 2121–4.

Hess, D. (1968). *Biochemische Genetik.* Springer-Verlag, Berlin.

Hijwegen, T. (1973). Autonomous and induced pterocarpanoid formation in the Leguminosae. *Phytochemistry,* **12**, 375–80.

Huber, H. (1969). Die Samenmerkmale und Verwandtschaftsverhältnisse der Liliifloren. *Mitt. Botan. Staatssammlung München,* **8**, 219–538.

Ingham, J. L. (1972). Phytoalexins and other natural products as factors in plant disease resistance. *Bot. Rev.*, **38**, 343–424.

Jacobson, M. & Crosby, D. G. (1971). *Naturally Occurring Insecticides.* Marcel Dekker, Inc., New York.

Jones, D. A. (1972*a*). On the polymorphism of cyanogenesis in *Lotus corniculatus* L. IV. The Netherlands. *Genetica*, **43**, 394–406.

Jones, D. A. (1972*b*). Cyanogenic glycosides and their function. In *Phytochemical Ecology* (ed. J. B. Harborne), pp. 103–24. Academic Press, London.

Jones, D. A. (1973). On the polymorphism of cyanogenesis in *Lotus corniculatus* L. v. Denmark. *Heredity*, **30**, 381–6.

Kaneko, Ch., Hayashi, S. & Ishikawa, M. (1964). On the structure of verticillol. *Chem. Pharm. Bull. Tokyo*, **12**, 1510–14.

Kircher, H. W. & Heed, W. B. (1970). Phytochemistry and host plant specificity in *Drosophila*. *Recent Advances in Phytochemistry*, vol. **3**, 191–209 (eds C. Steelink and V. Runeckles). Appelton-Century-Crofts, New York.

Kreig, M. B. (1964). *Green Medicine, the Search for Plants that Heal.* Rand McNally and Co., New York.

Kupchan, S. M. (1972). Recent advances in the chemistry of tumor inhibitors of plant origin. In *Plants in the Development of Modern Medicine* (ed. T. Swain), pp. 261–78. Harvard Univ. Press, Cambridge, Mass.

Leuenberger, H. (1972). *Gesund durch Gift*. Deutsche Verlags-Anstalt, Stuttgart.

Levin, D. A. (1971). Plant Phenolics: an ecological perspective. *Am. Natur.*, **105**, 157–81.

Luckner, M. (1969). *Der Sekundärstoffwechsel in Pflanze und Tier.* VEB G. Fischer, Jena.

Luckner, M. (1971). Was ist Sekundärstoffwechsel? *Pharmazie*, **26**, 717–24.

Mabry, T. J. (1972). Major frontiers in phytochemistry. *Recent Advances in Phytochemistry*, vol. **4**, 273–306 (eds V. C. Runeckles & J. E. Watkin). Appelton-Century-Crofts, New York.

Mabry, T. J. (1973). The chemistry of chemical races. In *Chemistry in Evolution and Systematics* (ed. T. Swain), pp. 377–400 (IUPAC International Symposium, Strasbourg, 1972). Butterworths, London.

Major, R. T. (1967). The Ginkgo, the most ancient living tree. *Science*, **157**, 1270–3.

Markham, K. R. (1971). A chemotaxonomic approach to the selection of opossum resistant willows and poplars for use in soil conservation. *N.Z. J. Sci.*, **14**, 179–86.

Mc Kern, H. H. G. (1965). Volatile oils and plant taxonomy. *J. Proc. R. Soc. N.S. Wales*, **98**, 1–10.

Miyasaka, T. (1964). Interrelation of methyl sciadopate with communic acid and daniellic acid. *Chem. Pharm. Bull.*, *Tokyo*, **12**, 744–7.

Morgan, E. D. & Thornton, M. D. (1973). Azadirachtin in the fruit of *Melia azedarach*. *Phytochemistry*, **12**, 391–2.

Evaluation problems

Moritz, O. & Frohne, D. (1967). *Einführung in die pharmazeutische Biologie*, 4 Aufl. G. Fischer Verlag, Stuttgart.

Nakanishi, N. (1967). *The Chemistry of Natural Products 4*, Proc. 4th intern. IUPAC symposium on the Chemistry of Natural Products, Stockholm 1966, pp. 89–113. Butterworths, London.

Naves, Y. R. & Mazuyer, G. (1939). *Les parfums naturelles*. Gauthiers-Villars, Paris.

Paech, K. (1950). *Biochemie und Physiologie der sekundären Pflanzenstoffe*. Springer-Verlag, Berlin.

Pickersgill, B. (1971). Relationships between weedy and some cultivated forms in some species of Chili peppers (genus *Capsicum*). *Evolution*, **25**, 683–91.

Rehr, S. S., Bell, E. A., Janzen, D. H. & Feeny, P. P. (1973). Insecticidal amino acids in legume seeds. *Biochemical Systematics*, **1**, 63–7.

Romeike, A. (1961). Scopolaminbildung in der Artkreuzung *Datura ferox* L. × *Datura stramonium* L. *Die Kulturpflanze*, **9**, 171–80.

Romeike, A. (1966). Ein Modell für die Entstehungsweise chemischer Rassen. *Die Kulturpflanze*, **14**, 129–35.

Rothschild, M. (1972). Some observations on the relationships between plants, toxic insects and birds. In *Phytochemical Ecology* (ed. J. E. Harborne), pp. 1–12. Academic Press, London.

Runeckles, V. C. & Watkin, J. E. (eds) (1972). *Recent Advances in Phytochemistry*, 4. Appleton-Century-Crofts, New York.

Schnepf, E. (1969). Sekretion und Exkretion bei Pflanzen. *Protoplasmatologia, Handbuch der Protoplasmaforschung*, Band VIII (8). Springer-Verlag, Berlin.

Schönbeck, F. & Schröder, C. (1972). Role of antimicrobial substances (tuliposides) in tulips attacked by *Botrytis* spp. *Physiol. Plant Pathol.*, **2**, 91–9.

Schröder, C. (1972*a*). Die Bedeutung der γ-Hydroxysäuren für das Wirt-Parasit-Verhältnis von Tulpe und *Botrytis* spp. *Phytopath. Z.*, **74**, 175–81.

Schröder, C. (1972*b*). Untersuchungen zum Wirt-Parasit-Verhältnis von Tulpe und *Botrytis* spp. I. Stabilität und Aktivität der Tuliposide. *Z. Pflanzenkrankh. u. Pflanzenschutz*, **79**, 1–9.

Sengbusch, R. von (1953). Ein Beitrag zur Entstehungsgeschichte unserer Nahrungs-Kulturpflanzen unter besonderer Berücksichtigung der Individualauslese. *Der Züchter*, **23**, 353–64.

Slob, A. (1973). Tulip allergenes in *Alstroemeria* and some other Liliiflorae. *Phytochemistry*, **12**, 811–15.

Sondheimer, E. & Simeone, J. B. (eds) (1970). *Chemical Ecology*. Academic Press, New York.

Stahl, E. (1888). Pflanzen und Schnecken, eine biologische Studie über die Schutzmittel der Pflanzen gegen Schneckenfrass. *Jena. Z. Naturwiss. u. Medizin*, **22** (N.F. 15), 557–684.

Stahl, E. (1971). Chemische Rassen bei Arzneipflanzen mit ätherischem Öl. *Pharm. Weekblad*, **106**, 237–44.

Steelink, C. & Runeckles, V. C. (eds) (1970). *Recent Advances in Phytochemistry*, **3**, *Phytochemistry and the plant environment*. Appleton-Century-Crofts, New York.

Stoessl, A. (1970). Antifungal compounds produced by higher plants. *Recent Advances in Phytochemistry*, **3**, 143–80 (eds C. Steelink & V. C. Runeckles). Appleton-Century-Crofts, New York.

Swain, T. (ed.) (1972). *Plants in the Development of Modern Medicine*. Harvard University Press, Cambridge, Mass.

Swain, T. (ed.) (1973). *Chemistry in Evolution and Systematics* [IUPAC International Symposium, Strasbourg 1972, reprinted from *Pure Appl. Chem.*, **34**, 353–672 (1973)].

Taniguchi, T. & Oshima, Y. (1972*a*). Phrymarolin-I, a novel lignan from *Phryma leptostachya* L. *Agric. Biol. Chem., Tokyo*, **36**, 1013–25.

Taniguchi, T. & Oshima, Y. (1972*b*). Structure of Phrymarolin-II. *Agric. Biol. Chem., Tokyo*, **36**, 1489–96.

Taniguchi, T. & Oshima, Y. (1972*c*). New gemlinol-type lignan, leptostachyol acetate. *Tetrahedron Letters*, 1972, pp. 653–6.

Tétényi, P. (1970). *Infraspecific chemical taxa of medicinal plants*. Akadémiai Kiadó, Budapest.

Wada, K. & Munakata, K. (1971). Insect feeding inhibitors in plants III. *Agric. Biol. Chem., Tokyo*, **35**, 115–18.

Whittaker, R. H. (1970). The biochemical ecology of higher plants. In *Chemical Ecology* (eds E. Sondheimer & J. B. Simeone), pp. 43–70. Academic Press, New York.

Whittaker, R. H. & Feeny, P. P. (1971). Allelochemics: chemical interactions between species. *Science*, **171**, 757–70.

Wilde, J. de & Schoonhoven, L. M. (eds) (1969). *Insect and Host plant*. (Proc. 2nd International Symposium, Wageningen, Netherlands, 1969, reprint from *Entomologia Experiment. et Appl.*, **12**, 471–810.) North Holland Publ. Co., Amsterdam, London.

Zenk, M. H. (1967). Biochemie und Physiologie der sekundären Pflanzenstoffe. *Ber. dt. bot. Ges.*, **80**, 573–91.

PART IV

Conservation and storage

22. Problems of long-term storage of seed and pollen for genetic resources conservation

E. H. ROBERTS

The vast majority of species have seeds whose period of viability may be extended by lowering their temperature and moisture content during storage. In such seeds the moisture content may be reduced to 2–5 per cent or even lower before further drying ceases to increase the viability period and the reverse process tends to set in (Roberts, H., 1959; Ching, Parker & Hill, 1959; Evans, 1957; Harrington & Crocker, 1918; Nutile, 1964). There is no evidence that there is any limitation to the minimum temperature which may be used, with the exception, which would not normally be important in practice, that freezing injury may occur if the moisture content is above 15 per cent (Roberts, 1972b). I have described such viability behaviour as orthodox (Roberts, 1973a). The recent very interesting work by T. A. Villiers, which is reported in Chapter 23, has now shown that within such orthodox species the typical smooth relationship between moisture content and viability period may show a discontinuity when the seed is fully imbibed; in this condition the viability is extended enormously beyond that which would be predicted by extrapolation from the pattern of decreasing viability period with increase in moisture content up to the point of full imbibition.

There is a second but much smaller group of species which does not show the simple orthodox viability behaviour at moisture contents below fully imbibed which has just been described. In this group, which I have called 'recalcitrant' (Roberts, 1973a), a decrease in moisture content below some relatively high value – anything between 12 and 31 per cent moisture content, depending on the species – tends to decrease the period of viability. Examples of recalcitrant seeds include many of the large-seeded trees, e.g. *Taxus* spp., *Carya* spp., *Juglans* spp., *Carpinus caroliniana*, *Corylus* spp., *Castanea* spp., *Quercus* spp., *Artocarpus* spp., *Citrus* spp., *Swietenia* spp., *Malpighia glabra*, *Aleurites fordii*, *Hevea brasiliensis*, *Acer saccharinum*, *Aesculus* spp., *Blighia sapida*, *Litchi chinensis*, *Cola nitida*, *Theobroma cacao*, *Dovyalis hebe-*

carpa, *Flacourtia indica*, *Nyssa* spp., *Eugenia uniflora*, *Myriciara cauliflora*, *Trapa natans*, *Achras zapota*, *Calocarpum sapota*, *Chrysophyllum cainito*, *Diospyros virginicum*, *Coffea* spp., and many Palmaceae (Harrington, 1972) including *Elaeis guineensis* (Rees, 1963). In addition to forestry species, it will be noticed that the list includes many tropical fruits and important plantation crops, such as rubber, oil palm, coffee and cocoa. Although they are of less economic importance, it is also worth mentioning that a number of aquatic species also show recalcitrant behaviour (Harrington, 1972).

Little work has been done on defining the quantitative relationships between environmental parameters and period of viability in recalcitrant species; consequently in this account I shall concentrate entirely on orthodox species, about which there is much more information. However, it is perhaps worth mentioning that since such recalcitrant seeds appear to be averse to drying, future research into the practical problems of long-term storage in such species might not only make use of the discoveries of Villiers on orthodox species, which have already been mentioned, but also of techniques developed from research into the low-temperature storage of sperms (Mann, 1964; Polge, 1972) and, more recently, the storage of mouse embryos (Whittingham, Leibo & Mazur, 1972) which have now been stored at low temperatures for at least 23 weeks (Whittingham, personal communication, 1973) and can probably be stored for much longer. Such animal systems are like recalcitrant seeds in the sense that they are normally damaged by desiccation.

In attempts to apply such cryogenic techniques to recalcitrant species, the work of Sakai & Noshiro, reported in Chapter 24, may be helpful – even though their work has been carried out on orthodox species.

The long-term storage of the orthodox species, however, presents no serious problem: the principles are now well understood – though not as well known as one would hope – and the aim of future research should be to define the quantitative relationships more precisely for more species so that the conditions necessary for given periods of storage can be specified with more authority. The establishment of the principal relationships between environmental parameters and viability period has been slow to develop, even though it has been known for a long time that the important factors are temperature and moisture content and, to a lesser extent, oxygen pressure: lowering any of these increases the period of viability. The quantitative relationship between oxygen pressure and seed viability is still not defined, but the quantitative effects of temperature and moisture are now reasonably clear.

Earlier approaches to the problem of mathematically defining the pattern of loss of seed viability in relation to the environment include attempts to describe seed survival curves (Gane, 1948; Watson, 1970), descriptions of the relationship between temperature and the period of viability (Groves, 1917; Touzard, 1961; Watson, 1970), and attempts to quantify the relationship between both temperature and moisture content and some measure of viability (Roberts, 1960; Hutchinson, 1944; Harrington, 1963; Hukill, 1963). These individual approaches have recently been discussed in detail (Roberts, 1972a) and will not be dwelt on here; instead I shall concentrate on the system, which arose out of some initial work on results for wheat (Roberts, 1960), which enables the complete pattern of loss of viability to be described in relation to storage temperature and moisture content so that percentage viability may be predicted after any period under any given combination of temperature and moisture content.

Under any given set of constant conditions, a given sample of seeds has a particular mean viability period and there is a random distribution of the viability periods of the individual seeds around this mean value. Thus the survival curves can be described as negative cumulative normal distributions, or ogives. At least this is so in the six species whose survival curves have been examined in detail – wheat (*Triticum aestivum* L.) (Roberts, 1961a), rice (*Oryza sativa* L.) (Roberts, 1961b), broad beans (*Vicia faba* L.), peas (*Pisum sativum* L.), barley (*Hordeum vulgare* L.) (Roberts & Abdalla, 1968) and tomatoes (*Lycopersicon esculentum* (L.) Mill.) (Rees, 1970). A slight modification of this distribution may occur when the storage conditions are sufficiently bad to result in a mean viability period of a week or less (Roberts & Abdalla, 1968); but such conditions are hardly likely to be a limiting factor in long-term seed storage.

It turns out that, although the spread of the distribution (of which the standard deviation, σ, is a measure) increases if the mean viability period of the sample is increased by improving the storage conditions, the Coefficient of Variation (which is the standard deviation expressed as a percentage of the mean value of the distribution) remains the same (Roberts, 1960; 1961a; 1961b; Roberts & Abdalla, 1968). The log of the mean viability period shows a negative linear relationship with both temperature and moisture in which the temperature and moisture content terms do not interact (Roberts, 1960; 1961b; Roberts & Abdalla,

271

1968) in all five species which have been investigated – wheat, rice, barley, broad beans, peas. Consequently it is possible to predict the percentage viability after any given period under any combination of temperature and moisture content using what I have termed the three basic viability equations (Roberts, 1972a).

These are as follows:

(1) The distribution of viability periods of the individual seeds in a population is described by normal distribution,

$$y = \frac{1}{\sigma\sqrt{(2\pi)}}\, e^{-(p-\bar{p})^2/2\sigma^2}$$

where y is the relative frequency of deaths occurring at time p, $\bar{p}$ is the mean viability period, and σ is the standard deviation of deaths in time.

(2) The spread of the distribution in time is directly proportional to the mean viability period, i.e.

$$\sigma = K_\sigma \bar{p}$$

where K_σ is a constant.

(3) The relationship between temperature, moisture content and mean viability period is described by the following equation:

$$\log \bar{p} = K_v - C_1 m - C_2 t$$

where m is per cent moisture content (wet weight basis), t is temperature (°C), and K_v, C_1 and C_2 are constants.

These basic viability equations have been shown to be reasonably accurate for predicting percentage viability from a few days to several years at least (Roberts, 1972a); and there is some evidence (Roberts, 1972a), based on the germination of a very small sample (25 seeds) of 123-year-old barley seeds found in the foundation stone of the Nuremberg City Theatre (Aufhammer & Simon, 1957), to suggest that these equations are at least not wildly inaccurate for predicting viability for extremely long periods, and consequently it has been argued that they may be used with reasonable confidence for predicting the conditions necessary for long-term storage.

The application of the basic viability equations to long-term seed storage for genetic conservation

More evidence is needed but the indications are that the basic viability equations can probably be applied to a very large number of species: it is difficult not to be too impressed by the fact that they have been found to apply to all of the first five species which have been examined in

272

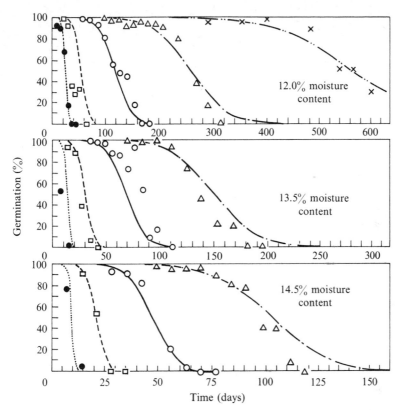

Fig. 22.1. Survival curves for seed of *Oryza sativa* hermetically sealed in air at different temperatures and moisture contents. Note that the time scale is different for each moisture content: reading from top to bottom each scale is a 2× magnification of the previous one. Storage temperatures were as follows: 27 °C (×); 32 °C (△); 37 °C (○); 42 °C (□); 47 °C (●). The curves were fitted according to the three basic viability equations from which the following constants were determined for viability period in days: $K_v = 6.531$, $C_1 = 0.159$, $C_2 = 0.069$, and $K_\sigma = 0.210$. (From Roberts, 1961 *b*.)

detail (Roberts, 1972*a*). An example of the application of the equations to some results obtained with rice is given in Fig. 22.1. The values of the four viability constants (K_σ, K_v, C_1 and C_2), however, are different for different species. But once the constants have been determined, the equations can be applied to describe the pattern of loss of viability for a species in the form of a nomograph, which is a convenient way of providing the information in an easily accessible form (Roberts, 1972*a*). Viability nomographs have been published for the five species for which there are sufficient data to calculate the four viability constants (Roberts & Roberts, 1972*a*), and an example for one of these species, *Vicia faba*, is shown in Fig. 22.2.

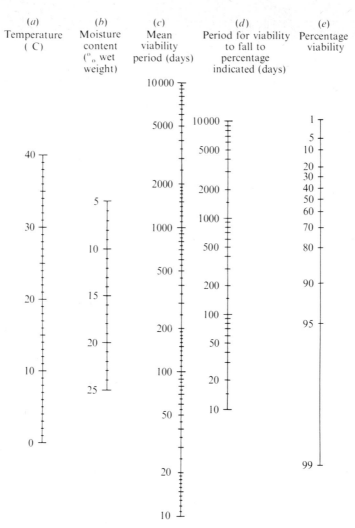

Fig. 22.2. Viability nomograph for *Vicia faba*. Viability nomographs may be used in various ways; two of the most useful are as follows. (1) To estimate the time taken for viability to fall to any given percentage at any given temperature and moisture content, put a ruler on the required temperature (scale *a*) and moisture content (scale *b*). Note the value on scale *c* (this gives the mean viability period). Using this point on scale *c* as a pivot, move the ruler to indicate the required percentage viability on scale *e*. The value now indicated on scale *d* is the time taken for viability to drop to the percentage chosen. (2) To find the alternative combinations of temperature and moisture content necessary to prevent viability falling below a given level after a given period, select the minimum percentage viability required on scale *e*. Select required storage period on scale *d*. Put a ruler through both points and note value on scale *c*. Using this point on scale *c* as a pivot, move the ruler through scales *a* and *b*. Any position of the ruler indicates a combination of temperature (scale *a*) and moisture content (scale *b*) which should achieve the object. (From Roberts & Roberts, 1972*a*.)

One point which such nomographs strongly emphasise is that it is inappropriate to discuss seed storage and viability in terms of time or temperature or moisture content alone: all three have to be considered simultaneously (strictly speaking one should also add oxygen pressure to this list, but for simplification it can usually be ignored as a non-variable factor, because under open storage conditions it is more or less constant at 21 per cent, and in hermetic storage it soon becomes constant at near 0 per cent through respiration). A second point, which follows from this, is that the commonly held concept of a 'safe moisture content' for storage is misleading: the relationship between time, temperature, moisture content and period of viability is smooth (apart from the exceptions noted in the introductory paragraph); there is no discontinuity. It follows that it is only legitimate to use the term 'safe moisture content' if, at the same time, the temperature, the required period of storage, and the required minimum percentage viability are also specified.

The amount of experimental work needed to calculate the four viability constants for a species is relatively small. It has been pointed out that, in theory, only four germination tests are required. The theory, which is explained in greater detail elsewhere (Roberts 1972a; 1973a), is illustrated in Fig. 22.3. Of course it would be statistically unwise to base the calculations on four germination tests; nevertheless, the main point to be emphasised is that in a carefully planned investigation it would be possible to obtain reliable results on the basis of a modest amount of practical work. It is worth emphasising that the four viability constants have some real meaning in terms of the viability behaviour. The value K_v indicates the potential viability under ideal conditions; the value of C_1 indicates the sensitivity to moisture content and C_2 the sensitivity to temperature; and the value of K_σ indicates the extent of the seed-to-seed variability in viability period.

So far in this discussion, in order to establish the main points clearly, I have not raised the difficulties which may be met in trying to apply the basic viability equations to the practical problems of long-term seed storage. These are discussed in detail elsewhere (Roberts, 1972a; 1973a); here it will be sufficient to outline the main problems.

(1) The extent of intraspecific genotypic differences in viability are not known in most cases: sometimes the differences appear to be small – e.g. in tomato, bean, pea, water-melon, cucumber, sweet corn (James, Bass & Clark, 1967) and rice (Roberts, 1963); but in other cases there may be up to a threefold difference in viability period between cultivars

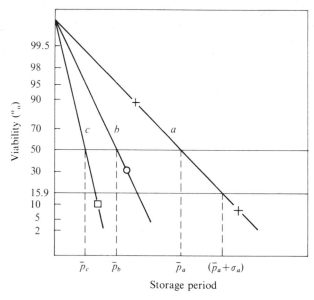

Fig. 22.3. Diagram illustrating that theoretically only four germination tests are necessary to determine the four seed viability constants. The graph shows some hypothetical results for percentage viability plotted on a probability scale against time (cumulative normal survival curves then appear as straight lines). It is assumed that three batches of the initial sample have been stored under different sets of conditions (environments *a*, *b* and *c* in order of their increasingly deleterious effect on viability); and these conditions include two temperatures and two moisture contents. Two germination tests (+) carried out at different times for seed stored in environment *a* would determine the slope of the survival curve from which the values of the mean viability period, $\bar{p}_a$ (time taken for viability to drop to 50 per cent), and the standard deviation, σ_a (time taken for viability to drop from the 50 per cent level to the 15.9 per cent level), could be calculated for this storage environment; substituting these values in equation (2) the value K_σ could be determined. Two further germination tests ($\bigcirc$, $\square$) carried out after any reasonable time on seed stored in environments *b* and *c* respectively would fix the positions of the survival curves for these environments, since these curves would also be straight-line and have a common intercept with the survival curve determined for environment *a*. From these curves two additional values ($\bar{p}_b$ and $\bar{p}_c$) for mean viability could be obtained. Finally, using the three values for the mean viability period ($\bar{p}_a$, $\bar{p}_b$, $\bar{p}_c$) obtained under three sets of conditions, three equations (similar to equation 3) can be solved simultaneously, thus determining the values K_v, C_1, and C_2 (from Roberts, 1973*a*).

when stored under uniform conditions, e.g. in lettuce (Harrison, 1966). In most cases, however, there is no critical information.

(2) Although only extreme environmental conditions before harvest are likely to affect subsequent seed viability to any marked extent (Austin, 1972), the conditions at or after harvest can have some effect (Moore, 1972; Austin, 1972; Roberts, 1972*a*), particularly mechanical damage in large-seeded legumes and the application of too high a temperature during drying. There are also undefined effects of

provenance but which are probably related to climate at the time of ripening and harvest (Andersen & Andersen, 1972; Roberts, 1973*a*).

(3) The values of the viability constants calculated so far were determined under hermetic storage. Application of the basic equations using these values to 'open' storage is likely to overestimate the period of viability because of the higher partial pressure of oxygen in open storage (Roberts, 1972*a*; 1973*a*). Nevertheless, it is felt that the error involved will not be serious in practice, particularly if rather better conditions are provided than the minimum specified, so that the storage system includes some margin of safety.

With regard to all three sources of variation described above, it is important to remember that these factors can only affect the pattern of loss of viability by affecting the value of one or more of the four viability constants. The problem of defining further the effects of these sources of variation on viability therefore should not be difficult. In summary, there is an urgent need to study the effects of genotype, pre-storage environment (mechanical damage, drying damage and the various aspects of provenance), and oxygen pressure on the values of the viability constants. Not all the constants, of course, may be affected by all these factors.

The genetic stability of stored seeds

There have been a number of recent reviews of the chromosome damage which occurs in seeds as they age and the associated genetic changes (Ashton, 1956; D'Amato & Hoffman-Ostenhof, 1956; Barton, 1961; Roberts, Abdalla & Owen, 1967; Roberts, 1972*b*); accordingly there is no call here to cite all the voluminous literature on this topic. The development of our knowledge of the genetic instability of seeds during storage has been summarised into four phases as follows (Roberts, 1973*b*).

At the beginning of this century mutant phenotypes were observed with a greater frequency in plants produced from old seed as compared with plants produced from young seed. At this time it was assumed, quite wrongly, that the mutant genotypes were present from the beginning of storage and that ageing selected out the mutants since the mutations in some way conferred greater longevity.

Secondly, there was a great deal of work starting in the early 1930s involving observations on the frequency of chromosome aberrations, pollen abortion and mutant phenotypes which showed that chromosome damage was a general phenomenon induced by ageing in seeds.

Thirdly, in the late 1930s it became apparent that the age of the seed could not be considered in isolation: the temperature and moisture content during storage were also important. Increases in time of storage, and the temperature and moisture content during storage all lead to an increase in chromosome damage and mutation.

Fourthly, in the 1960s it was shown that any combination of time, temperature, moisture content and oxygen not only leads to predictable loss of viability but also to a predictable amount of chromosome damage as well (Roberts, Abdalla & Owen, 1967; Abdalla & Roberts, 1968; 1969). The quantitative evidence so far as the production of recessive mutations is concerned is not as precise, but, nevertheless, the same rule appears to apply (Abdalla & Roberts, 1969). In other words it does not matter how quickly seeds lose viability or what conditions lead to that loss, for a given loss of viability there will be a given amount of genetic damage in the surviving seeds. Even if the past history of a sample of seeds is unknown, a knowledge of the percentage viability will enable a reasonably accurate prediction to be made about the amount of genetic damage in the surviving seeds.

Fig. 22.4a illustrates the accumulation with time of aberrant cells (cells containing chromosome breakages) in surviving *Vicia faba* seeds stored under different conditions and Fig. 22.4 b the concomitant loss of viability. In Fig. 22.4c the same data are replotted to show percentage aberrant cells against percentage viability. It will be seen that for all storage conditions – except for the most severe where viability was lost within a week – the same curve describes the relationship between the percentage of aberrant cells in the embryo of the surviving seeds and percentage viability. Under very severe storage conditions, this rule breaks down and a separate curve is necessary to describe the relationship; but this is of little consequence in the present context since such poor storage conditions would obviously not be used when an attempt is being made to preserve viability for long periods.

Very similar relationships to those just described have also been shown in the two other species, *Pisum sativum* and *Hordeum distichon*, which have been subjected to quantitative investigation in the same way (Abdalla & Roberts, 1968; 1969). Similar relationships have also been shown between percentage viability and frequency of abnormal ana-phases in *Lactuca sativa*, but in this case the asymptotic level of abnormal anaphases per root reached the very high value of 90 per cent when

278

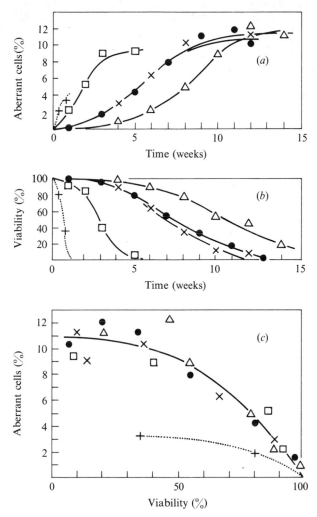

Fig. 22.4. (*a*) The increase in the mean frequency of aberrant cells in the surviving popula-tion of seeds of *Vicia faba* stored under various combinations of temperature and moisture content (m.c.).

(*b*) Seed survival curves for the same treatments.

(*c*) The relationship between percentage viability and the mean frequency of aberrant cells for the same treatments.

Conditions were as follows:

45 °C, 18 per cent m.c. (+); 45 °C, 11 per cent m.c. (●); 35 °C, 18 per cent m.c. (□); 35 °C, 15.3 per cent m.c. (×); 25 °C, 18 per cent m.c. (△). (From Roberts, Abdalla & Owen, 1967).

germination had dropped to 50 per cent (Harrison & McLeish, 1954; Harrison, 1966).

Although it has been possible to assess reasonably accurately the amount of visible chromosome damage associated with loss of viability there is no corresponding estimate of the total amount of mutation. But in the three species which have been investigated it was found that, in general, any storage treatment which leads to a loss of viability of about 50 per cent induces recessive chlorophyll mutations in about 1–4 per cent of the surviving seeds (Abdalla & Roberts, 1969). This represents a considerable amount of mutation and, it has been argued, is equivalent to that produced by a treatment of fresh seeds with 10000 r of X-rays (Roberts, 1973b). If it is assumed that the genes affecting chlorophyll synthesis are not outstandingly more unstable than all the other genes, then there seems no avoidance of the conclusion that the *total* amount of mutation present in the surviving seeds in samples which have lost 50 per cent viability must be extremely high.

In normal seed storage practice it would be hoped that conditions are such that loss of viability does not occur to the extent of 50 per cent. We have no direct information concerning the amount of recessive mutation associated with smaller losses of viability. However, if we make the assumption, which is not unreasonable (see, for example, Caldecott, 1961), that there is a more or less linear relationship between the frequency of chromosome aberrations and the frequency of recessive mutations, then it follows from the curve shown in Fig. 22.4c (and similar curves for peas and barley) that almost half as many mutations may be found in the surviving seeds when the viability is still 80 per cent as compared with samples showing 50 per cent viability. Consequently the available evidence would suggest that conditions leading to relatively small losses of viability can result in a very considerable accumulation of heritable damage.

So far, in spite of a considerable amount of work which has been reviewed by D'Amato & Hoffman-Ostenhof (1956) and Roberts (1972b; 1973b), it is still not clear what causes the damage to chromosomes during storage. In some cases the evidence suggests the accumulation of mutagens with time, but in other cases there is little evidence of mutagenic substances. Certainly the phenomenon of the accumulation of chromosome damage with time is not peculiar to seeds but also seems to be a characteristic of those animal tissues in which there is also no cell division and consequently in which aberrations cannot be removed by diplontic selection (Curtis, 1963; Price, Modak & Makinodan, 1971).

Neither do we know what causes loss of viability in seeds, although the major current theories have been discussed recently (Roberts, 1972*b*). Nevertheless, it is not necessary to know the mechanism of mutation or loss of viability in order to apply what we do know to the practical problems of the long-term storage of genetic stocks of seeds.

Practical approaches to the long-term storage of genetic stocks of seeds

The main objectives in the storage of genetic stocks of seeds should be two-fold. First, seeds need to be stored for long periods in order to avoid the costs, complications and risks involved in growing plants at frequent intervals in order to replace stocks of seeds as they begin to lose viability. Secondly, it is important that the stocks held in storage show the minimum amount of genetic alteration.

It will now be clear that the second objective can be met providing conditions are such that the loss of viability is small. Accordingly, it is recommended that in designing storage systems one should adopt the principle that viability should not be allowed to fall below some relatively high value – say 85–90 per cent. Starting then, say, from the 90 per cent value on scale *e* of nomographs of the type shown in Fig. 22.2, one then decides what would be a reasonable period of storage on scale *d*. These two values determine the mean viability period on scale *c*. From this point may be derived all the alternative pairs of conditions of temperature and moisture content on scales *a* and *b* which will give 90 per cent viability for the chosen storage period.

One can now proceed to the engineering and mechanical problems as to how any of these alternative pairs of conditions can be achieved and which would be the most desirable in terms of cost, convenience and reliability. So far as the control of moisture content is concerned, two main methods are available. First, seeds can be dried to a given moisture content and then sealed in hermetic containers. Alternatively, the relative humidity of the air surrounding the seeds can be controlled, say, by air-conditioning apparatus.

The second method depends on the well-known fact that the moisture content of seeds shows an equilibrium hygroscopic relationship with the relative humidity of the inter-seed atmosphere. There are two complications concerning this relationship which have to be recognised in the design of seed stores. The first is that the equilibrium relationship

281

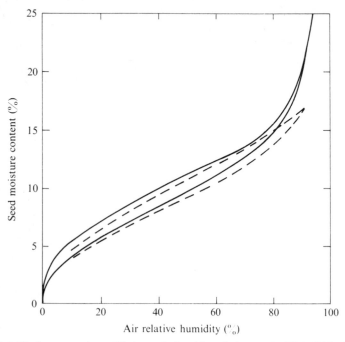

Fig. 22.5. The hygroscopic equilibrium relationships of wheat at 35 °C (solid line) and paddy (rice in husk) at 25 °C (dashed line). In both cases the upper curve represents the desorption relationship and the lower curve represents the absorption relationship. Data for wheat from Hubbard, Earle & Senti (1957) and for paddy from Breese (1955). Diagram reproduced from Roberts (1972*a*).

shows hysteresis, i.e. the relationship is not the same on absorption of water by the seed as it is on desorption. Typically the seed moisture content on desorption is 1–1.5 per cent higher than it is on absorption (see Fig. 22.5), although some of the hygroscopic equilibrium curves published earlier showed greater discrepancies than this between desorption and absorption. Fortunately, from the point of view of calculating what may happen in practice, the effect of temperature at high humidities is minimal; at low humidities a reduction of temperature by 10 °C around the value 20 °C increases the equilibrium moisture content at a given relative humidity by about 1 per cent, but the increase in moisture content for a similar drop in temperature over a lower range of temperature is less (Roberts, 1972*a*). As a safety precaution, in seed storage practice it would be advisable to base calculations of the intended seed moisture content on the desorption hygroscopic relationship, since this will indicate the seed moisture content

282

at which the seeds can be expected to equilibrate if the seeds are put in store at a higher moisture content than intended.

The second complication is that different seeds show different equilibrium relationships. However this complication does not present many difficulties from the practical point of view once it is recognised that in this respect seeds fall largely into two categories, the oily and the non-oily seeds. For a given relative humidity the oily seeds show a lower equilibrium moisture content: e.g. at 40 per cent relative humidity non-oily seeds like cereals or beans show moisture contents of about 9 or 10 per cent, whereas oily seeds like groundnuts or soya beans have moisture contents of about 6 or 7 per cent (Roberts, 1972*a*). Tables summarising published data on the hygroscopic equilibrium of seeds of a wide range of species have recently been published (Roberts & Roberts, 1972*b*).

Which storage strategy to apply – hermetic storage or air-conditioning – will depend, as already suggested, on relative costs, convenience and reliability. Problems of cost will not be dealt with here but would lend themselves to solution by standard systems engineering procedures. With regard to reliability, one advantage of air-conditioned rooms or cold rooms is that it is immediately apparent when the system fails, and something can be done about it. On the other hand, systems which depend on hermetic storage do not necessarily show when the seal of a container is no longer perfect. Nevertheless, particularly for small volumes of seeds, near foolproof container designs can be used. For example in the Japanese National Seed Storage Laboratory for Genetic Resources the seeds are first dried to 4–6 per cent moisture content and then stored in standard food cans which are then placed in cold rooms (Ito, 1972). It would now also be worth considering packages made of plastic-laminated aluminium foil which are used increasingly for commercial packaging of seeds. Such packages are now generally reliable and have the advantage that they occupy little space. However they cannot easily be resealed after opening so that it would be necessary to use several packages per sample if seeds are to be distributed from the long-term store or used for germination tests.

In the United States National Seed Storage Laboratory the system depends on air-conditioning; the standard storage conditions employed are 4 °C with an average relative humidity of 32 per cent (James, 1972). For many non-oily crop seeds this should give an equilibrium seed moisture content of about 9 per cent. Rough calculations show that for the five common crop species mentioned earlier for which nomographs have been constructed, such conditions should maintain 90 per cent viability for 12 to 135 years, depending on the species; *Vicia faba*, for example, the species illustrated in Fig. 22.2, would be expected to store satisfactorily for about 27 years under such conditions. Many wild species seem to have intrinsically greater longevity than crop plants (Harrington, 1970) and might well last very much longer under these conditions.

Obviously, from what has been said earlier, one should not take these figures as being accurate prognostications of behaviour, but they do give some indication of the order of magnitude of the satisfactory storage periods involved. Any storage system, however, should include some programme of regular routine tests in order to check that the seeds are behaving as anticipated. If, for some reason, any sample begins to lose viability sooner than expected, then a sub-sample should be taken for multiplication. Such routine tests could, of course, be quite infrequent, say every five to ten years, in systems designed for retention of 85–90 per cent viability for 10–100 years.

In view of the somewhat complicated and time-consuming arrangements that have to be made for replenishment of stocks because of reductions in viability (see, for example, Ito, 1972 and Allard, 1970), it would be advisable for seed storage laboratories to aim at very long storage periods and consequently to consider using subzero temperatures, such as those found in commercial deep-freeze cabinets, which operate at about −20 °C. Under such conditions it would probably be preferable to maintain low seed moisture content by storing seed in hermetic containers, although this may not be necessary with an appropriate design of refrigeration equipment which maintains low water vapour pressure in the store.

There is no direct experimental evidence to show how long seeds would last under such conditions, but extrapolation from the basic viability equations would suggest satisfactory storage periods of the order of centuries, at least·for the five species to which the equations have been applied. There is certainly plenty of evidence to show that extremely low temperatures are not damaging to orthodox seeds, providing their moisture content is less than 15 per cent (Roberts, 1972*a*). Seeds of many species have been reduced to a temperature below −271 °C without any harmful effects (Lipman, 1936), and there is also evidence that very low temperatures, −18 °C or less, are more satisfactory than higher temperatures and allow storage for many years with very little deterioration in many species (Barton, 1966; von Senbusch, 1955;

Weibull, 1953; 1955). Some comments on the economics of very low-temperature storage facilities are included as an appendix to this chapter.

The storage of pollen

Harrington (1970) pointed out that probably the easiest and least expensive way of preserving plant genetic resources is by seed storage. There is also some potential for storing pollen for this purpose but he emphasised that in this case the practical maximum period of storage appears to be a few years – much less than for seeds. The longest period of pollen storage so far recorded appears to be nine years – for apple pollen when stored at about $-20\ ^\circ$C (Ushirozawa & Shibukawa, 1951). This genus, *Malus*, belongs to the group of families noted for the longevity of their pollen – Rosaceae, Primulaceae, Saxifragaceae and Pinaceae; families showing intermediate longevity are Amaryllidaceae, Liliaceae, Ranunculaceae, Salicaceae and Scrophulariaceae; short-lived families, where recorded maximum periods of viability are a matter of days, include Alismataceae, Cyperaceae, Commelinaceae, Juncaceae and especially Gramineae (Holman & Brubaker, 1922).

At the present time, then, pollen storage appears to be a much less promising method for the long-term conservation of genetic resources because, in addition to the more difficult storage problems, there are also the complications involved in the genetical strategy of how the pollen could be used to reconstitute the diploid material from whence it originated. However, such diploid reconstitution would not be necessary in all applications and there is still much fundamental work to be done on the long-term storage of pollen; thus it would be premature to exclude pollen storage as a method of genetic conservation at this stage.

There have been a number of good reviews on pollen storage, e.g., Visser (1955), Johri & Vasil (1961), Linskens (1964) and Harrington (1970). Not a great deal of significance has been added to the literature since the more recent of these reviews, and no attempt will be made here to provide yet another survey of the literature. Instead, I intend to underline the similarity of pollen storage problems to those of seeds and suggest areas of investigation which may be profitable.

There seems to be little doubt that the most important environmental factors so far as pollen viability is concerned are the same as for seed storage: viz temperature, moisture content and oxygen pressure; lowering any of these tends to increase the period of viability. However, with regard to moisture content (in the case of pollen normally measured indirectly as the relative humidity of the atmosphere in contact with the grains), there is in many species a limit below which a further decrease in moisture content no longer leads to an increase in viability. For many species this limit lies somewhere between moisture contents in equilibrium with 8–25 per cent relative humidity (Holman & Brubaker, 1926; Visser, 1955; Johri & Vasil, 1961; Linskens, 1964), though in some species, especially those showing particularly short periods of pollen viability, the pollen does not seem to be able to withstand even slight desiccation; for example relative humidities below 50 per cent are very detrimental to the pollen of *Zea mays* (Daniel, 1955).

In general this classification into two groups, namely, the larger orthodox group where there is a continual improvement in viability period with reduction of moisture content to very low levels, and a smaller recalcitrant group which cannot withstand desiccation, is very reminiscent of the situation we find in seeds. Although there are these strong parallels in behaviour between seeds and pollen, it is also interesting to note that a species showing orthodox seed behaviour may have recalcitrant pollen, e.g. the cereals.

In spite of the fact that pollen and seeds show many similarities with regard to their response to storage conditions, it might be argued that the survival curves in each case are of a different form and thus the two processes must be fundamentally different. As we have seen, the survival curves of seeds are typically negative cumulative normal distributions. Little critical work seems to have been done on the survival curves of pollen, but from some recently published data on two species (Heslop-Harrison & Heslop-Harrison, 1970), it can be seen that pollen probably tends to behave like other unicellular non-dividing organisms (Roberts, 1972b), in that log of the number of survivors plotted against time is approximately straight-line (Fig. 22.6). (At least the data for *Tagetes* conformed to this pattern, although the data for *Zebrina* is somewhat anomalous: the initial loss of viability conformed to this pattern but then the probability of grain survival increased.) Such curves describe the situation when a constant proportion of the survivors die during each unit interval of time. They imply that, however long a pollen grain has survived, the probability of its death in the following unit interval of time remains the same. In other words, in common with many unicellular organisms, the loss of viability of pollen grains appears to be a random process and does not show the typical ageing characteristic of many multicellular organisms where the probability of death in-

286

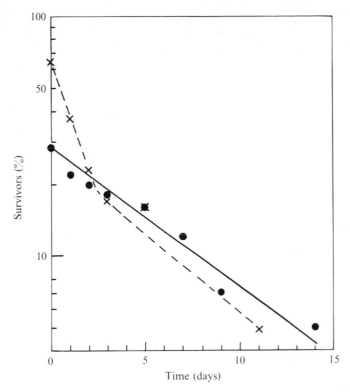

Fig. 22.6. Survival curves of two species of pollen, *Tagetes patula* (● —) and *Naegelia zebrina* (× — —), as indicated by intracellular retention of fluoroscein produced by enzymic hydrolysis of fluoroscein diacetate. Log per cent viable pollen is plotted against time. (Re-drawn from the original data of Heslop-Harrison & Heslop-Harrison, 1970.)

creases with age. This point will not be laboured here but it is worth noting, as has been discussed in detail elsewhere (Roberts, 1972*b*), that it is possible to reconcile such survival curves of unicellular organisms with those of seeds in terms of a common stochastic viability theory and thus there is no basis at the present time for considering loss of viability in pollen as an intrinsically different process from loss of viability in seeds.

Although there have been a number of systematic investigations of the effects of temperature on pollen viability period, and other investigations on the effects of moisture content on pollen viability period, there do not appear to have been any systematic multifactorial experiments which include an examination of the effects of both environmental parameters at the same time. Such an investigation would be particularly helpful since, if a simple relationship could be found between these environmental factors and period of viability, as has been found for seeds, then extrapolation from the relationship discovered would at least indicate what practical treatments might lead to the most promising and economical methods of extending the viability period.

At the beginning of this section, it was mentioned that the potential for long-term pollen storage does not appear to be as promising as for seeds because, under similar conditions, pollen has a far shorter viability period. Nevertheless, it should be recognised that the whole range of feasible storage conditions has not yet been thoroughly explored. For example, although it is known that extremely low temperatures do not deleteriously affect the pollen of many species, there do not appear to have been any long-term experiments using such conditions. Linskens (1964) has reported several cases where pollen has been kept for short periods at -180 to -190 °C without any deleterious effects. And so far as extremely low moisture content is concerned, King (1965) freeze-dried pollen and, storing it in nitrogen or under vacuum, found that it remained viable throughout the three-year experiments. There is obviously a need for more long-term experiments using both extremely low temperatures and moisture contents: it is not sufficient merely to know that such treatments are not deleterious; it is necessary to know how beneficial they are. Visser (1955) stored pollen of *Pyrus malus* for two years in liquid air and found that it was 'as good as fresh pollen' believing that it would keep indefinitely under such conditions. Although we shall never know if his prediction of immortality is correct, it would be very useful to know whether such conditions would at least allow pollen to be kept for very long periods.

Conclusions

(1) Seed storage in general has more potential value than pollen storage for the long-term preservation of genetic resources, but the full potential of pollen storage has not yet been investigated.

(2) In spite of the fact that seeds and pollen have different types of survival curve, they appear to respond very similarly to the main factors which control viability period – temperature, moisture content and oxygen pressure. In neither case is it clear why viability is lost, but the fundamental causes could be very similar.

(3) On the basis of their response to moisture content, species fall into two groups in respect of both seeds and pollen. The majority show 'orthodox' behaviour in which decreases in moisture content down to

very low values cause corresponding increases in viability period; in these species, lowering the temperature and oxygen pressure also increases viability period. In the remaining 'recalcitrant' species, desiccation below relatively high critical levels (often 12 per cent or more in seeds, or moisture content in equilibrium with 50 per cent relative humidity or more in pollen) results in rapid loss of viability. In both seeds and pollen it has not been possible to store recalcitrant species for long periods. Species having orthodox seeds may have recalcitrant pollen.

(4) In the case of recalcitrant seeds it has not been possible so far to perceive any general quantitatively expressed laws with regard to storage treatments. In the case of the orthodox species, however, it has been found possible to develop three basic viability equations which predict percentage viability after any period under any combination of temperature and moisture content. Nomographs based on these equations provide a convenient basis on which to design seed stores for long-term genetic conservation.

(5) Even in orthodox species T. A. Villiers has recently shown that the effect of moisture content suddenly reverses when the seed becomes fully imbibed: the viability period is now greater than would be predicted by extrapolation from lower moisture contents. This discovery may be particularly useful if it can be applied to the problems of storing recalcitrant species.

(6) Chromosome damage and genetic mutation is associated with loss of seed viability during storage. In this respect the age of the seed is of little consequence; providing the seeds have been stored in conditions which maintain high viability there will be little genetical deterioration. On the other hand if storage conditions are such as to lead even to quite small losses of viability, so that germination falls below say, 85 per cent, the seeds will have accumulated a considerable amount of mutation, irrespective of whether they have been stored for a short or long time.

(7) Major subjects requiring further study include the following: (*a*) the application of the three basic viability equations to seeds of more species in addition to the five to which they have already been applied; (*b*) further studies on the effects of genotype and pre-storage conditions in terms of their effects on the four viability constants applicable to the basic viability equations; (*c*) studies on recalcitrant seeds applying, if possible, Villiers' discovery of increased longevity of fully imbibed seeds and making use of the approaches adopted for the storage of

mammalian sperms and embryos which are also easily damaged by desiccation; (*d*) quantitative multifactorial investigations on the effects of temperature and moisture content on the viability period of pollen in order to discover whether the viability period can be described in simple quantitative terms; (*e*) preferably as part of such studies there is a need for long-term investigations into the effects of extremely low temperatures and moisture contents on pollen viability.

References

Abdalla, F. H. & Roberts, E. H. (1968). Effects of temperature, moisture and oxygen on the induction of chromosome damage in seeds of barley, broad beans, and peas during storage. *Ann. Bot.*, **32**, 119–36.

Abdalla, F. H. & Roberts, E. H. (1969). The effects of temperature and moisture on the induction of genetic changes in seeds of barley, broad beans, and peas during storage. *Ann. Bot.*, **33**, 153–67.

Allard, R. W. (1970). Problems of maintenance. In *Genetic Resources in Plants – their Exploration and Conservation* (eds O. H. Frankel and E. Bennett), pp. 491–4. Blackwell, Oxford and Edinburgh.

Andersen, S. & Andersen, K. (1972). Longevity of seeds of cereals and flax. *Acta agric. scand.* **22**, 3–10.

Ashton, T. (1956). Genetical aspects of seed storage. In *The Storage of Seeds for Maintenance of Viability* (by E. B. Owen), pp. 34–8. Commonwealth Agricultural Bureaux, Farnham Royal, England.

Aufhammer, G. & Simon, U. (1957). Die Samen landwirtshaftlicher Kulturpflanzen im Grundstein des chamaligen Nürnberger Stadttheaters und ihre Keimfähigkeit. *Z. Acker- u. PflBau*, **103**, 454–72.

Austin, R. B. (1972). Effects of environment before harvesting on viability. In *Viability of Seeds* (ed. E. H. Roberts), pp. 114–49. Chapman and Hall, London.

Barton, L. V. (1961). *Seed Preservation and Longevity*. Leonard Hill, London.

Barton, L. V. (1966). Effects of temperature and moisture on viability of stored lettuce, onion and tomato seeds. *Contrib. Boyce Thompson Inst.*, **23**, 285–90.

Breese, M. H. (1955). Hysteresis curves in the hygroscopic equilibria of rough rice at 25 °C. *Cereal Chem.*, **32**, 481–7.

Caldecott, R. S. (1961). Seedling height, oxygen availability, storage and temperature: their relation to radiation-induced genetic and seedling injury in barley. In *Effects of Ionising Radiations in Seeds*, pp. 3–24. International Atomic Energy Agency, Austria.

Ching, T. M., Parker, M. C. & Hill, D. D. (1959). Interaction of moisture and temperature on viability of forage seeds stored in hermetically sealed cans. *Agron. J.*, **51**, 680–4.

Curtis, H. J. (1963). Biological mechanisms underlying the ageing process. *Science*, **141**, 686–94.

D'Amato, F. & Hoffman-Ostenhof, O. (1956). Metabolism and spontaneous mutations in plants. *Adv. Genet.*, **8**, 1–28.

Daniel, L. (1955). Polleneltartas esírázóképes állapothan. *Növénytermeles*, **1**, 133–52.

Evans, G. (1957). The viability over a period of fifteen years of severely dried ryegrass seed. *J. Brit. Grassld Soc.*, **12**, 286–9.

Gane, R. (1948). The effect of temperature, humidity, and atmosphere on the viability of Chewing's Fescue grass seed in storage. *J. Agric. Sci.*, **38**, 90–2.

Groves, J. F. (1917). Temperature and life duration of seeds. *Bot. Gaz.*, **63**, 169–89.

Harrington, G. T. & Crocker, W. (1918). Resistance of seeds to desiccation. *J. Agric. Res.*, **14**, 525–32.

Harrington, J. F. (1963). Practical advice and instructions on seed storage. *Proc. Int. Seed Test. Ass.*, **28**, 989–94.

Harrington, J. F. (1970). Seed and pollen storage for conservation of plant gene resources. In *Genetic Resources in Plants – their Exploration and Conservation* (eds O. H. Frankel and E. Bennett), pp. 501–21. Blackwell, Oxford and Edinburgh.

Harrington, J. F. (1972). Seed storage and longevity. In *Seed Biology* (ed. T. T. Kozlowski), vol. III, pp. 145–245. Academic Press, New York and London.

Harrison, B. J. (1966). Seed deterioration in relation to storage conditions and its influence upon germination, chromosomal damage and plant performance. *J. Nat. Inst. Agric. Bot.*, **10**, 644–63.

Harrison, B. J. & McLeish, J. (1954). Abnormalities of stored seeds. *Nature, Lond.*, **173**, 593–4.

Heslop-Harrison, J. & Heslop-Harrison, Y. (1970). Evaluation of pollen viability by enzymatically induced fluorescence; intracellular hydrolysis of fluoroscein diacetate. *Stain Technol.*, **45**, 115–20.

Holman, R. M. & Brubaker, F. (1926). On the longevity of pollen. *Univ. Calif. Publs. Bot.*, **13**, 179–204.

Hubbard, J. E., Earle, F. R. & Senti, F. R. (1957). Moisture relations in wheat and corn. *Cereal Chem.*, **34**, 422–33.

Hukill, W. V. (1963). Storage of seeds. *Proc. Int. Seed Test Ass.*, **28**, 871–3.

Hutchinson, J. B. (1944). The drying of wheat. III. The effect of temperature on germination capacity. *J. Soc. Chem. Ind.*, **63**, 104–7.

Ito, H. (1972). Organisation of the National Seed Storage Laboratory for Genetic Resources in Japan. In *Viability of Seeds* (ed. E. H. Roberts), pp. 405–16. Chapman and Hall, London.

James, E. (1972). Organisation of the United States National Seed Storage Laboratory. In *Viability of Seeds* (ed. E. H. Roberts), pp. 397–404. Chapman and Hall, London.

James, E., Bass, L. N. & Clark, D. C. (1967). Varietal differences in longevity of vegetable seeds and their response to various storage conditions. *Am. Soc. Hort. Sci.*, **91**, 521–8.

Johri, B. M. & Vasil, I. K. (1961). Physiology of pollen. *Bot. Rev.*, **27**, 326–81.

King, J. R. (1965). The storage of pollen – particularly by the freeze-drying method. *Bull. Torrey bot. Club*, **92**, 270–87.

Linskens, H. F. (1964). Pollen physiology. *Ann. Rev. Pl. Physiol.*, **15**, 255–70.

Lipman, C. B. (1936). Normal viability of seeds and bacterial spores after exposure to temperatures near the absolute zero. *Pl. Physiol.*, **11**, 201–5.

Mann, T. (1964). *The Biochemistry of Semen*, pp. 350–63, 2nd edn. Methuen, London.

Moore, R. P. (1972). Effects of mechanical injuries on viability. In *Viability of Seeds* (ed. E. H. Roberts), pp. 94–113. Chapman and Hall, London.

Nutile, G. E. (1964). Effect of desiccation on viability of seeds. *Crop Sci.*, **4**, 325–28.

Polge, C. (1972). Artificial control of reproduction. In *Reproduction of Mammals*, vol. v (eds C. R. Austin and R. V. Short). Cambridge University Press, London.

Price, G. B., Modak, S. P. & Makinodan, T. (1971). Age-associated changes in the DNA of mouse tissue. *Science*, **171**, 917–20.

Rees, A. R. (1963). A large-scale test of storage methods for oil palm seed. *J. West African Inst. Oil Palm Res.*, **4** (13), 46–51.

Rees, A. R. (1970). Effect of heat-treatment for virus attenuation on tomato seed viability. *J. Hort. Sci.*, **45**, 33–40.

Roberts, E. H. (1960). The viability of cereal seed in relation to temperature and moisture. *Ann. Bot.*, **24**, 12–31.

Roberts, E. H. (1961a). Viability of cereal seed for brief and extended periods. *Ann. Bot.*, **25**, 373–80.

Roberts, E. H. (1961b). The viability of rice seed in relation to temperature, moisture content, and gaseous environment. *Ann. Bot.*, **25**, 381–90.

Roberts, E. H. (1963). An investigation of inter-varietal differences in dormancy and viability of rice seed. *Ann. Bot.*, **27**, 365–9.

Roberts, E. H. (1972a). Storage environment and the control of viability. In *Viability of Seeds* (ed. E. H. Roberts), pp. 14–58. Chapman and Hall, London.

Roberts, E. H. (1972b). Cytological, genetical and metabolic changes associated with loss of viability. In *Viability of Seeds* (ed. E. H. Roberts), pp. 253–306. Chapman and Hall, London.

Roberts, E. H. (1973a). Predicting the storage life of seeds. *Seed Sci. Tech.*, **1**, 499–514.

Roberts, E. H. (1973b). Loss of seed viability: chromosomal and genetical aspects. *Seed Sci. Tech.*, **1**, 515–27.

Roberts, E. H. & Abdalla, F. H. (1968). The influence of temperature, moisture and oxygen on period of seed viability in barley, broad beans, and peas. *Ann. Bot.*, **32**, 97–117.

Roberts, E. H., Abdalla, F. H. & Owen, R. J. (1967). Nuclear damage and the ageing of seeds, with a model for seed survival curves. *Symp. Soc. exp. Biol.*, **21**, 65–100.

Roberts, E. H. & Roberts, D. L. (1972a). Viability nomographs. In *Viability of Seeds* (ed. E. H. Roberts), pp. 417–23. Chapman and Hall, London.

Roberts, E. H. & Roberts, D. L. (1972*b*). Moisture content of seeds. In *Viability of Seeds* (ed. E. H. Roberts), pp. 424–37. Chapman and Hall, London.

Roberts, H. M. (1959). The effect of seed storage conditions on the viability of grass seeds. *Proc. Int. Seed Test Ass.*, **24**, 184–213.

Touzard, J. (1961). Influences de diverses conditions constantes de température et d'humidité sur la longévité des graines de quelques espèces cultivées. Adv. Hort. Sci. and their Applications. *Proc. 15th Int. Hort. Congr.*, *Nice*, **1**, 339–47. Pergamon, Oxford.

Ushirozawa, K. & Shibukawa, J. (1951). Studies on the germination and fertilization of long-preserved apple pollen. *Aomori Apple Research Station*, **4**.

Visser, T. (1955). Germination and storage of pollen. *Meded. van de Landb. Hoogesch.*, *Wageningen*, **55**, 1–66.

von Senbusch (1955). Die Erhaltung der Keimfähigkeit von Samen bei tiefen Temperature. *Züchter*, **25**, 168–9.

Watson, E. L. (1970). Effect of heat treatment upon the germination of wheat. *Can. J. Plant Sci.*, **50**, 107–14.

Weibull, G. (1953). The cold storage of vegetable seed and its significance for plant breeding and the seed trade. *Rep. 13th Int. Hort. Congr.*, *1952*.

Weibull, G. (1955). The cold storage of vegetable seeds, further studies. *Rep. 14th Int. Hort. Congr.*, *1954*, pp. 647–67.

Whittingham, D. G., Leibo, S. P. & Mazur, P. (1972). Survival of mouse embryos frozen to $-196°$ and $-269\,°C$. *Science*, **178**, 411–14.

Appendix

The economics of long-term storage of seeds which show orthodox viability behaviour

If it is recognised that, from the biological point of view, the lower the storage temperature the better, then the major factor determining the temperature to be used in practice is the capital cost of the store and associated refrigeration equipment necessary to maintain a given temperature. There is little difference in the cost of a storage plant designed to maintain any temperature within the range $-10\,°C$ to $-20\,°C$, but there would be an increased cost of about 15 per cent if the plant were designed to maintain a temperature of $-23\,°C$ for example. There would be little difference in the running costs of operating storage facilities at say $-10\,°C$ or $-20\,°C$, providing the store was not being entered very frequently. Consequently it is suggested that the main storage facility at Genetic Resources Centres should be run at $-20\,°C$ and that orthodox seeds should be stored at about 5 per cent moisture content in moisture-proof containers at this temperature.

The cost of storage facilities designed to operate at $-20\,°C$ is not very great but varies with the storage capacity required. The relationship between volume and cost is given approximately by the equation, cost $= K \times \text{vol}^{\frac{2}{3}}$ (for cost in £ sterling and volume in cubic metres the value of K in the UK is currently about 270 for 'off-the-peg' units). Thus there is a worthwhile economy of scale.

In order to estimate roughly the cost of storing various numbers of samples, current prices of various sizes of 'off-the-peg' low-temperature storage units have been taken. Fifteen per cent has been added to the basic price in order to allow for provision of stand-by refrigeration equipment which, it is felt, should be standard for long-term seed storage facilities. If the units on which the estimates are based are surrounded by a brick or similar permanent structure it is estimated that their life should be greater than 100 years. The refrigeration equipment would need renewing every 12–15 years at about 20 per cent of the current capital cost.

The following assumptions have been made in estimating the number of samples which can be held in a store of given volume. (*a*) The samples of seeds will be stored in cylindrical containers of 2000 cm³ average volume (this would allow for the storage of at least a kilogram of seeds of most species), and each cylinder will occupy a square-sectional prismatic space of 2600 cm³. (*b*) The containers will be placed on movable shelving of the type found in library stacks so that in a rectangular store, the only working space that is necessary would be two passages 90 cm wide running at right-angles. Table 22.1 has been compiled incorporating these assumptions.

It is impossible to make comparisons between different practices of long-term seed conservation which are based on experience. There may be considerable errors in extrapolating what happens at very low temperatures and moisture contents from what happens at higher values and there may be a tendency to over-estimate the period of viability (Roberts, 1972*a*; 1973*a*), but it is the only method which is available. Nevertheless, such estimates will give some indication of the order of magnitude of the multiplication intervals to be expected, and the relative comparison between different storage systems is probably reasonably valid.

Table 22.2 shows the conditions employed in three seed-storage banks and compares them with the conditions proposed here in terms of the expected periods for viability of three common crop species to drop to 95 per cent of their initial viability.

Table 22.1. *The capital cost of storing various numbers of seed samples at* − 20 °C

Capacity of 'off-the-peg' storage unit		Basic cost in UK*		15 % addition to basic cost for stand-by refrigeration unit		Useful storage space		No. of 1 kg samples	Capital cost per sample	
ft³	m³	£	$	£	$	m³	% of total space		£	$
3000	85	5200	11960	5980	13750	60	70	22800	0.26	0.60
4000	113	6000	13800	6900	15870	84	74	31900	0.22	0.50
6500	183	8000	18400	9200	21160	148	80	56200	0.16	0.38
10000	282	10400	23920	11960	27500	240	85	91200	0.13	0.30

* At December 1973 prices and exchange rates £1.00 = $2.30 US.

23. Genetic maintenance of seeds in imbibed storage

T. A. VILLIERS

The special ability of seeds to exist in a state of suspended animation in an air-dry condition is a striking biological phenomenon. However, it is well known that such seeds deteriorate, more or less rapidly according to the conditions of storage, and this becomes especially important if they are to be used for long-term storage.

Because of the great importance of storage conditions in controlling the rate of seed deterioration, a vast body of practical information has accumulated concerning the effects of moisture, temperature, gaseous environment and microbial activities, on the longevity of seeds. In general, it is recognized that an increase in seed moisture content, warm temperatures and the presence of oxygen all tend to shorten the life spans of the majority of seeds.

However, although it has been possible to formulate general rules which allow manipulation of the ambient conditions to give a long storage life, the physiological and biochemical factors which cause loss of seed viability are not known (Roberts, 1972, Chapter 2). A knowledge of the mechanism of deterioration should bring about at the least a better understanding of the relative importance of the various storage conditions, and perhaps suggest new or improved methods.

A different approach to the problem is possible if we consider the seed in its natural habitat. It is evident that in the field, seeds seldom meet with optimum storage conditions. After drying on the parent plant, seeds may lie on the soil surface for a long period, during which time they are subject to fluctuating atmospheric humidity and high temperatures. If they should become buried under surface litter or in the soil, they may lie in a fully imbibed state for very many years, apparently held in a state of induced or secondary dormancy. Wareing (1966) has quoted several ecological studies which show that seeds of weed species may lie viable in the soil for many years, and concluded that considering the known requirements of seeds for retention of viability when they are stored dry, it is surprising that many seeds retain their viability longer in moist soil than in dry storage. Barton (1961) stated that although it may not be surprising that seeds which are dry,

297

and therefore have their metabolic processes greatly reduced, are still capable of germination after years of storage, the extended life spans of some species in the soil, where they are presumably fully imbibed, is more difficult to explain.

The only suggestion which has previously been offered to account for the longevity of seeds in the soil is that because they are held in a state of dormancy, their metabolic rates are reduced to a minimum and they are therefore able to survive for long periods (Toole & Toole, 1953; Roberts, 1972). However, this does not explain why seeds in the soil should live *longer* than seeds stored air-dry, nor does it explain the difficulty that in air-dry storage an increase in seed moisture greatly shortens the period of viability. Extrapolation of the results of dry storage experiments would predict a very short life span for seeds in the soil. Further, it is known that dormant seeds are in fact capable of a wide range of metabolic activities, including food material interconversion, membrane synthesis and organelle production (Villiers, 1971).

In air-dry seeds the metabolic rate is very low. An increase in water content would be expected to increase the rate of metabolism. However, this drastically reduces longevity. Together with the reduction in longevity which occurs with rise in temperature, and in the presence of oxygen, this tends to support the prevalent idea that air-dry seeds are able to survive for long periods because of their low metabolic rate.

I consider that a satisfactory explanation of the loss of viability in seeds should take into account not only the information obtained from dry-storage experiments, but also the apparently contradictory observations on the behaviour of seeds in the field. It is therefore pertinent to ask what is the difference between the physiology of seeds stored dry and seeds in the soil, which could account for the apparent anomaly of long life at high moisture contents.

It is possible to reconcile these differences once we accept that all the macromolecular components of cells are subject to deterioration, and that in normal tissues these components are maintained structurally and functionally in good order by repair mechanisms, or are replaced as whole units by new organelles.

The existence of repair and maintenance systems in living organisms is now generally accepted. In addition to the repair of spontaneous deterioration, damage to DNA by irradiation and chemical mutagens can be repaired by various enzyme-controlled mechanisms. This is firmly established for micro-organisms, and much evidence is being obtained for its occurrence in the cells of higher organisms (Hanawalt, 1972).

Complex patterns of enzyme activity, involving the manipulation of large structural units, would be unlikely to operate in air-dry seeds, but could take place in fully imbibed tissues. Therefore damage to macro-molecules, membranes and entire organelles must accumulate in air-dry tissues without the possibility of repair, until the amount of damage sustained becomes too great for recovery when the seed eventually imbibes water. On the other hand, in seeds fully imbibed during the storage period, the damage could be corrected as and when it occurs.

It is therefore postulated that seeds which are imbibed during storage are able to maintain macromolecules in good order, whereas those having a moisture content in equilibrium with the surrounding atmosphere are unable to do so. From this hypothesis the following predictions may be made. Firstly, that within the range of moisture contents achieved by seeds in equilibrium with atmospheric humidity, any conditions which prolong viability should be those which tend to stabilize complex macromolecules, and secondly, that the storage of seeds for long periods in a fully imbibed state should be possible with little genetic or structural damage accruing. The remainder of this contribution will be concerned with confirming the second prediction, and with a consideration of evidence for the activity of repair and maintenance systems in seeds.

Results

Moisture content and seed longevity

Toole & Toole (1953) maintained lettuce seeds at a range of air-dry moisture contents and also fully imbibed in the laboratory. The imbibed seeds were kept at 30 °C in the dark in order to prevent germination. A high germination capacity was retained by the fully imbibed seeds, while over similar periods of time, seeds stored air-dry had shorter life spans, and those stored at high relative humidities lost viability within a few weeks.

These experiments were repeated in my laboratory to serve as a basis for further investigation. Seeds of lettuce, varieties Arctic King, Grand Rapids, Big Boston and Webb's Wonderful were each brought to a range of moisture contents by storing over various concentrations of sulphuric acid (Hale, 1958). They were then sealed into jars and stored in a dark incubator at 30 °C. Seeds of each variety from the same

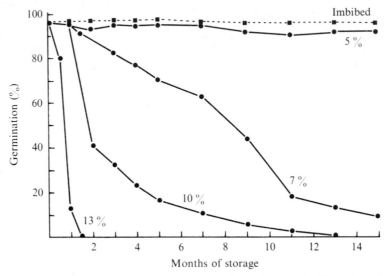

Fig. 23.1. Germination of samples of lettuce seeds, variety Arctic King, from batches stored at 30 °C with 5 per cent, 7 per cent, 10 per cent and 13 per cent seed moisture content, together with a sample stored fully imbibed. The imbibed seeds were maintained on moist filter paper in the dark to prevent germination occurring. Germination tests were conducted at 22 °C under fluorescent lighting.

provenances were imbibed on wet filter paper in Petri dishes also stored in the dark at 30 °C. Water was added from time to time as necessary to keep the seeds from drying.

With increase in moisture content from 5 per cent to 13 per cent there occurred a drastic shortening of the period of viability such that at 13 per cent moisture content the seeds were all dead by the sixth week (Fig. 23.1). The fully imbibed seeds, however, showed a very high germination capacity which has now been maintained in the same experiments for 18 months. A variable but small percentage of the imbibed seeds germinated during this time in wet storage, but in most cases more than 80 per cent of the original population remained dormant and could be stimulated to germinate at any time by placing under fluorescent lighting at 22 °C.

Similar experiments were set up with seeds of the forest trees *Fraxinus excelsior* and *F. americana*. These behaved in exactly the same manner, with the imbibed-seed samples remaining fully viable after all the dry-stored seeds had died. In previous experiments on dormancy states, which were not planned as experiments on seed ageing, seeds of *F. excelsior* have been maintained for eight years fully imbibed in the laboratory, and have retained their viability.

Chromosome damage

Cell division cannot occur in seeds during dry storage, and therefore in this respect at least, cells in stored seeds are similar to postmitotic cells, which are considered to be of great importance in the ageing of normal tissues. Ageing damage to macromolecules and structures must therefore accumulate since there is no possibility of the selection of undamaged cell lines, which may take place in a population of dividing cells.

It is well known that the incidence of chromosome damage increases during the ageing of seeds in dry storage. Roberts, Abdalla & Owen (1967 and Chapter 22) reviewed this subject and came to the conclusion that ageing in seeds can be explained on the basis of an accumulation of genetic damage seen as chromosomal aberrations in the first mitotic divisions of early germination. They point out that this visible chromosome damage is only a small aspect of the total genetic damage which has probably occurred. Gross damage would never be seen by this technique because mitosis would not take place at all, and many other mutations could also occur which might not result in structural alterations visible in the chromosomes.

Root-tip squashes stained by the aceto-carmine method were examined at the time when the first cell divisions were taking place in germinating seeds. The seeds were selected from the samples stored at the range of moisture contents described above. Chromosome bridges and deletions were scored as percentages of the total number of cells in anaphase and early telophase. Forty root tips from each germinating seed sample were examined, giving between 300 and 400 cells in anaphase or telophase.

It was found that chromosome damage increased with time of storage and especially so at higher seed moisture contents. However, there was no increase in the number of chromosome aberrations in the seeds which had been stored fully hydrated. On the contrary, the number of aberrations appeared to *decrease* slightly from the level in the original sample (Fig. 23.2).

The most striking difference lies in the comparison between the sample stored at 5 per cent moisture content, and that stored fully hydrated for the same period of time. A seed moisture content of 5 per cent would generally be considered excellent for storage, and, as can be seen from Fig. 23.1, a high percentage germination was retained by the seeds under such storage conditions. However, 13 per cent of the dividing

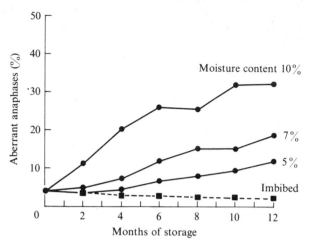

Fig. 23.2. Chromosome aberrations, expressed as a percentage of all cells in late anaphase, in radicle cells of lettuce seeds germinating after storage under the conditions described in the caption to Fig. 23.1.

cells were aberrant compared with only 2 per cent in the seeds stored imbibed.

It is important from the point of view of practical seed storage that, even though a high germination percentage can be maintained, the seeds can accumulate a large number of mutations. It thus appears that there are at least two components in the ageing of seeds as has been suggested previously by Abdalla & Roberts (1968) and also for mammalian tissues by Curtis (1966). One appears to be concerned with genetic damage whereas the other component is probably extrachromosomal.

Mutations appearing in the phenotype

If germination (as assessed by the emergence of the radicle) can occur in spite of the accumulation of genetic damage, it is necessary to show whether normal plants can be produced from seeds stored under the various conditions described above. Fifty seedlings were potted, where possible, from each of the samples germinated for the assessment of chromosomal aberration.

In general it was found that although some 'germination' occurred after three months' storage with moisture contents higher than 7 per cent, very few of the seedlings grew into normal plants. A large number of deformities was observed, including stunted growth, subdivided and twisted cotyledons and leaves, aberrant pigmentation, swollen

302

roots and necrosis of the root meristems. Examples of the extreme variability in development of the few remaining plants from a large number of dry-stored seeds can be seen in Plate 23.1.

However, seeds stored fully imbibed for more than a year grew vigorously and uniformly when stimulated to germinate by illumination at 22 °C. Under a 16-hour-long day regime they bolted and flowered, producing viable pollen and a good set of viable seeds (Plate 23.2).

State of seeds obtained commercially

In view of the above findings it was considered of interest to determine the moisture contents, germination percentages and degree of chromosome damage in commercial lettuce seeds. Seeds were purchased from large, reputable companies in the USA and Europe. Only seeds which had been sealed in cans, and therefore stored in reasonably good conditions, were used. Table 23.1 shows that of five lettuce varieties chosen, the moisture contents lay within the range 4–6 per cent, and whereas the percentage germination was high, and the seeds therefore considered saleable, they had in several cases accumulated considerable genetic damage. The unknown factor was the age of the seed. Even the dates given could not be confirmed beyond any doubt.

Reversal of ageing damage

From Figure 23.2 it can be concluded that when seeds which have already accumulated a certain amount of genetic damage are imbibed and replaced in storage, no further damage occurs during the period of the experiment, but rather a slight reversal of the original damage may be seen. If damage is continually sustained by the chromatin, but is repaired in the course of maintenance and repair processes during the normal metabolism of all cells, then it may be that the steadily accumulating chromosomal damage in dry-stored seeds can be arrested when the seeds are placed in imbibed storage, and that damage already sustained might be reversed by repair mechanisms. Confirmation of this was sought in the following ways.

Chromosome damage. A large sample of lettuce seeds of the variety Big Boston, with a moisture content of 7 per cent, was stored at 30 °C. Samples were withdrawn and germinated at intervals and the percentage of aberrant cell divisions recorded. After five months a sample of these

Plate 23.1. Lettuce seedlings, variety Arctic King, showing the extreme variability in growth vigour of the few seedlings produced from a single germination test of a batch of seeds after 3 months' storage with a moisture content of 10 per cent (germination 30 per cent).

Plate 23.2. Flowering lettuce plants grown from seeds stored fully imbibed for 14 months at 30 °C in the dark (germination 92 %).

Table 23.1. *Relationship between viability, expressed by percentage germination, and degree of chromosome damage in commercially available crop seeds of* Lactuca sativa. *Seed lots were sealed in cans by the producer*

Variety	Age of seed (months)	Moisture content (%)	Chromosome aberrations (%)	Germination (%)
New York	?	5.5	9.5	94
Big Boston	20	6	13.0	93
Arctic King	?	4.2	4	95
Grand Rapids	?	5	8	96
Webb's Wonderful	14	5.3	3	96

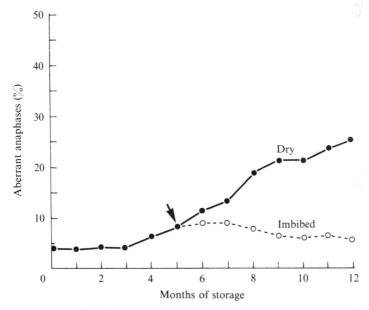

Fig. 23.3. Percentage chromosome aberrations in lettuce seeds, variety Big Boston, stored at 30 °C with a moisture content of 7 per cent. After 5 months (arrow) a portion of these seeds was imbibed on moist filter paper and returned to store at 30 °C in the dark. Determinations of percentage chromosome aberrations were continued at monthly intervals on both the air-dry and the imbibed seeds.

seeds was fully imbibed with water on wet filter paper and replaced in the dark at 30 °C. Determinations of the percentage of aberrant cell divisions in germination tests of the original dry-stored batch and also the newly-imbibed seeds were continued. The results are presented in Fig. 23.3, where it can be seen that the trend of increasing chromosome damage was arrested when the seeds were imbibed, and a decrease in the degree of damage demonstrated.

Attempts to demonstrate a more dramatic repair of severe genetic damage have not been successful as it was found that when seeds were imbibed which had accumulated damage to the extent of 15 per cent aberrant divisions, many became subject to microbial attack and died. It was considered that this degree of structural damage to the chromosomes represented a degree of genetic or general cellular damage too extensive to be repaired.

Irradiation experiments. Although the working hypothesis is that, when seeds are dry, the repair systems cannot act and that the damage therefore accumulates, it could be argued that, whereas the drying of seeds damages the chromatin, when hydrated seeds are stored no damage occurs; in other words, the *status quo* is maintained rather than that active repair occurs.

As it had proved difficult to demonstrate that cellular repair processes occur after imbibition of seeds which had accumulated damage in previous dry-storage periods, it was decided to inflict a controlled amount of damage by means of irradiation. Preliminary experiments using gamma irradiation from a cobalt-60 source showed that in lettuce seeds chromosome aberrations increased with increase in the irradiation dose from 3 per cent in unirradiated seeds to approximately 60 per cent at 2000 r.

In subsequent experiments a dose of 600 r was chosen, expected to give approximately 8 per cent aberrant cell divisions. Four dry and four dark-imbibed samples of Grand Rapids lettuce seeds were irradiated in the dark with 600 r at a dose-rate of 200 r/min. Immediately after irradiation, water was added to one batch of the dry-irradiated seeds, and these, together with one dish of the imbibed-irradiated seeds, were germinated in the light. The remaining three dishes of dry seeds and three dishes of imbibed seeds were stored in the dark.

After two days' storage, water was added to the second batch of dry-stored seeds and this was germinated in the light, together with the second batch of the imbibed-stored seeds. This was repeated after four days, and again after eight days. Chromosome aberrations were recorded, and the lengths of the fully grown hypocotyls measured at the time when the first true leaves were expanding. Fig. 23.4 shows that following an initial increase, the percentage of aberrant cells in seeds stored imbibed had decreased after 8 days.

306

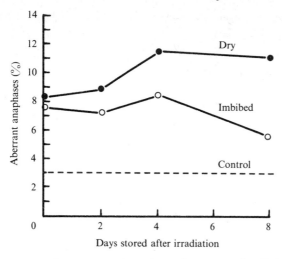

Fig. 23.4. Percentage chromosome aberrations after gamma-irradiation of air-dry and imbibed seeds. Following irradiation at 600 r, samples of the seeds were germinated immediately, and the remainder stored and tested after 2, 4 and 8 days in air-dry and imbibed storage.

Fig. 23.5 shows that when germination was forced immediately after irradiation, the length of the seedling hypocotyls was reduced by 50 per cent whether the seeds were irradiated dry, or after imbibition. However, whereas the seeds stored dry following irradiation tended to deteriorate even further, those stored imbibed without being allowed

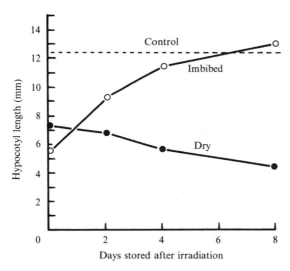

Fig. 23.5. Hypocotyl lengths of lettuce seedlings grown from seeds gamma-irradiated and stored under the conditions described in Fig. 23.4.

to germinate until illuminated gradually recovered and produced the normal seedling height after about six days in storage. It is considered that this is good evidence for the repair of damage in imbibed seeds as opposed to dry-stored seeds.

Electron microscopy of damaged tissues

Comparisons were made of the appearance of embryo cells from seeds after periods of storage at various moisture contents. A large amount of information has been collected on the sequence of cytological events occurring during the ageing process in dry storage, but it is merely intended here to demonstrate some of the differences between dry-stored and imbibed-stored seeds, and to suggest how damage to cytoplasmic structures could initiate or amplify damage to the genetic material.

In electron microscopical studies of aged embryos, changes in the reaction of the nuclei to electron stains such as uranyl acetate and lead citrate are very striking. In embryo root-tips from normal, unaged embryos, the nuclei stain homogeneously except for a fine network identified as the heterochromatin. During ageing, the heterochromatin stains more deeply until, in embryos from seeds of low viability, very dense clumps of chromatin occur. The staining reaction of the nuclei in seeds stored fully imbibed for similar periods appears to be quite normal (Plate 23.3).

It was previously suggested that evidence exists for the operation of at least two factors in ageing, one genetic and one not directly connected with the genome. It is of interest to examine briefly the type of damage which accumulates in the cytoplasm of stored seeds.

Within the cytoplasm, damage to the cellular membrane systems becomes apparent during dry storage of seeds. Mitochondria, plastids and the nucleus all sustain damage to the membranes, and the vacuoles of the embryo cells appear to play an important part in the process of autolysis which occurs in aged embryos during imbibition.

During early imbibition, the cells of the embryo appear normal except for the dense staining reaction of the nuclei described above. However, as imbibition proceeds, the vacuolar membranes appear to disperse, and this is followed by general autolysis of the cytoplasm. First the mitochondria and plastids disintegrate, followed by the nuclear envelope; then the plasma membrane becomes dissociated from the cell wall and disperses. The lipid-body membranes also disperse and the

308

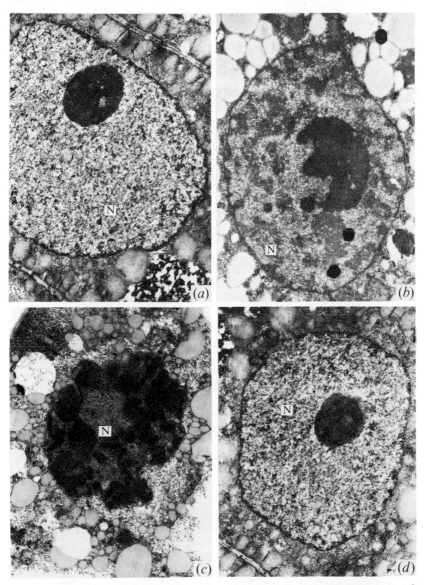

Plate 23.3. Nuclei from radicle meristem cells of lettuce (variety Arctic King), 24 hr. after placing in germination test conditions. Glutaraldehyde/osmium fixation, followed by lead citrate and uranyl acetate staining: N, nucleus. Magnification: × 16000. Seeds stored as follows before germination test: (*a*) Fresh seed immediately after receiving from supplier, chromatin stains as fine network; (*b*) after 2 weeks at 30 °C with 13 per cent seed moisture content, chromatin more heavily stained; (*c*) after 4 weeks at 30 °C with 13 per cent moisture content, chromatin clumped and deeply stained, embryo probably non-viable; (*d*) after 4 weeks' storage at 30 °C but fully imbibed, chromatin gives normal staining reaction; embryo viable.

309

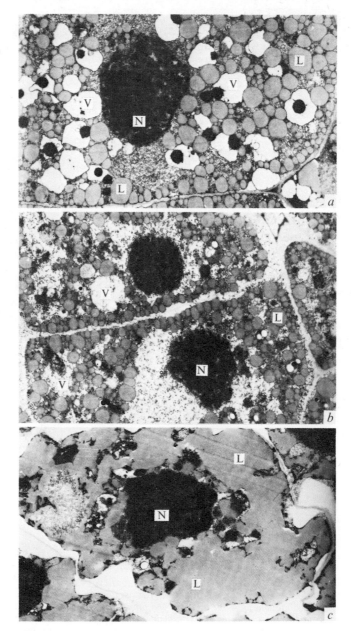

Plate 23.4. (For legend see p. 311 opposite.)

lipid coalesces into large, confluent masses. Some of these stages are illustrated in Plate 23.4.

It may be that even a small amount of damage to the membrane systems could have a significant effect within the cell. Membranes and macromolecules may be damaged by peroxidation reactions caused by free radicals, possibly produced by the spontaneous ionization of water molecules, in turn affecting the polyunsaturated lipids of the cytoplasmic membranes (Barber & Bernheim, 1967). Such peroxidation reactions are greatly enhanced by the presence of oxygen.

This can result in the destruction of the lipid itself and the formation of insoluble lipid–protein complexes by cross-linking, making the membranes less efficient and inactivating enzymes. As one of the primary functions of membranes is to allow the compartmentation of cytoplasmic activities, enzymes and substrates may leak out from organelles and amplify the original damage. Simon & Harun (1972) have shown that in the very early stages of water uptake by dry seeds, the plasma membranes do not act efficiently as differentially permeable membranes until a certain degree of hydration has been achieved, before which substances are able to leak across the membranes. If this is also true of the inner cytoplasmic membranes, it could have far-reaching effects in ageing seeds. In the case of the vacuoles, which have been suggested to be the lysosomes of the plant cell (Matile, 1968), the leakage of hydrolytic enzymes could cause a wide range of damage. Proteolytic enzymes, ribonuclease and deoxyribonuclease could therefore damage the chromatin, among other things, during the imbibition process in seeds stored air-dry for long periods.

It should be noted that the cytological organization of the embryo cells of seeds stored fully imbibed for similar periods appeared to be normal in all respects even when they were kept without germinating until seeds from the same harvest, but stored air-dry at moisture contents of 7 to 13 per cent were all dead.

Legend to Plate 23.4. (See p. 310 opposite.)

Plate 23.4. Radicle meristem cells of lettuce (variety Arctic King) at increasing times after beginning of imbibition. Seeds stored air-dry at 30 °C with 13 per cent moisture content for 4 weeks before test. L, lipid; N, nucleus; V, vacuole. (*a*) Eight hours after beginning of imbibition. Cytoplasm still organized, but nucleus dense and abnormal. Magnification: × 8000. (*b*) Sixteen hours later, the vacuoles have disintegrated, the cytoplasm appears disorganized. Magnification: × 5700. (*c*) Forty-eight hours after beginning of imbibition, cytoplasm totally disorganized, lipid body membranes dispersed and lipid has become confluent. Magnification: × 8000.

Discussion

Genetic and structural damage occurs and accumulates in seeds stored air-dry at rates determined by the moisture content of the seeds and by environmental factors such as temperature and the availability of oxygen. If we consider possible reasons for the greater longevity of seeds lying in moist soil, it appears possible that air-dry seeds may gradually accumulate damage and so deteriorate because of the inability of repair enzymes to function in air-dry cells, whereas continuing repair and maintenance may be possible in fully hydrated tissues.

The results show that maintaining seeds imbibed without germinating reversed the trend of increasingly rapid deterioration found when the moisture content was increased within the range achieved by seeds in equilibrium with atmospheric humidity. Seeds kept imbibed were able to germinate and produce a full crop of vigorous plants which eventually flowered and in turn set new seed. Imbibed-stored seed showed no cytological deterioration when the tissues were viewed in the electron microscope and very few chromosome aberrations were seen by light microscopy.

When seeds, previously dry-stored and showing an increasing degree of damage, were fully imbibed and then returned to store, there was an immediate arrest of the trend of deterioration, and evidence was obtained for a reversal of the damage already accumulated. This was considered to be due to the activity of repair mechanisms in moist-stored seeds rather than a mere stabilization of the macromolecular structures by hydration.

When damage was deliberately inflicted on the seed tissues by gamma-irradiation, it could be demonstrated that while seeds irradiated and stored dry showed no recovery, but rather a further deterioration, seeds which had been irradiated when imbibed, and subsequently stored imbibed, showed a positive recovery from the damage.

The ageing damage in dry-stored seeds is considered to be due to peroxidation reactions, causing deterioration of the lipids and proteins of the membranes. This is autocatalytic and could cause general damage throughout the cells. The fact that stored seed tissues are air-dry would not prevent such peroxidation reactions. Water in the immediate vicinity of macromolecules has been suggested to be in a semicrystalline state (Verzar, 1968). As tissues become air-dry, the remaining water becomes tenaciously bound to the cell colloids (Askenov, Askochenskaya &

Petinov, 1969), and many enzyme-controlled reactions, especially those involving large structural units, would be unlikely to occur under these conditions. Being bound to the seed colloids, however, even a small amount of water would still be able to permit peroxidation reactions which require only the proximity of water and oxygen to the double bonds of the unsaturated lipids of the cytoplasmic membranes.

Oxygen tension was not varied during these investigations, but it is known that increased availability of oxygen increases the rate of seed deterioration (Harrison, 1966). If the mechanism of damage to the macromolecules is by free-radical peroxidation, then vacuum storage or replacement of the air by an inert gas such as nitrogen would greatly reduce such damage and would be expected to lead to a greater life-span of the seeds. Tappel (1965) states that an enhancing effect of damage by oxygen is in itself an indication of the involvement of lipid peroxidation reactions.

In the study of ageing, somatic mutation theories have the most convincing experimental support. Curtis (1967) has shown that mutations increase during ageing and that species of animals with short lifespans have higher mutation rates. An organism becomes gradually less efficient as it accumulates a large number of mutations, and postmitotic cells eventually die. In addition to strand breakages, deletions and point mutations, damage can also be caused by cross-linkage within and between DNA molecules and also between the DNA and its associated protein in the chromatin. However, I believe that there are two separate but connected factors concerned in the loss of viability of seeds. These are damage to the membrane systems and enzymes, and damage to the genome. Omura, Siekevitz & Palade (1968) suggested that a continuing turnover of membrane proteins and lipids occurs, and Luzzati *et al.* (1969) have stated that any given portion of a membrane may be in a transient state.

Although genetic damage increases during dry storage, this may not actually affect the early germination process. The initial stages of germination are concerned with cell enlargement and protrusion of the radicle, and may therefore be prevented because of damage to the membrane systems of the cytoplasm. Cell division does not begin until the radicle has extended beyond the seed coats (Haber & Luippold, 1960), and there is evidence that the early germination process in quiescent seeds does not directly involve the genetic material (Marcus & Feeley, 1964; Dure & Waters, 1965). In addition, the delay frequently encountered before germination in old seed samples is possibly due to

313

the necessity for a period during which repairs are effected and organelles replaced (Berjak & Villiers, 1972).

However, following early germination, development of the seedling becomes controlled by the genome, and aberrant cell divisions, visible deformities, loss of vigour, and pollen abortion may be due in part to damage accumulated in the genome during dry storage of the seed.

One of the obvious deficiencies of any theory of ageing damage applied to metabolizing cells and tissues is that a system capable of repair, or of being replaced, is unlikely to be of importance in ageing. However, in the case of air-dry tissues such as seeds, spores and pollen grains this cannot be an objection. In dry seeds, repair systems probably do not function, and this offers a reasonable explanation for the observed differences in longevity between seeds stored dry and those stored fully imbibed.

Although these experiments were designed to show why seeds lose viability and accumulate genetic damage in dry storage, the results suggest that wet storage may be useful where the maintenance of a high degree of genetic stability is required. Lettuce seeds were used in the present work because they can be caused to become dormant at high temperatures. With other species of crop plants, most of which do not show dormancy characteristics, artificial means of preventing germination during wet storage would have to be found. One such method would be to keep the seeds imbibed in a non-toxic solution of high osmotic pressure.

This work was supported by funds from the South African Council for Scientific and Industrial Research, and the South African Atomic Energy Board.

References

Abdalla, F. H. & Roberts, E. H. (1968). Effects of temperature, moisture, and oxygen on the induction of chromosome damage in seeds of barley, broad beans, and peas during storage. *Ann. Bot.*, **32**, 119–36.

Askenov, S. I., Askochenskaya, N. A. & Petinov, N. S. (1969). The fractions of water in wheat seeds. *Fiz. Rast.*, **16**, 58–62.

Barber, A. A. & Bernheim, F. (1967). Lipid peroxidation: its measurement, occurrence, and significance in animal tissue. In *Advances in Gerontological Research*, vol. II (ed. B. L. Strehler), pp. 355–403. Academic Press, New York.

Barton, L. V. (1961). *Seed Preservation and Longevity*. Leonard Hill, London.

Berjak, P. & Villiers, T. A. (1972). Ageing in plant embryos. II. Age-induced damage and its repair during early germination. *New Phytol.*, **71**, 135–44.

Curtis, H. J. (1966). A composite theory of ageing. *Gerontologist*, **6**, 143–9.

Curtis, H. J. (1967). Radiation and ageing. In *Aspects of the Biology of Ageing*, SEB Symposium XXI (ed. H. W. Woolhouse), pp. 51–63. Cambridge University Press, London.

Dure, L. S. & Waters, L. C. (1965). Long-lived messenger RNA: evidence from cotton seed germination. *Science*, **147**, 410–12.

Haber, A. H. & Luippold, H. J. (1960). Separation of mechanisms initiating cell division and cell expansion in lettuce seed germination. *Pl. Physiol., Lancaster*, **35**, 168–73.

Hale, L. J. (1958). *Biological Laboratory Data*. Methuen, London.

Hanawalt, P. C. (1972). Repair of genetic material in living cells. *Endeavour*, **31**, 83–7.

Harrison, H. J. (1966). Seed deterioration in relation to storage conditions and its influence upon germination, chromosomal damage and plant performance. *J. natn. Inst. agric. Bot.*, **10**, 644–63.

Luzzati, V., Gulik-Krzywicki, T., Tardieu, A., Rivas, E. & Reiss-Husson, F. (1969). Lipids and membranes. In *The Molecular Basis of Membrane Function* (ed. D. C. Tosteson), pp. 79–93. Prentice-Hall Inc., Englewood Cliffs, N.J.

Marcus, A. & Feeley, J. (1964). Activation of protein synthesis in the imbibition phase of seed germination. *Proc. natn. Acad. Sci. USA*, **51**, 1075–9.

Matile, Ph. (1968). Vacuoles as lysosomes of plant cells. *Biochem. J.*, **111**, 26–7.

Omura, I., Siekevitz, P. & Palade, G. E. (1968). Turnover of constituents of the endoplasmic reticulum membranes of rat hepatocytes. *J. biol. Chem.*, **242**, 2389–96.

Roberts, E. H., Abdalla, F. H. & Owen, R. J. (1967). Nuclear damage and the ageing of seeds. In *Aspects of the Biology of Ageing*, SEB Symposium XXI (ed. H. W. Woolhouse), pp. 65–99. Cambridge University Press, London.

Roberts, E. H. (ed.) (1972). *Viability of Seeds*. Chapman and Hall, London.

Simon, E. W. & Harun, R. M. R. (1972). Leakage during seed imbibition. *J. exp. Bot.*, **23**, 1076–85.

Tappel, A. L. (1965). Free-radical peroxidation damage and its inhibition by vitamin E and selenium. *Fedn Proc.*, **24**, 73–8.

Toole, V. K. & Toole, E. H. (1953). Seed dormancy in relation to seed longevity. *Proc. int. Seed Test. Ass.*, **18**, 325–8.

Verzar, F. (1968). Intrinsic and extrinsic factors of molecular ageing. *Exp. Geront.*, **3**, 69–75.

Villiers, T. A. (1971). Cytological studies in dormancy. I. Embryo maturation during dormancy. *New Phytol.*, **70**, 751–60.

Wareing, P. F. (1966). Ecological aspects of seed dormancy and germination. In *Reproductive Biology and Taxonomy of Vascular Plants*. BSBI Conf. Reports, No. 9 (ed. J. G. Hawkes), pp. 103–21. Pergamon Press, Oxford.

Table 22.2. *Expected multiplication intervals under various systems of storage*

Organisation	Temperature	Moisture content	Estimated multiplication interval, years (time taken for viability to drop by 5%)*		
			Triticum aestivum	Hordeum† distichon	Vicia faba
National Seed Storage Laboratory, Fort Collins, USA	4 °C‡	32% R.H. (i.e. about 9% moisture content for non-oily seeds; about 4–6% for oily seeds)	9	109	20
Plant Genetic Resources Centre, Izmir, Turkey	0 °C	6% moisture content for cereals, 7–8% for oily seeds	30	710	90
National Seed Storage Laboratory, Hiratsuka, Japan	−10 °C§	4–6% moisture content	120	5900	440
Conditions proposed here	−20 °C	5% moisture content	390	33500	1600

* Calculated from seed viability nomographs (Roberts & Roberts, 1972).
† Recent evidence suggests that the cultivar used to produce these figures may have a viability period which is greater by a factor of 10 than other barley cultivars.
‡ Three rooms are also available at −12 °C.
§ Seed for distribution is held at −1 °C.

24. Some factors contributing to the survival of crop seeds cooled to the temperature of liquid nitrogen

A. SAKAI & M. NOSHIRO

It is well known that plants which resist intensive dehydration or are extremely frost-hardy retain their viability after exposure to temperatures of liquid nitrogen ($-196\ °C$) or even down to several degrees above absolute zero (Ching & Slabaugh, 1966; Ichikawa & Shidei, 1971; Levitt, 1972; Lipman, 1934; 1936; Luyet & Gehenio, 1938; Sakai, 1960a, b; 1962). Fully-hydrated seeds, except those that over-winter in cold climates, are generally killed by slight freezing. Luyet (1937) has shown that even in the normally hydrated state, plant cells can survive immersion in liquid nitrogen if both the cooling and rewarming rates are ultra-rapid. Little information is available on the factors contributing to the survival of crop seeds cooled to the temperature of liquid nitrogen.

The present study was undertaken to clarify the effects of cooling and rewarming rates on the survival of seeds at relatively high moisture contents, and to develop a technique to preserve seeds of crop plants in liquid nitrogen. The work was carried out on seeds which can be stored normally under low moisture content and consequently do not present any insuperable difficulties with regard to long-term storage. Nevertheless, the findings may be relevant to the storage of 'racalcitrant' seeds, which cannot be dried to low moisture content and thus cannot at present be stored for long periods.

Materials and methods

The species used in these experiments were rice (*Oryza sativa* L. cv. Hóryu), winter wheat (*Triticum aestivum* L. cv. Mukakomugi), soybean (*Glycine max* Merill cv. Hórei), alfalfa (*Medicago sativa* L. cv. Du Puits) and Italian ryegrass (*Lolium multiflorum* Lam. cv. Billion).

The water content was adjusted to the desired levels by moistening

317

the seeds in a saturated atmosphere at 10 or 20 °C. Their water content as percentage dry weight was determined by oven drying at 75 °C for 16 hours. In making comparisons with other work, it is important to note that water contents in seed technology are usually quoted as percentage moisture content on a wet weight basis. Table 24.5 indicates the relationship between the two methods of expressing water contents. Viability was estimated by germinating at 26 °C in the dark alongside untreated control seeds. Germination was measured by counting the seeds in which both radicles and coleoptiles or plumules were exposed through the split testas within periods which varied from two to seven days according to species. In every experiment 30 seeds were used.

Seeds were placed in a wire mesh basket (2 cm diameter, 5 cm long) or in a plastic vessel (2 cm diameter, 3 cm long), and were cooled by immersion in liquid nitrogen. In some experiments, seeds placed in petri dishes were cooled in 5 °C steps at hourly intervals to successively colder temperatures down to −30 °C and then held at those temperatures for 16 hours. Soybeans immersed directly in liquid nitrogen from room temperature cracked into several pieces due to a rapid change in temperature. They were therefore placed in a plastic vessel (2 cm diameter, 3 cm long) with a sinker and were then immersed in liquid nitrogen (cooling rate: 0.4 degC/sec). Seeds cooled to the temperatures of liquid nitrogen were rewarmed rapidly (rewarming rate: 25 degC/sec) in water at 36 °C or slowly (0.5 degC/sec) in air at 0 °C. The temperature of the material was determined with 0.1 mm copper–constantan thermocouples and was recorded with an oscilloscope. The cooling rate was calculated from the time required for the temperature to cool to within the range of 0 to 5 °C above the final temperature. The rewarming rate was represented as the time required for the temperature to rise from the final temperature to −5 °C.

Results

The effect of water content on the germination of crop seeds which had been rewarmed rapidly or slowly following immersion in liquid nitrogen was determined for the following species: rice, wheat, soybean, alfalfa and Italian ryegrass. The result obtained with unhulled rice is shown in Fig. 24.1. With water content below 15 per cent (dry weight), almost all the rice germinated after immersion in liquid nitrogen regardless of rewarming rates, while with water content above 21 per cent, no germination was observed when rewarmed rapidly or

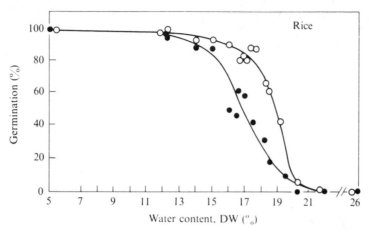

Fig. 24.1. Effect of water content on the germination of unhulled rice seeds (cv. Hóryu) rewarmed rapidly (25 degC/sec) or slowly (0.5 degC/sec) following immersion in liquid nitrogen (cooled at 50 degC/sec). O, rapid rewarming; ●, slow rewarming.

slowly. In contrast, in the intermediate water content, 16 to 19 per cent, the rewarming rate remarkably influenced the germination; a higher rate of germination was observed in the seeds rewarmed rapidly.

Similar results were obtained in wheat seeds (Fig. 24.2) and Italian ryegrass (Fig. 24.3). However, when wheat seeds with water content below 18 per cent were rewarmed rapidly following immersion in liquid nitrogen, some of them cracked into a few pieces due to a rapid change in temperature. Accordingly the maximum germination of

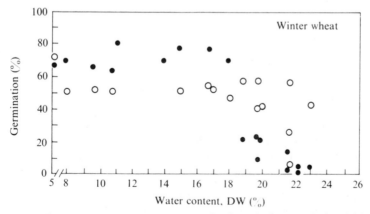

Fig. 24.2. Effect of water content on the germination of wheat seeds (cv. Mukakomugi) rewarmed rapidly (25 degC/sec) or slowly (0.5 degC/sec) following immersion in liquid nitrogen (cooled at 50 degC/sec). O, rapid rewarming; ●, slow rewarming.

319

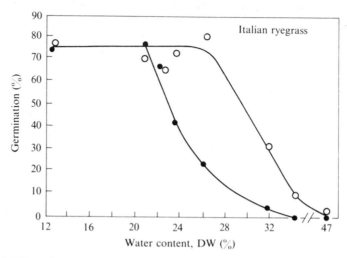

Fig. 24.3. Effect of water content on the germination of seeds of Italian ryegrass rewarmed rapidly (25 degC/sec) or slowly (0.5 degC/sec) following immersion in liquid nitrogen (cooled at 50 degC/sec). ○, rapid rewarming; ●, slow rewarming.

wheat seeds rewarmed rapidly was always much lower than those rewarmed slowly except for the seeds with a water content of 19 to 23 per cent, where the percentage germination of the seeds rewarmed rapidly (25 degC/sec) was much higher than those rewarmed slowly (0.5 degC/sec). These results suggest that little or no injury occurs during a rapid cooling to the temperature of liquid nitrogen, but it easily occurs during slow rewarming.

We also observed that the germination rate of the wheat seeds with a water content of 16 to 18 per cent decreased considerably when held at $-30\,°C$ for 20 minutes before being rewarmed rapidly in water at $36\,°C$ following removal from liquid nitrogen, while in seeds with 11.3 per cent water, no appreciable decrease in the germination, even when held at $-30\,°C$, occurred (Table 24.1). An abrupt decrease in the germination of seeds with 15.9 per cent water was further observed at temperatures between -40 and $-30\,°C$ during rewarming following removal from liquid nitrogen (Table 24.2).

To determine the effect of cooling rate on the survival of hydrated seeds, wheat seeds with different water contents were cooled rapidly (50 degC/sec) or slowly (1.7 degC/sec) to the temperature of liquid nitrogen and then rewarmed rapidly (25 degC/sec) in water at $36\,°C$ or slowly (0.5 degC/sec) in air at $0\,°C$. As shown in Table 24.3, in the wheat seeds with 21.5 per cent water, the germination rate of seeds

Table 24.1. *The decrease in germination rate of hydrated wheat seeds (cv. Mukakomugi) held at −30 °C, following removal from liquid nitrogen*

Rewarming conditions	Germination percentages at the following water contents (DW %)			
	11.3	16.0	17.0	18.2
Rewarmed rapidly in water at 36 °C	73	59	54	41
Held at −30 °C for 20 min before rewarming rapidly in water at 36 °C	67	21	4	5

Seeds were first placed in a wire mesh basket at room temperature and then immersed in liquid nitrogen.

Table 24.2. *Temperature range causing damage to wheat seed at 15.9 per cent (DW) moisture content in the process of rewarming, following removal from liquid nitrogen*

Temperature at which seeds were held for 1 hr following removal from liquid nitrogen (°C)	Germination (%)
−20	38
−30	55
−40	94
−50	100
−70	100
−90	100
−120	100

Seeds were immersed in liquid nitrogen and held at different temperatures for one hour before being rewarmed rapidly in water at 36 °C.

cooled rapidly was much higher than those cooled slowly. This fact implies that in the seeds with 21.5 per cent water, slow cooling to the temperature of liquid nitrogen causes injury. When the water content was 24.5 per cent or above, all the seeds were killed after rapid cooling even when rewarmed rapidly.

The germination of wheat seeds cooled very slowly to different temperatures decreased with increasing water content (Table 24.4). Fully-hydrated wheat seeds (40 per cent water) did not survive freezing even to −5 °C. The same results were obtained in other crop seeds tested.

Table 24.3. *Effects of rates of cooling and rewarming on the germination of wheat seeds (cv. Mukakomugi) hydrated to different levels*

Conditions of cooling and rewarming	Germination percentages at the following water contents (DW %):			
	17.5	21.5	24.5	26.0
Rapid cooling*–rapid rewarming	92	57	0	4
Rapid cooling–slow rewarming	77	27	0	0
Slow cooling†–rapid rewarming	95	39	25	9
Slow cooling–slow rewarming	96	22	4	0

* Seeds placed in a wire mesh basket were directly immersed in liquid nitrogen from room temperature (cooling rate: 50 degC/sec).
† Seeds in aluminium foil were enclosed in a plastic vessel (2 cm diameter, 3 cm long) and the vessel was then transferred to air at $-150\,°C$ (cooling rate: 1.7 degC/sec) and held there for one hour. Seeds were rewarmed rapidly (rewarming rate: 25 degC/sec) in water at 36 °C or slowly (0.5 degC/sec) in air at 0 °C.

Table 24.4. *Effect of water content on the germination of wheat seeds (cv. Mukakomugi) cooled slowly to different temperatures*

Temperature (°C)	Germination percentages at the following water contents (DW %):					
	16.0	17.0	17.7	18.5	19.5	40.0
− 5	—	—	95	—	86	9
− 10	95	90	95	90	93	0
− 15	88	85	97	90	91	0
− 20	81	85	97	100	100	0
− 30	90	70	74	80	56	0
− 70	85	52	40	38	24	0
− 120	84	—	—	—	—	0

Wheat seeds hydrated to different levels were cooled in 5 degC steps at hourly intervals to successively colder temperatures down to $-30\,°C$ and then held at those temperatures for 16 hours respectively. Some seeds were further cooled to -70 or $-120\,°C$ and held at those temperatures for three hours respectively. These seeds were all rewarmed slowly in air at 0 °C.

In contrast, the wheat seeds with a water content of 16 per cent survived slow cooling down to at least $-120\,°C$, or slow rewarming following rapid cooling in liquid nitrogen as shown previously. Interestingly we recently found that seeds of some weed species native to cold climates (e.g. *Lepidium* sp., *Hieracium* sp., *Oenothera lamarckiana*) survived freezing to -20 or $-25\,°C$ even in a fully hydrated state.

In alfalfa seeds, only a slight difference was observed in the germina-

Table 24.5. *Effect of rewarming rates on the germination of alfalfa seeds (cv. Du Puits) rewarmed rapidly or slowly after immersion in liquid nitrogen*

Water content (%)		Germination percentages	
Dry weight	Fresh weight	Rapid rewarming	Slow rewarming
4.8	4.6	84	81
8.7	8.0	82	86
13.2	11.7	80	78
17.7	15.0	76	74
23.3	18.9	64	68
26.8	21.1	54	47
30.1	23.1	47	38
31.3	23.8	33	23
39.6	28.3	8	3
61.0	37.7	3	6

Seeds placed in a wire mesh basket were directly immersed in liquid nitrogen from room temperature. Seeds immersed in liquid nitrogen were rewarmed rapidly in water at 36 °C or slowly in air at 0 °C.

tion of seeds rewarmed either rapidly or slowly (Table 24.5). The reason for this is as yet unknown. When soybeans were immersed directly in liquid nitrogen from room temperature (cooling rate: 9 degC/sec), all cracked into several pieces due to a rapid change in temperature. Thus, slow cooling is required for maintaining viability of soybeans if they are to be cooled to the temperature of liquid nitrogen. The same mechanical injury was also observed in the soybeans with about 8 per cent water when rewarmed rapidly in water at 36 °C after cooling to the temperature of liquid nitrogen (Table 24.6).

Conditions of cooling and rewarming which enable viability to be maintained in crop seeds hydrated to different levels after immersion in liquid nitrogen are summarised in Table 24.7.

As a safe general practice it is concluded that the highest survival of crop seeds after immersion in liquid nitrogen may be obtained by the following procedures:

1. Dehydrate to the range of 8 to 15 per cent (DW) water.

2. Enclose in aluminium foil or an aluminium or plastic vessel with a screw cap.

3. The foil or vessel in which seeds are enclosed should be immersed in liquid nitrogen from room temperature.

4. After storage, the foil or vessel should be transferred to air at 0 °C or room temperature (slow rewarming).

323

Table 24.6. *Effect of water content on the germination of soybeans (cv. Hórei) rewarmed rapidly or slowly following immersion in liquid nitrogen*

Water content (DW %)	Germination percentages	
	Rapid rewarming	Slow rewarming
6.7	0*	80
11.9	67	72
14.6	100	88
14.9	95	95
15.8	87	100
18.8	0	0
21.8	0	0

Soybeans wrapped in aluminium foil were placed in a plastic vessel, and were then immersed in liquid nitrogen (0.4 degC/sec) for 30 minutes. Soybeans wrapped in aluminium foil were taken out of the plastic vessel and were then rewarmed rapidly (9 degC/sec) in water at 36 °C or slowly (0.5 degC/sec) in air at 0 °C.
* Cracked into several pieces.

Table 24.7. *Conditions of cooling and rewarming which enable viability to be maintained in seeds hydrated to different levels after immersion in liquid nitrogen*

Species	Water content (%)		Conditions of cooling and rewarming	Minimum germination (%)
	Dry weight	Fresh weight		
Rice (*Oryza sativa* cv. Hóryu)	5–17	5–15	R–R	80
	5–17	5–15	R–S	80
Wheat (*Triticum aestivum* cv. Mukakomugi)	5–18	5–16	R–S	80
Soybean (*Glycine max* cv. Hórei)	7–16	6–14	S–S	80
Italian ryegrass (*Lolium multiflorum* cv. Billion)	12–26	11–21	R–R	90
	12–20	11–17	R–S	90
Alfalfa (*Medicago sativa* cv. Du Puits)	5–20	5–17	R–R	80
			R–S	80

R–R: rapid cooling–rapid rewarming; R–S: rapid cooling–slow rewarming.
Seeds placed in a wire mesh basket (2 cm diameter, 5 cm long) or in a plastic vessel (2 cm diameter, 3 cm long) were cooled rapidly or slowly by immersion in liquid nitrogen. Seeds immersed in liquid nitrogen were rewarmed rapidly in water at 36 °C or slowly in air at 0 °C.

Discussion

All the wheat seeds with a water content above 24.5 per cent were killed even when rewarmed rapidly. This suggests that intracellular ice crystals formed during rapid cooling are large enough to be immediately damaging. In contrast, in the wheat seeds with water contents in the range of 19–22 per cent, the germination rate of the seeds that were rewarmed rapidly was much higher than those rewarmed slowly. One possible interpretation is that some freezable water remains in these seeds, but the ice crystals formed in the cells during rapid cooling may be small enough to be innocuous (Luyet, 1937; Sakai & Otsuka, 1967; 1972; Sakai, Otsuka & Yoshida, 1968; Sakai, 1970). When rewarming is carried out rapidly, the crystals may melt before they have time to grow to a damaging size, and the cells will then remain viable.

In sufficiently dehydrated seeds no harmful effect is observed except for mechanical injury, regardless of rates of cooling to liquid nitrogen and subsequent rewarming. If any freezable water remains in the seeds, the germination rate of the seeds cooled to the temperature of liquid nitrogen is very much affected by the cooling and rewarming rates used. It is therefore unreasonable to conclude on the basis of the germination rate that almost all of the freezable water in seeds will be lost by desiccation at some specific water content. It may however be possible to determine roughly what this water content is on the basis of germination percentage, provided that the seeds immersed in liquid nitrogen are rewarmed slowly by keeping them at − 30 °C for at least one hour. This water content lies around 16 per cent (DW) in the wheat seeds used. To check the validity of this method, some investigations have been made with an electron microscope (Sakai & Otsuka, 1967), by X-ray diffraction analysis on pollen (Ching & Slabaugh, 1966) and by differential thermal analysis (unpublished). Ching & Slabaugh reported that death of pollen grains containing about 30 per cent water was observed following rapid freezing in a bath at − 70 °C. They used X-ray diffraction analysis to discern the relationship between intracellular ice formation and the death caused by rapid freezing in pollen grains containing water in variable amounts. As a result, ice crystals were detected by X-ray diffraction in pollens containing 36 per cent or more water at about − 25 °C. Pollen samples with detectable ice crystals were found to be killed. However, the viability of some pollen samples containing about 30 per cent water was reduced by freezing but without detectable ice crystals.

Conservation and storage

It is suggested that the main factor in this problem is not whether ice is present within the cells, but rather the size of the ice crystals (Sakai *et al.*, 1968). The results obtained may therefore be explicable in terms of crystal size. Some direct evidence supporting this suggestion has been obtained with the electron microscope, as mentioned above. One of the urgent problems for solution at present is to determine the damaging size of intracellular ice crystals in the seeds.

References

Ching, T. M. & Slabaugh, W. H. (1966). X-ray diffraction analysis of ice crystals in coniferous pollen. *Cryobiology*, **2**, 321–7.

Ichikawa, S. & Shidei, T. (1971). Fundamental studies on deep-freezing storage of tree pollen. *Bull. Kyoto Univ. Forests*, **42**, 51–82. (In Japanese with English summary.)

Levitt, J. (1972). *Responses of Plants to Environmental Stresses*. Academic Press, New York, 697 pp.

Lipman, C. B. (1934). Tolerance of liquid-air temperatures by seeds of higher plants for sixty days. *Pl. Physiol., Lancaster*, **9**, 392–4.

Lipman, C. B. (1936). Normal viability of seeds and bacterial spores after exposure to temperatures near the absolute zero. *Pl. Physiol., Lancaster*, **11**, 201–5.

Luyet, B. J. (1937). The vitrification of organic colloids and of protoplasms. *Biodyamica*, **1**, 1–14.

Luyet, B. G. & Gehenio, P. M. (1938). The survival of moss vitrified in liquid air and its relation to water content. *Biodynamica*, **2**, 1–7.

Sakai, A. (1960*a*). Survival of the twigs of woody plants at − 196 °C. *Nature, Lond.*, **185**, 393–4.

Sakai, A. (1960*b*). Survival of plant tissue at super-low temperatures. III. Relation between effective pre-freezing temperatures and the degree of frost hardiness. *Pl. Physiol., Lancaster*, **40**, 882–7.

Sakai, A. (1962). The frost-hardiness of bulbs and tubers. *J. Hort. Ass. Japan*, **29**, 233–8. (In Japanese with English summary.)

Sakai, A. (1970). Some factors contributing to the survival of rapidly cooled plant cells. *Cryobiology*, **8**, 225–34.

Sakai, A. & Otsuka, K. (1967). Survival of plant tissue at super-low temperatures. V. An electron microscope study of ice in cortical cells cooled rapidly. *Pl. Physiol., Lancaster*, **42**, 1680–94.

Sakai, A. & Otsuka, K. (1972). A method for maintaining the viability of less hardy plant cells after immersion in liquid nitrogen. *Plant and Cell Physiol.*, **13**, 1129–33.

Sakai, A., Otsuka, K. & Yoshida, S. (1968). Mechanism of survival in plant cells at super-low temperatures by rapid cooling and rewarming. *Cryobiology*, **4**, 165–73.

25. Meristem culture techniques for the long-term storage of cultivated plants

G. MOREL

A very special feature of higher plants is the mode of growth of their stems and roots. At the apices of these organs small aggregates of cells remain embryonic, and continue to differentiate new tissues and organs throughout the whole life of the plant on a well-organised plan, so long as the flowering process has not been initiated. The meristems can be compared to permanent embryos; in other words, embryogenesis in plants is unlimited.

In the present chapter we shall be concerned with apical stem meristems only. The apex is a self-perpetuating structure in which some cells keep dividing and remain embryonic, whilst others differentiate into organs. In the actively dividing zone of the apex mitotic activity varies a great deal from one point to another. According to the latest interpretation put forward by Buvat (1955) three main zones can be distinguished.

(1) An axial apical zone on top of the apical dome, with very little mitotic activity.

(2) Below and to the side of this is a lateral zone of small highly meristematic cells with dense cytoplasm. This is the initial ring where most of the mitoses take place.

(3) Inside and below the central zone lies the meristematic precursor of the pith.

Apical meristem culture

Although the stem apex appears to be a self-perpetuating unit, it is necessary to understand the factors underlying its growth and differentiation and to know the degree of autonomy of the apex. Is it able to function if it is provided solely with the essential nutrients needed by plant cells in culture or does it require stimuli or growth regulators from other parts of the plant?

To answer these questions a technique has been devised of excising the apex and growing it on a nutrient medium (see review by Morel, 1965). The answer, however, is not a simple one, since experiments have shown that the degree of apical autonomy varies a very great deal according to species.

The nutritional requirements of the apex of certain lower plants such as the fern *Adiantum* are very simple. When excised and planted on one of the mineral media used for tissue cultures, such as White's or Knop's media supplemented with sugar, the apex will resume its growth, differentiate new leaves, make a new rhizome and regenerate a small plant.

On the other hand, the apices of most of the higher plants are not able to develop on such a medium. They produce callus-like outgrowths or very distorted and chlorotic stems. Since the undifferentiated tissues of these plants grow normally on such media, it means that the differentiation of the apical structure needs some factors normally provided by the plants but lacking in the medium. Numerous experiments with various herbaceous dicotyledons such as potato, dahlia, carnation and chrysanthemum, have shown that a gibberellin, namely gibberellic acid, is absolutely necessary for the normal development of the apex. When gibberellic acid is incorporated into the medium at a concentration of 0.1 mg/l, instead of callus and distorted growth the meristem differentiates into a perfectly normal stem.

Another rather critical requirement is the concentration of potassium ions in the medium. Most of the media used for plant tissue culture, such as White's or Gautheret's, have a fairly low concentration of potassium, e.g. White's medium: 0.8 meq/l; Gautheret's medium: 2.1 meq/l. Nevertheless they support a very good growth of undifferentiated tissue cells. When these media are supplemented with gibberellic acid the meristem starts to elongate but it remains chlorotic and soon stops growing. We have shown (Morel & Muller, 1964) that in order to obtain a normal development of the apex it is necessary to increase the potassium level of these media up to 10 meq/l. This can be simply done by the addition of potassium chloride or nitrate. The requirement for potassium is absolutely specific; potassium ions cannot be replaced by any other cations such as sodium or ammonium. Any well-balanced culture medium with high potassium content such as that of Murashige & Skoog (1962) will support a good growth of the apex.

The stem development of most woody plants is a complex phenomenon, periods of rapid elongation alternating with very long rest periods. In the stem of the lilac, for example, all the elongation takes place in

April–May in a few weeks and the apical meristem then dies. Further growth will be resumed by lateral buds after almost a year of dormancy. One must bear in mind that the cultivation of such a meristem means forcing it to grow almost one year ahead of its normal time. Preliminary results indicate that this can be done only after a very long vernalisation. Boxus & Quoirin (1973) have shown that for various *Prunus* species the stems must be kept in a refrigerator at 4 °C for almost six months before the meristem excision. The growth requirements of the apices of these plants are high. Besides a gibberellin they also need a cytokinin. So far, a strong cytokinin such as benzylaminopurine has given the best results.

On a medium provided with gibberellin (1 mg/l) and benzylamino-purine (1 mg/l) the formation of a new bud from the meristem can be observed. This bud might form a rosette of leaves but there is no stem elongation. To obtain the elongation of the stem the explant must be transferred into a new medium deprived of growth regulator. A new stem then elongates quickly but usually remains rootless. To promote formation this stem must be transferred into another medium with a fairly high auxin content of 1 mg/l.

So with most woody meristems a new plant can be obtained only by duplicating successively the conditions of plant growth: long dormancy – bud formation – bud elongation and then rooting.

Long-term storage of meristems

The three main features of plantlets grown *in vitro* which make them very valuable for long-term storage are: (1) their small size and very slow development; (2) their very high speed of multiplication; (3) the fact that they can easily be kept free from viruses, insect parasites, fungi or bacteria.

(1) The new plant obtained from a meristem is very small, the leaves are minute, and the stem very thin. As long as it is cultivated on agar, the plant develops very little. The growth, which is very rapid at first, slows down when the medium becomes depleted of nutrients. The plant stays alive for months without much growth and it can be kept in a test tube on nutrient medium for indefinite periods of time by transferring every six months or every year cuttings of nodes with the corresponding leaf.

Let us take the grape as an example of the saving in space, time and money that can be effected by meristem cultures. In the field the vine is very vigorous and needs a great deal of space to keep a large collection. The cultivation is very expensive, the vineyards have to be weeded several times during the growing season and fertilised and sprayed during the whole summer against mildew and other parasites. The vines need to be pruned several times also. The cost of maintenance of a large collection is therefore very high.

According to Galzy (1969) the apex of the grape stem can easily be cultivated on a medium with a high potassium concentration and a small dose of indole acetic acid (IAA) (0.1 mg/l). The tubes are kept at 20 °C with a light intensity of 3000 lux for 12 hours a day. The stems elongate and roots are formed after about 20 days. The plantlets are then transferred into a medium with a lower potassium concentration, such as Knop's solution, and deprived of auxin. They reach about 10 cm in three months.

They can then be multiplied by node cuttings. For this the plant is taken out of the test tube and cut in a Petri dish into stem fragments with a node and the corresponding leaf. Each fragment is then transferred to a new medium at 20 °C and new plants are produced after 50 days. The plants are then transferred to a temperature of 9 °C and at this temperature the speed of growth slows down and finally stops. However, the plant stays alive and if cuttings are taken after 290 days and placed at 20 °C they start growing at once. It is then possible to keep the plants alive with only one transfer a year.

So with the grape 800 cultivars with six replicates can easily be stored on a surface of only two square metres for about one year. We now have more than 15 years of experience and can show that grapes can be kept for unlimited periods of time in these conditions with one transfer per year. There is thus no comparison between the expenses needed to keep the cultures *in vitro* and those of cultivating the same amount in the fields. It would require about one hectare to accommodate the same number of plants.

(2) In most cases the speed of multiplication of the stock *in vitro* is very high. Meristem culture affords an entirely new means of clonal multiplication and has brought a real revolution in the cultivation of many plants. The cultivation of orchids affords a particularly good example.

Orchids have always been a luxury flower because it was not possible in the past to apply modern methods of industrial cultivation to them. In some countries the cut-flower business has become very large. In France, for example, it has become more important than large crops such as sugar beet. Carnation culture is developing fast in South American countries like Colombia and Ecuador. The orchid business is a real asset for tropical countries, such as Thailand and Singapore.

The management of cut-flower cultivation on such a large scale is not a simple matter and it needs careful programming. The grower has to be sure to produce at the right time, such as Christmas or Mother's Day, the exact number of flowers of the right quality. So far, this had been impossible with orchids. Cultivated plants are complex, hybrids are very heterozygous and the only way of propagation was by seed. The so-called back-bulb propagation is too slow, since it only allows the doubling of the number of plants every year. This explains the very high prices quoted for orchid plants up to now.

We have found (Morel, 1964) that meristem culture of orchids permits a five- to ten-fold increase in the number of plants every month. It is therefore possible to obtain several million plants in the span of one year. This technique is also being applied to other flower cultures, such as chrysanthemum.

In the case of the grape five cuttings can be obtained every month. Thus a single meristem culture could produce more than 10 million cuttings a year.

(3) The third advantage of meristem culture that is worth emphasising is the aspect of plant protection. As long as plants are grown in sterile culture there are no problems with pests or diseases. Not only can the stocks be kept virus-free with very great ease but as we found many years ago (Morel & Martin, 1952) meristem culture affords a new means of producing virus-free plants.

Later, Morel & Martin (1955) established that when plants were inoculated with a wide range of viruses the meristematic part of the apex stayed virus-free.

A technique has now been developed, based on this observation, of curing diseased plants. With potato, for example, the apical dome of young sprouts from tubers germinating at 37 °C, excised and transferred into a special medium, will produce in about two months' time small plants looking like seedlings. According to the variety and the disease, from 50 to 80 per cent of these plants will be virus-free. This technique has been applied to many other crops also, such as sweet potato, sugar cane, strawberries, etc. (Mori, 1971).

In conclusion, apical meristem cultures possess great advantages for the long-term storage of many economic plants. It should be possible to conserve a large collection in a small space and the costs of maintenance will be very low. Furthermore, the material can be kept free of virus and other diseases and can be propagated quickly when it becomes necessary to provide large quantities of plants for evaluation or breeding purposes. It can therefore be stated with confidence that the techniques outlined above could well be applied, with suitable modifications, to other plants. They would be particularly valuable for conserving species with 'recalcitrant' seeds (see Roberts, Chapter 22), which cannot at present be conserved in seed banks.

References

Boxus, P. & Quoirin, M. (1973). La culture de méristèmes apicaux de *Prunus*. *Physiol. veget.* (in press).

Buvat, R. (1955). Le méristème apical de la tige. *Ann. Biol.*, **31**, 596–656.

Galzy, R. (1969). Recherches sur la croissance de *Vitis rupestris* Scheele sain et court noué cultivé *in vitro* à différentes températures. *Ann. de Phytopathol.*, **1**, 149–66.

Morel, G. M. (1964). Tissue culture. A new means of clonal propagation of orchids. *Bull. Am. Orchid Soc.*, pp. 472–8.

Morel, G. M. (1965). La culture du méristème caulinaire. *Bull. Soc. fse Phys. Vég.*, **11**, 213–24.

Morel, G. M. & Martin, C. (1952). Guérison de Dahlia atteints d'une maladie à virus. *C. r. Acad. Sci.*, **235**, 1324–5.

Morel, G. M. & Martin, C. (1955). Guérison de plantes atteintes de maladies à virus par culture de méristèmes apicaux. *Rep. XIVth Int. Hort. Cong.*, **1**, 303–10.

Morel, G. M. & Muller, J. F. (1964). La culture *in vitro* du méristème apical de la pomme de terre. *C. r. Acad. Sci.*, **258**, 5470–3.

Mori, K. (1971). Production of virus free plants by means of meristem culture. *Japan Agric. Res. Quart.*

Murashige, T. & Skoog, F. (1962). A revised medium for rapid growth and bioassay with tobacco tissue cultures. *Physiologia Pl.*, **15**, 473–97.

26. The problem of genetic stability in plant tissue and cell cultures

F. D'AMATO

In 1939, working independently of each other, White, Gautheret and Nobécourt almost simultaneously announced the successful unlimited in-vitro growth of plant tissues. Whilst White achieved this goal by culturing tissue explants of the tumour-producing hybrid *Nicotiana glauca* × *N. langsdorfii* (an auxin autotroph), on an agar medium containing mineral salts, sucrose and yeast extract, the French workers succeeded in growing carrot root tissue on a wholly synthetic medium consisting of mineral salts, glucose and growth promoting factors [aneurine, cysteine and indole acetic acid (IAA)].

Since this initial discovery, the most varied types of tissues and cells from a large number of plant species have been brought into culture and grown indefinitely both in liquid and in solid media. In phanerogams, in-vitro cultures have been obtained from virtually all types of cells, although not necessarily all from the same species. For the long-term maintenance of cultures the most convenient and economic method is the subculture of calluses on solid media.

With the progressive improvement of techniques, methods for the regeneration of plants from in-vitro cultures are being increasingly applied to species of dicotyledons and monocotyledons, including important crops. Consequently, in-vitro cultures may be considered possible for long-term storage of genetic variability in plant breeding research. In this respect, plant tissue and cell cultures raise a series of questions, of which the following are of special significance:

(1) What are the nuclear conditions in cells of in-vitro cultures and how do they compare with the nuclear conditions of cells in the primary explants, i.e. in the in-vivo situation?

(2) Are the nuclear conditions occurring in cultures and the nuclear conditions of the plants which are regenerated from them related?

(3) Are methods available, or to be expected, for the long-term conservation of genetic stocks in culture?

Nuclear conditions in plant tissue and cell cultures

In the majority (more than 90 per cent) of the plant species investigated so far, the cells of differentiated tissues *in vivo* contain endoreduplicated (endopolyploid) nuclei, that is, nuclei whose chromosomes at interphase have undergone one or more duplications. In an extreme case, as in the largest cells of the embryo suspensor of *Phaseolus coccineus*, it is believed that eleven duplication cycles take place (2*n* polytene chromosomes with 2048 chromatids each) (Nagl, 1962). The degree of endopolyploidy may vary from tissue to tissue and in a given tissue may not be uniform; in other words, a variable proportion of cells may not endoreduplicate. Moreover, meristematic cell lines in differentiated plant tissues (procambium, pericycle) remain diploid, being practically immune from the endoreduplication process. In the remaining species (less than 10 per cent) histological differentiation occurs in a diploid condition; thus, in the absence of chromosome endoreduplication the nuclei are left in the same condition as at the end of the last mitosis before differentiation. The DNA content of these nuclei is 2*C*, which is typical of the pre-DNA synthesis phase (i.e. G_1 phase) of the diploid cell cycle. For reviews on the subject, see D'Amato (1952, 1965) and Partanen (1963).

Histological differentiation in a diploid condition is encountered with some frequency among the Asteraceae. Examples of plants having differentiated tissues in the 2*C* condition are lettuce, Jerusalem artichoke and *Crepis capillaris*. It seems possible that the mechanisms controlling the sequence from DNA synthesis to mitosis, which ensures histological differentiation *in vivo* in the diploid condition in these species, may still control cytological stability *in vitro* (D'Amato, 1972*a*). Thus, in *Crepis capillaris*, callus obtained from leaf tissue and grown for one year showed only diploid mitoses (2*n* = 6) and its interphase nuclei possessed DNA contents typical of the diploid level (Reinert & Küster, 1966). If, however, the leaf callus was grown for a longer time, polyploidization occurred whose frequency increased with time (after 20 months, polyploidization is seen in 28 per cent of calluses); moreover, the tendency to chromosome doubling was clearly enhanced in calluses derived from a haploid plant of the same species (after 12–18 weeks only, about 40 per cent of the cultures had become diploid) (Sacristan, 1971). The clear loss of cytological stability in the haploid *Crepis capillaris* raises the doubt that, even starting from haploids of species which do not endopolyploidize *in vivo*, it might not be possible to isolate permanently haploid tissue and cell cultures to be used in biochemical genetics, mutagenesis, etc. (see further below).

In the species showing endopolyploidy in their differentiated tissues the first mitoses which occur when an explant is brought into culture reflect, essentially if not totally, the pre-existing situation *in vivo*. Diploid nuclei give mitoses with $2n$ chromosomes and endopolyploid nuclei give mitoses with either $2n$ diplochromosomes or $2n$ quadruplo-chromosomes. Sixteen-ploid and higher-ploid nuclei are generally not stimulated to division; if they are, they do not contribute, to any significant extent, to the proliferating cell population. Mitoses with diplo-chromosomes and quadruplochromosomes occurring in the first phase of callus induction are a clear cytological evidence that endopolyploid nuclei are already present in the in-vivo explant (Bennici, Buiatti, D'Amato & Pagliai, 1971).

Other evidence comes from cytophotometric determination of DNA content in individual nuclei. Thus, in tobacco the pith tissue and the majority of the derived calluses showed nuclei with the same DNA contents ($2C$, $4C$, $8C$, $16C$). However, in a callus grown on a medium containing IAA and kinetin $32C$ nuclei were measured (Devreux, Saccardo & Brunori, 1971; Pätau, Das & Skoog, 1957). The possibility indicated by this tobacco data, that in-vitro culture brings endopolyploidy in a callus to a higher level than in the initial explant, is now well documented for the haploid cultivar Kleine Liebling of *Pelargonium* (Bennici, Buiatti & D'Amato, 1968). A DNA cytophotometric analysis demonstrated that in the primary explant (stem internode) the nuclei were $1C$, $2C$, $4C$ and $8C$ and the mitoses were haploid ($2C$) with $n = 9$ chromosomes, whilst in the callus some $16C$ and $32C$ nuclei were formed and the mitoses were $2C$ (haploid), $4C$ (diploid) and $8C$ (tetraploid) (Fig. 26.1). Therefore, in *Pelargonium*, as in the tobacco pith *in vitro* (Pätau *et al.*, 1957), the culture conditions were not able to stimulate nuclei with the higher endopolyploid levels to divide mitotically. Endopolyploidization is also induced by non-synthetic culture media containing coconut milk, as seen in *Haplopappus gracilis* (Mitra & Steward, 1961) and in haploid *Antirrhinum majus* (Melchers & Bergmann, 1959). In this last material, during the first eight transfers *in vitro* in a medium containing IAA the callus showed haploid mitoses only ($n = 8$). However, following a passage on a medium containing 2,4-dichlorophenoxyacetic acid (2,4-D) there occurred a progressive increase in polyploidy up to $8x$ and $16x$ (where x is the base number).

That the composition of the culture medium plays an important role

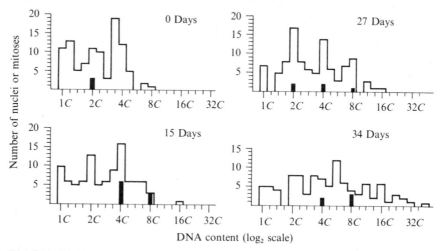

Fig. 26.1. DNA content in interphase nuclei (empty bars) and mitoses (solid bars) in the haploid *Pelargonium* cultivar Kleine Liebling.

Material: internode at the moment of explant (zero days) and in callus derived from it after 15, 27 and 34 days. Transfer of the primary explant on fresh medium was made on the fifteenth day. Note the selection in favour of diploid (4C) and tetraploid (8C) mitoses and the progressive increase in degree of endopolyploidy in the interphase nuclei (from Bennici *et al.*, 1968).

in the induction and maintenance of mitosis in cells of different ploidy was first demonstrated by Torrey (1961; 1967; Matthysse & Torrey, 1967). In the differentiated portions of the pea root, the only diploid cell line is the pericycle; the differentiated cells are endopolyploid, at least in their majority. If root segments are grown *in vitro* on Shigemura's synthetic medium, only the diploid cells proliferate; if kinetin or yeast extract plus 2,4-D are added to the culture medium, the tetraploid cells are selectively stimulated to mitosis and the proliferating cell population will then consist exclusively of tetraploid cells. Since a combination of an auxin and a cytokinin is often essential for DNA synthesis and mitosis (Skoog & Miller, 1957), their quantitative ratios in a culture medium can greatly influence the composition of the proliferating cell population. This situation was clearly demonstrated by a recent investigation on callus subcultures in *Haplopappus gracilis* (Fig. 26.2).

Cell selection by artificial means might help in future to reduce or overcome karyological instability in plant tissue and cell cultures. Gupta & Carlson (1972) found that the addition of parafluorophenylalanine (PFP) in the culture medium caused preferential growth of calluses derived from pith tissue of haploid plants and inhibited growth

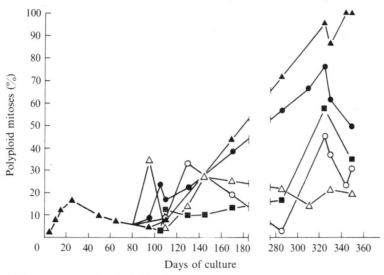

Fig. 26.2. Frequency of polyploid mitoses at different culture durations of calluses of *Haplopappus gracilis* grown on Linsmayer and Skoog's medium containing five different concentrations (in ppm) of kinetin (K) and naphthalene acetic acid (NAA). (▲, K 0.02 ppm, NAA 1 ppm; ○, 0.02 ppm, 2 ppm; △, 0.02 ppm, 4 ppm; ●, 0.2 ppm, 1 ppm; ■, 1 ppm, 1 ppm.) The initial callus was grown for 80 days (4 transfers at 20-day intervals) on the medium containing K 0.02 ppm and NAA 1 ppm and then subdivided into 5 sub-cultures. After 350 days in the medium with the lowest K and NAA content, only polyploid mitoses were seen (from Bennici *et al.*, 1971).

of callus derived from pith tissue of diploid plants of tobacco. In this work, no attention was paid to the nuclear conditions of the primary explants. Consequently it is not known whether in PFP-treated cultures, cells with nuclear conditions other than haploid may be carried forward in the culture and thus be available for regenerative processes. As will be seen later (Table 26.6), this possibility is well established for the 'haploid' callus of rice. It is, therefore, apparent that an adequate knowledge of the selective cellular effects in culture of some treatments (such as particular auxin–cytokinin ratios, PFP treatment and other treatments which may be tried in future) can be obtained only when chromosome counts are supplemented with a determination of the nuclear conditions of non-dividing cells by DNA cytophotometry and – when feasible – with a cytological analysis of plants regenerated *in vitro*.

Another genomic change of common occurrence in plant tissue cultures is aneuploidy, which is frequently expressed with a great variation in chromosome number. Examples of this are tobacco (Tables 26.1,

337

Table 26.1. *Chromosome numbers in stem pith tissue of tobacco* (Nicotiana tabacum, *2n = 4x = 48*) *and in the callus derived from it. The number in brackets denotes the mitotic figures with the indicated chromosome number (from Murashige & Nakano, 1967)*

Approximate duration of culture (years)	Mitotic figures analysed (no.)	Chromosome numbers
0[a]	53	48 (25), 96 (28)
0[b]	44	48 (4), 96 (31), 192 (7) aneuploids (2): 182, 184
1	15	96 (4), 192 (3) and aneuploids (8): 54, 74, 86, 88, 92, 156, 158, 304
6	12	Aneuploids (12): 108, 122, 124, 140, 146, 148 (2), 152 (2), 154, 160, 174

a, b: Explants excised from stem segments at 2.5–10.5 cm (*a*) and at 15.5–22.5 cm (*b*) from the apex. An increase in degree and frequency of polyploidy in the older pith is evident.

26.2), *Saccharum* species hybrids (Table 26.3) and *Haplopappus gracilis* (Table 26.4). Aneuploidy as well as polyploidy increase in frequency with increasing age of the culture (Tables 26.1, 26.2, 26.4). In one and the same species, the frequency and extent of aneuploidy may be related to the nature of the primary explant, as in tobacco (Table 26.2) and in rice, in which the frequency of aneuploidy is greater in the callus derived from roots than in the callus derived from stem internodes (Yamada, Tanaka & Takahashi, 1967).

As to chromosome structural changes, they have often been observed in in-vitro culture (see reviews in Partanen, 1963; Sacristan, 1971). In a *Crepis capillaris* callus, Sacristan (1971) described several types of chromosome mutations – mostly translocations – and observed that, in some subcultures, all cells showed one and the same chromosome mutation. If sampling errors in making subcultures are excluded, this observation would give the first indication of a selective advantage of mutated over normal cells.

Nothing is so far known about the possible occurrence *in vitro* of small deficiencies and gene mutations, but it seems plausible that such changes may accumulate with time in tissue and cell cultures. If stable haploid tissue cultures become available, the problem of spontaneous mutation rate in in-vitro systems will be amenable to analysis.

Table 26.2. *Chromosome number in cells of calluses derived from different organs of tobacco and wheat (from Shimada, 1971)*

Species (2n)	Origin of callus	Duration of callus subculture	Number of mitoses observed	Chromosome number		Notes
				Mean	Range	
Tobacco (48)	Pith	1.5 year	49	81.1	38–153	
	Root tip	3–7 months	36	55.7	45–96	21 cells (64 %) with 2n = 48
	Root tip	2 years	10	116	46–198	only 1 cell with 2n = 48
Common wheat (42)	Root tip	1 year	40	41.8	26–84	
	Root tip	4 years	26	42.5	32–84	10 cells (40 %) with 2n = 42
	Ovary	2 years	28	41.9	28–84	
	Stem	1 month	10	42.3	40–44	
Emmer wheat (28)	Root tip 1*	4 years	50	27.4	14–56	25 cells (50 %) with 2n = 28
	Root tip 2*	4 years	50	26.1	9–56	28 cells (56 %) with 2n = 28

* 1, 2. In the last two years, the calluses were grown on two different media: (1) supplemented with 1 mg/l of 2,4-dichlorophenoxyacetic acid (2,4-D) and 1 g/l of casein hydrolysate; and (2) supplemented with 0.2 mg/l of kinetin, 40 mg/l of adenine sulphate and 1 mg/l of 2,4-D.

Table 26.3. *Chromosome numbers in tissue cultures derived from internode parenchyma of five interspecific Saccharum (sugar cane) hybrids and analysed after 6 years of suspension culture (based on data from Heinz, Mee & Nickell, 1969; Heinz & Mee, 1971)*

Hybrid	Chromosome numbers			
	In the original plants (2n)	In the culture	In 3 clones of single-cell origin	In the plants regenerated from the culture
H37-1933	106	—	—	106 (8)[c]
H50-7209	108–128	71–90, 101–110, 154–160, > 300	111–140 and polyploidy 71–90, 110–311[a], 51–90, 151–185[b]	Chromosome mosaicism in one and the same plant[d]: 94–120 (37), 17–118 (1)
H49-5	114	51–113	—	—
H49-3533	114	61–74, 126–145	—	—
H39-7028	122	High polyploidy	—	—
N. CO. 310	112	156–200	—	—

[a]: 110, 130, 137, 145, 148, 164, 181, 311.
[b]: in addition, 111, 113, 126, 147, 191.
[c]: in brackets, number of regenerated plants.
[d]: great variation in morphological characters and isozyme patterns of plants.

Table 26.4. *Distribution of chromosome numbers in cells of four strains of Haplopappus gracilis (2n = 4) derived from stem segments cultured in a medium containing coconut milk and naphthalene acetic acid. In-vitro transfers at monthly intervals (from Blakely & Steward, 1964)*

Strain	Months in culture	Chromosome number													
		4	5	6	7	8	9	10	11	12	13	14	15	16	>16
DS	8	46	—	—	—	—	—	—	—	—	—	—	—	—	—
	12	63	—	—	—	1	—	—	—	—	—	—	—	—	—
	25	40	—	—	—	10	—	1	—	—	—	—	—	1	—
WS	10	67	—	—	—	46	—	—	—	—	—	—	—	—	—
	12	112	—	—	—	25	—	—	—	—	—	—	—	—	—
	20	—	—	—	—	2	—	4	1	2	—	1	—	3	1
	25	—	—	—	—	—	—	5	12	12	5	4	1	2	7
G22 (subclone from DS)	7	12	—	—	—	1	—	—	—	—	—	—	—	—	—
	11	121	—	—	—	5	—	—	—	—	—	—	—	—	—
	12	59	—	—	—	—	—	—	—	—	—	—	—	—	—
F1 (subclone from DS)	7	—	—	—	12	1	3	1	—	—	—	—	—	—	—
	11	—	—	—	5	27	30	3	3	—	—	7	—	—	4
	12	—	—	1	8	26	4	3	1	—	—	3	—	1	1

Nuclear conditions of plants regenerated *in vitro*

Through comparative cytological analyses of tissue cultures and of the plants which are regenerated from them, the types of cells involved in the organogenetic process have been ascertained. In some materials, such as carrot (Steward, Blakely, Kent & Mapes, 1963), rice (Nishi, Yamada & Takahashi, 1968) and common wheat (Shimada, Sasakuma & Tsunewaki, 1969) in spite of the presence in the culture of a variety of chromosome numbers (diploid, polyploid and aneuploid) all regenerated plants have been found to be diploid. In many other materials, such as sugar cane and tobacco, plant regeneration *in vitro* occurs not only at the diploid, but also at the polyploid and aneuploid levels (see Tables 26.3 and 26.5).

Odd-number polyploid (i.e. triploid, pentaploid, etc.) plants are sometimes obtained *in vitro*. Triploid shoots and plants have been obtained not only from triploid tissue, as in the in-vitro cultured endosperm of *Dendrophthoe falcata* (Johri & Nag, 1968), but also from calluses originating from somatic tissue of tobacco (Sacristan & Melchers, 1969).

With the recent developments in the in-vitro culture of anthers, plants with haploid and other euploid chromosome numbers have been regenerated, either directly from pollen grains or indirectly from calluses originating from pollen grains (for reviews, see D'Amato, 1972*a*; Nitsch, 1972). Of particular interest is the whole series of euploid plants – from haploid to pentaploid – derived from the callus of pollen origin in rice (Table 26.6). Although this callus showed haploid mitoses only (Niizeki & Oono, 1971), it obviously also contained diploid, triploid, tetraploid and pentaploid cells. The most plausible mechanism for the formation of triploid and pentraploid cells in in-vitro cultures is nuclear fusion.

Plants regenerated *in vitro* probably originate from single cells. The origin of carrot embryoids (which later develop into plants) from an isolated somatic cell has been documented by time-lapse microcinematography (Backs-Hüsemann & Reinert, 1970). A single-cell origin can also be inferred from the observation that, in general, plants regenerated *in vitro* are not chimeras (see review in D'Amato, 1972*a*).

Table 26.5. *Chromosome number of shoots or plants regenerated from callus of tobacco* (Nicotiana tabacum, $2n = 4x = 48$) *of four different origins*

Tn_1 and Tn_2: calluses heterotrophic for auxin coming from cultures which were started by Prof. R. Gautheret in 1961; Tc and Ta: calluses autotrophic for auxin (resp. transformed by crown gall and habituated) coming from cultures which were started by Prof. Gautheret in 1946 (based on data from Sacristan & Melchers, 1969)

Strains	Chromosome numbers in the callus	Cultures used for regeneration (no.)	Regeneration of		Chromosome number in shoots or plants
			shoots	plants	
Tn_1	96 (88 %), 192 (10 %), 48 (1 %)	2690	Yes	1 only	65–71
Tn_2	As above	1777	Yes	None	82–90
Tc	68–82 with mode at 72 ($3x$)	1692	Yes	Yes	60–70[a,b]
Ta	As above	988	Yes	Yes	57–65[b]

[a]: The 138 plants analysed were not chimerical (identical chromosome number in all cells observed).
[b]: All plants were sterile.

Table 26.6. *Chromosome number of plants regenerated from calluses originated from pollen grains* (anther culture) *of rice* ($2n = 2x = 24$). *All dividing cells in the calluses were haploid* ($n = 12$)

Chromosome number of regenerated plants	Total numbers of regenerated plants	
	Nishi & Mitsuoka, 1969	Niizeki & Oono, 1971
x	4	8
$2x$	4	15
$3x$	1	2
$4x$	0	1
$5x$	7	0
Total	16	26

The genetic continuity of the meristem cell line

Vascular plants are characterized by a unique feature in development and growth, namely, the localization at the end of the axial organs of embryonic tissues, the apical meristems, which all through the life of the individual form new tissues and organs which are added to those formed during embryogenesis (see also Morel, Chapter 25). Because of this, plants have sometimes been defined as organisms with 'continued embryogenesis' or 'recurrent ontogenesis'. Recurrent ontogenesis in

343

plants makes separation of somatic cells and germ cells impossible. The shoot apical meristem acts as a store of the genetic information of the plant, which is directly delivered to the sporogenous tissue of the flower (germ line) at the time of change from the vegetative to the reproductive phase. Since the meristems are the forerunners of the germ line, genetic stability of the meristem cell line is the prerequisite for the genetic stability of the germ line. Important factors of genetic stability in meristems seem to be: (i) the strict control of the sequence of DNA-synthesis and mitosis which does not allow the extra duplications of DNA which are responsible for somatic polyploidy, and (ii) the continuous division, which eliminates at least part of the spontaneously occurring chromosome structural changes and genetic defects impairing the reproductive integrity of cells. Any device which in nature might ensure the long-term conservation of the genetic make-up of a species by maintaining meristems in a quiescent state should have had a selective advantage. In phanerogams, the seed is the most advanced device which has been developed so far for the storage of genetic information in plants. The nuclear cytology of the process of development and 'storage' of meristems in seed embryos favours this view (for further elucidation of the above discussion, see D'Amato, 1964*a, b*; 1972*b*).

From the above discussion it is apparent that for plants which can be propagated by seed the best method for the long-term conservation of a given genetic stock is seed storage. In other plants, especially those propagated vegetatively, the genetic stability of apical meristems can be, and indeed is, utilized with great advantage in in-vitro culture. Following the classical observation of Morel & Martin (1955) that virus-free dahlia plants could be obtained from apical meristems grown *in vitro*, meristem culture has been applied with success – by Morel himself and many other workers – to the extensive in-vitro propagation of many plant species. In orchids, in which propagation by meristem culture is now applied on an industrial scale, an initial plant can be maintained, by continuous propagation *in vitro*, in the juvenile state (as regeneration protocorms) for many years, probably indefinitely. The ability of meristem culture to free clones from viruses adds to the merits of this technique (reviews in Morel, 1964*a, b* and in Chapter 25).

Concluding remarks

From the above discussion it appears that when in-vitro cultures are established from non-meristem plant tissues great variation in nuclear conditions is generally found in the culture. The most common nuclear change is polyploidy which, in the largest majority of plant species (more than 90 per cent), pre-exists in the initial explant, that is, *in vivo*. However, polyploidization may be favoured by the in-vitro culture conditions. Since gene duplication in polyploid cells allows the survival of aneuploid derivatives, the cultures may often comprise an array of cells with very different chromosome numbers. Even in species which do not endopolyploidize *in vivo*, long-term in-vitro culture favours polyploidization and derived genomic changes. In addition, structural chromosomal changes and probably more minute genetic changes (small deficiencies, gene mutations) may accumulate in culture.

In many plant species, great variation in chromosome number is obtained among plants regenerated from in-vitro cultures. In a few other species (e.g. carrot, rice, common wheat) plant regeneration *in vitro* occurs at the diploid level. If it were confirmed that in these and possibly other species, plant regeneration *in vitro* occurred at the diploid level only, this would indicate the possibility of conservation *in vitro* of the meristem cell line of the initial plant. In view of its great significance, such a phenomenon should be investigated extensively with appropriate cytological and genetic analyses aiming at exactly defining the degree of in-vitro stability of the meristem cell line.

Because of the great genetic variability which is present in plant tissue and cell cultures, these represent an extraordinary reservoir of genetic information to be exploited for genetic and breeding research. If, however, long-term conservation of given genetic stocks is required, plant tissue and cell cultures can serve this scope only when genetic stability *in vitro* is achieved. At present, the genetic stability of the meristematic cell line can be successfully utilized only by means of meristem culture. This technique is of the greatest value in vegetatively propagated plants, in which it permits the long-term preservation of clones – the so-called 'mericlones', i.e. meristem derived clones – in a virus-free condition. For in-vitro cultures other than meristem culture, genetic stability (or at least relative stability at a given ploidy level) is at present no more than a remote possibility. Nevertheless, efforts should be concentrated towards a better knowledge of the dynamics of cell populations in in-vitro systems. Such a knowledge is the essential pre-requisite for any serious progress towards the goal of genetic stability.

Conservation and storage

References

Backs-Hüsemann, D. & Reinert, J. (1970). Embryobildung durch isolierte Einzelzellen aus Gewebekulturen von *Daucus carota*. *Protoplasma*, **70**, 49–60.

Bennici, A., Buiatti, M. & D'Amato, F. (1968). Nuclear conditions in haploid *Pelargonium in vivo* and *in vitro*. *Chromosoma*, **24**, 194–201.

Bennici, A., Buiatti, M., D'Amato, F. & Pagliai, M. (1971). Nuclear behaviour in *Haplopappus gracilis* grown *in vitro* on different culture media. *Les Cultures de Tissus de Plantes*, *Coll. Internat. C.N.R.S.*, **193**, 245–50.

Blakely, L. M. & Steward, F. C. (1964). Growth and organized development of cultured cells. VII. Cellular variation. *Am. J. Bot.*, **51**, 809–20.

D'Amato, F. (1952). Polyploidy in the differentiation and function of tissues and cells in plants. A critical examination of the literature. *Caryologia*, **4**, 311–58.

D'Amato, F. (1964a). Cytological and genetic aspects of ageing. In *Genetics Today*, *Proc. XI Int. Congr. Genetics*, pp. 285–95. Pergamon Press, Oxford.

D'Amato, F. (1964b). Nuclear changes and their relationships to histological differentiation. *Caryologia*, **17**, 317–25.

D'Amato, F. (1965). Endopolyplody as a factor in plant tissue development. In *Proc. Int. Conf. Plant Tissue Cult.* (eds P. R. White and A. R. Grove), pp. 449–62. McCutchan Publ. Corp., Berkeley.

D'Amato, F. (1972a). Significato teorico e pratico delle colture *in vitro* di tessuti e cellule vegetali. Accad. Naz. Lincei, Roma, *Problemi Attuali di Scienza e di Cultura*, Quaderno No. **173**, 55 pp.

D'Amato, F. (1972b). Morphogenetic aspects of the development of meristems in seed embryos. In *The Dynamics of Meristem Cell Populations* (eds M. W. Miller and C. C. Kuehnert), pp. 149–63. Plenum Publish. Corp., New York.

Devreux, M., Saccardo, F. & Brunori, A. (1971). Plantes haploïdes et lignes isogéniques de *Nicotiana tabacum* obtenues per culture d'anthères et de tiges *in vitro*. *Caryologia*, **24**, 141–8.

Gautheret, R. J. (1939). Sur la possibilité de réaliser la culture indefinie des tissus de tubercules de carotte. *C. r. Acad. Sci.*, **208**, 218–20.

Gupta, N. & Carlson, P. S. (1972). Preferential growth of haploid plant cells *in vitro*. *Nature New Biol.*, **239**, 86.

Heinz, D. J. & Mee, G. W. P. (1971). Morphologic, cytogenetic and enzymatic variation in *Saccharum* species hybrid clones derived from callus tissue. *Am. J. Bot.*, **58**, 257–62.

Heinz, P. J., Mee, G. W. & Nickell, L. G. (1969). Chromosome number of some *Saccharum* species hybrids and their cell suspension cultures. *Am. J. Bot.*, **56**, 450–6.

Johri, B. M. & Nag, K. K. (1968). Experimental induction of triploid shoots *in vitro* from endosperm of *Dendrophthoe falcata* (L.f.) Ettings. *Cur. Sci.*, **37**, 606–7.

346

Matthysse, A. G. & Torrey, J. G. (1967). Nutritional requirements for polyploid mitoses in cultured pea root segments. *Physiologia Pl.*, **20**, 661–72.

Melchers, G. & Bergmann, L. (1959). Untersuchungen an Kulturen von haploiden Geweben von *Antirrhinum majus*. *Ber. deutsch bot. Ges.*, **71**, 459–63.

Mitra, J. & Steward, F. C. (1961). Growth induction in cultures of *Haplopappus gracilis*. II. The behaviour of the nucleus. *Am. J. Bot.*, **48**, 358–68.

Morel, G. M. (1964*a*). Tissue culture. A new means of clonal propagation of orchids. *Bull. Am. Orchid Society*, pp. 473–78.

Morel, G. M. (1964*b*). La culture *in vitro* du méristème apical. *Rev. Cytol. Biol. Vég.*, **27**, 307–14.

Morel, G. M. & Martin, C. (1955). Guérison de plantes atteintes de maladies à virus par cultures de méristèmes apicaux. *Rep. XIV Int. Hort. Congr.*, **1**, 303–10.

Murashige, T. & Nakano, R. (1967). Chromosome complements as a determinant of the morphogenic potential of tobacco cells. *Am. J. Bot.*, **54**, 963–70.

Nagl, W. (1962). 4096-Ploidie und 'Riesenchromosomen' im Suspensor von *Phaseolus coccineus. Naturwiss.*, **49**, 261–2.

Niizeki, H. & Oono, K. (1971). Rice plants obtained by anther culture. *Les Cultures de Tissus de Plantes, Coll. Int. C.N.R.S.*, **193**, 251–7.

Nishi, T. & Mitsuoka, S. (1969). Occurrence of various ploidy plants from anther and ovary culture of rice plants. *Jap. J. Genet.*, **44**, 341–6.

Nishi, T., Yamada, Y. & Takahashi, E. (1968). Organ redifferentiation and plant restoration in rice callus. *Nature, Lond.*, **219**, 508–9.

Nitsch, J. P. (1972). Haploid plants from pollen. *Z. PflZücht.*, **67**, 3–18.

Nobécourt, P. (1939). Sur la perennité et l'augmentation de volume des cultures de tissus végétaux. *C. R. Soc. Biol.*, **130**, 1270–1.

Partanen, C. R. (1963). Plant tissue culture in relation to developmental cytology. *Int. Rev. Cytol.*, **15**, 215–43.

Pätau, K., Das, N. K. & Skoog, F. (1957). Induction of DNA synthesis by kinetin and indoleacetic acid in excised Tobacco tissue. *Physiologia Pl.*, **10**, 949–66.

Reinert, J. & Küster, H. J. (1966). Diploide, chlorophylhaltige Gewebekulturen aus Blättern von *Crepis capillaris* (L.) Wallr. *Z. PflPhysiol.*, **54**, 213–22.

Sacristan, M. D. (1971). Karyotypic changes in callus cultures from haploid plants of *Crepis capillaris* (L.) Wallr. *Chromosoma*, **33**, 273–83.

Sacristan, M. D. & Melchers, G. (1969). The caryological analysis of plants regenerated from tumorous and other callus cultures of Tobacco. *Molec. gen. Genet.*, **105**, 317–33.

Shimada, T. (1971). Chromosome constitution of tobacco and wheat callus cells. *Jap. J. Genet.*, **46**, 235–41.

Shimada, T., Sasakuma, T. & Tsunewaki, K. (1969). *In vitro* culture of wheat tissues. I. Callus formation, organ redifferentiation and single cell culture. *Can. J. Genet. Cytol.*, **11**, 293–304.

Skoog, F. & Miller, C. O. (1957). Chemical regulation of growth and organ formation in plant tissues cultured *in vitro*. SEB Symposium, XI, *The Biological Action of Growth Substances*, pp. 118–31.

Steward, F. C., Blakely, L. M., Kent, A. & Mapes, M. O. (1963). Growth and organisation in free cell cultures. *Brookhaven Symposia in Biology*, **16**, 73–88.

Torrey, J. G. (1961). Kinetin as trigger for mitosis in mature endomitotic cells. *Exper. Cell Res.*, **23**, 281–99.

Torrey, J. G. (1967). Morphogenesis in relation to chromosomal constitution in long term plant tissue cultures. *Physiologia Pl.*, **20**, 265–75.

White, P. R. (1939). Potentially unlimited growth of excised plant callus in an artificial nutrient. *Am. J. Bot.*, **26**, 59–64.

Yamada, Y., Tanaka, K. & Takahashi, E. (1967). Callus induction in rice, *Oryza sativa* L. *Proc. Jap. Acad.*, **43**, 156–9.

27. Technical aspects of tissue culture storage for genetic conservation

G. G. HENSHAW

In the botanical context the term 'tissue culture' is generally applied rather loosely to the whole range of culture systems, including organ, callus, suspension, cell and protoplast cultures. Techniques for the initiation of such cultures are now well established for a wide range of monocotyledonous and dicotyledonous plant species and valuable contributions are already beginning to be made in the commercial plant breeding area, particularly as a highly efficient means of vegetative propagation ('micro-propagation'). In the future, other culture techniques will be used to accelerate the production of homozygous lines of certain species, and protoplast cultures will probably be used for non-sexual hybridisation programmes.

Tissue culture storage techniques might also be valuable for the conservation of plant genetic resources since in principle the cultures could be maintained indefinitely by regular medium replenishment, plants being regenerated in large numbers when required. Such a procedure would have the following advantages:

(1) The space requirement is relatively small compared with the requirement for field cultivation.

(2) The maintenance procedure is relatively simple and cheap.

(3) The propagation potential of the cultures can be very high.

(4) The problems of genetic erosion in stocks, which can be serious under field conditions, are completely avoided.

(5) The techniques can also be used to produce and maintain pathogen-free stocks of plant material.

In practice, however, there are certain difficulties which would almost certainly ensure that tissue culture storage techniques would only be used where the more orthodox genetic conservation procedures are unsatisfactory because of limited seed viability, unavailability of seed, inefficient vegetative techniques, etc. The difficulties can include the following:

(1) With certain species difficulties can be encountered in establishing suitable cultures. These problems have become less formidable as new

media have been devised; the medium of Murashige & Skoog (1962) has now been used with a wide variety of species and the recent medium of Schenk & Hildebrandt (1972) was used successfully by these authors with an extensive range of monocotyledonous and dicotyledonous crop plants. Nevertheless, some species do not readily respond, and suitable cultures may only be established after a considerable amount of painstaking empirical work.

(2) Regeneration of plants from cultures of certain species may be difficult and the reasons are not always readily established. It is likely that the precise environmental and nutritional requirements for releasing the morphogenetic potential of these species may not easily be identified, although there is always the possibility that the original cultures are derived from cells which are no longer totipotent.

(3) The morphogenetic potential of cultures, which rarely seems to be a stable property, especially in callus and suspension cultures, can be lost after growth under in-vitro conditions for an extended period of time. There are some indications that this can be the result of physiological changes which are reversible (Steward, Kent & Mapes, 1966; Reinert, Backs-Husemann & Zerban, 1971) but it has often been suggested that the loss of potential is the result of genetic changes in the cultures (e.g. Torrey, 1967).

(4) Plant tissue cultures, like many other tissue and cell cultures, frequently contain a proportion of cells which are genetically abnormal in the sense that they exhibit various chromosomal aberrations (D'Amato, 1965). It is also likely that less readily observable mutations at the gene level accumulate, but these have not yet been extensively studied.

Genetic instability

Since the details and implications of the genetic instability of plant tissue cultures have already been fully discussed by D'Amato (Chapter 26), it is sufficient to say here that this certainly constitutes the most serious problem with regard to the use of tissue cultures for long-term genetic conservation. This problem will only be surmounted by establishing conditions which stabilise the genetic constitution of tissue culture cells or by using cultures which are inherently more stable. The latter property is more likely to be found in meristem cultures, since some of the plant's normal control mechanisms might still operate (see Morel, Chapter 25).

There can be two approaches to the problem of stabilising the genetic constitution of callus or cell cultures:

(1) The cultures can be examined to see whether particular medium constituents or culture regimes cause genetic aberrations or selectively favour the growth of aberrant cells. It is already known (Torrey, 1967) that the genetic composition of tissue cultures can be influenced by the hormonal balance in the medium but it is not known whether long-term stabilisation could ever be achieved by adjustment of the culture conditions.

(2) The processes of DNA synthesis and cell division, probably the most genetically unstable stages in the cell cycle, might be suppressed. There is evidence that parenchymatous cells in certain plants (e.g. the cactus *Cereus giganteus*) remain alive, although not necessarily without genetic change, for more than 100 years and yet retain the ability to divide in response to wounding (MacDougal, 1926; Popham, 1958 as reported by Cutter, 1969).

Low-temperature storage of tissue cultures

Unfortunately, there is little information about the mechanisms re-straining mitosis *in vivo* and it is difficult to suppress cell division for any appreciable length of time in tissue cultures, using inhibitors or minimal media, without causing the death of the cells. Low temperature would almost certainly be a more satisfactory way of suppressing cell division and DNA synthesis in tissue cultures for the purpose of long-term genetic stabilisation.

Although it has been common practice for the last twenty years to store micro-organisms, animal cell cultures, tissues and spermatozoa at the temperature of liquid nitrogen, there have been surprisingly few attempts to apply such techniques to plant cells and particularly to plant tissue cultures. Since there has been no evidence of genetic change in micro-organisms or cells frozen and preserved at very low tempera-tures (Ashwood-Smith, 1965; Ashwood-Smith, Trevino & Warby, 1972; Shikama, 1964), this would seem to be an ideal method for the long-term conservation of genetic materials in gene banks.

Some of the most extensive studies of plant cells at very low tempera-tures have been carried out by Krasavtsev (1967), Sakai (1960; 1965; 1971; Sakai & Yoshida, 1967; Sakai & Otsuka, 1972) and their respec-tive associates, and these have shown that at least some of the principles established with animal cells and micro-organisms are applicable to plant cells. It is now clear, for example, that survival of cells at very low temperature is dependent on precisely controlled rates of cooling and warming, and that various substances (the so-called 'cryoprotectants') are capable of protecting cells from freezing damage.

When a cell is subjected to subzero temperatures it initially super-cools, regaining thermodynamic equilibrium either by transferring water to the exterior (if the cooling rate is sufficiently slow or if the permeability of the cell is sufficiently high) or by intracellular freezing (see review by Mazur, 1969). A number of factors could cause the damage at low temperature, including intracellular ice-crystal formation, increases in the concentrations of intra- and extracellular solutes (so-called 'solution effects') and denaturation of proteins as a result of dehydration when the water leaves the cell. It has been suggested by Mazur (1965; 1969) that the optimal cooling rate, which causes least cell damage, has to be slow enough to allow water to leave the cell before freezing, but fast enough to prevent the increasing concentrations of salts from damaging the cell. The optimal warming rate seems to be one that prevents ice-formation by re-crystallisation during the warming period.

It has been difficult to produce a unified theory to explain the mech-anism of action of cryoprotectants since they fall into two classes: compounds which penetrate the cell (low molecular weight substances such as glycerol and dimethylsulphoxide) and compounds which cannot penetrate the cell or which penetrate the cell slowly (low molecular weight substances such as sucrose and high molecular weight sub-stances such as polyvinylpyrrolidone, dextran and certain proteins). It has been suggested that the penetrating compounds prevent extreme 'solution effects' from occurring (Lovelock, 1953) but it has also been postulated that both types of compounds might work in the same way by protecting the surface membranes (Mazur, 1970).

In addition to the cooling and warming of cells at the optimal rates in the presence of suitable cryoprotectants, it is essential to provide correct refrigeration conditions for long-term storage. There can be a steady loss of viability unless ice-crystal growth, which can occur at temperatures down to $-130\,°C$, is completely suppressed. Since this temperature is below the range of electrical refrigerators, the cells must be stored in liquid nitrogen ($-196\,°C$) refrigerators which are readily available commercially and relatively cheap to buy and operate in countries where supplies of liquid nitrogen are obtainable.

Unfortunately the rules for successful low-temperature storage can only be stated in these very general terms and the precise conditions

352

for any particular cell type can only be determined empirically. Almost certainly callus and suspension cultures will prove to be easier experimental subjects in this context than organs or organ cultures because of the smaller range of cell types which they contain and because of the relative ease with which cryoprotectants can be administered.

There has been a small number of reports of survival of cultured plant cells at very low temperatures. Quatrano (1968) reported that 14 per cent of flax (*Linum usitatissimum*) cells were still capable of reducing a tetrazolium compound (sometimes taken to indicate viability) after being frozen at a relatively slow rate (between 5 degC and 10 degC per minute) to − 50 °C, in the presence of 10 per cent dimethylsulphoxide, and then thawed rapidly. Latta (1971) showed that at least some carrot (*Daucus carota*) cells survived and grew after a short period of storage at − 40 °C or − 196 °C in the presence of a suitable cryoprotectant (5 per cent glycerol or 10 per cent dimethylsulphoxide). On the other hand, *Ipomoea* cells only survived these conditions if they had been grown in a medium containing 6.5 per cent sucrose for a few days before freezing and then only in the presence of other cryoprotectants (2.5 per cent glycerol and 2.5 per cent dimethylsulphoxide).

These studies can be taken as an indication that tissue culture cells can survive being frozen to liquid nitrogen temperatures under suitable conditions. It is now essential to determine systematically the optimum conditions for freeze-storage, varying the rates of cooling and thawing, the type and concentration of cryoprotectant, the type of culture (callus, suspension, etc.) and the physiological state of the cells (dividing, non-dividing, etc.). Until recently this type of work has been slowed down by the problem of rapidly determining the proportion of viable cells in a sample which has been subjected to a freezing treatment. A demonstration of an ability to divide has been the only satisfactory way of assessing viability and this procedure is both time-consuming and technically exacting with tissue cultures. Further, it does not give an accurate estimation of the proportion of cells which have survived.

Cell survival, however, can now be assessed quite reliably by the use of two staining techniques which act in a complementary manner. Evan's blue stain is strongly absorbed by the cytoplasm of dead cells, but excluded by the plasmalemma of living cells (Gaff & Okong' O-ogola, 1971); fluorescein diacetate, which does not fluoresce in ultraviolet light, is absorbed by living cells and then hydrolysed by esterases to form fluorescein which is detectable with the ultraviolet fluorescence microscope (Heslop-Harrison & Heslop-Harrison, 1970; Widholm,

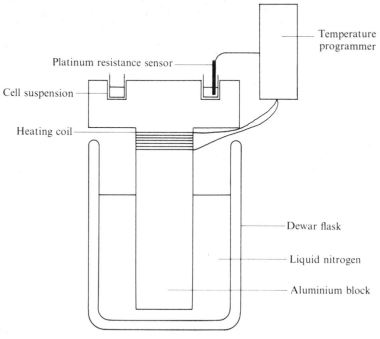

Fig. 27.1. Programmed freezing apparatus. The aluminium block is cooled by. partial immersion in liquid nitrogen and the temperature in the cell suspension is continuously monitored so that the rate of cooling can be controlled by a programmed reduction in the power supplied to the heating coil. Temperature programmer: Stanton Redcroft Ltd., Copper Mill Lane, London, S.W.7.

1972). Widholm showed that, out of almost thirty reagents which have been used at various times for assessing viability in other systems, these together with phenosafranine, were the only reagents which could be used reliably with plant tissue cultures, and this has been confirmed in our own work.

For our own work we have constructed an apparatus (Fig. 27.1) in which cells can be frozen at accurately controlled rates. Thin-walled glass vials, containing aliquots of a suspension culture, are placed in close-fitting holes drilled in a circular aluminium block attached to the top of an aluminium rod, the base of which is immersed in liquid nitrogen. The rate of cooling is controlled electronically by steadily reducing the power to a heating tape wrapped around the top of the aluminium rod, at a rate determined by a temperature programmer which monitors the actual temperature in a replicate vial with a platinum temperature-sensor. The advantage of this technique lies in the

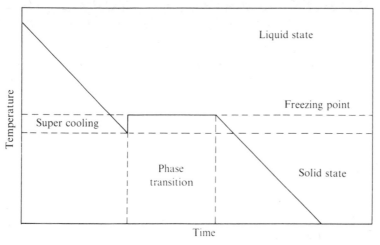

Fig. 27.2. Cooling curve for biological materials (diagrammatic). Biological materials cool to a point below their freezing point (i.e. supercool) before freezing occurs. When freezing begins the temperature rises to the freezing point, and remains steady until the phase transition is complete, when cooling then continues.

electronic feedback control which can maintain a constant cooling rate even during the period of phase transition when the medium around the cells would normally supercool and then warm up to freezing point before freezing (Fig. 27.2). Warming up is achieved, either by reversing the controls of the programmer, for the slower rates, or by plunging the vials into water at 30 °C for faster rates.

The results have shown that *Acer pseudoplatanus* cells are killed if cooled rapidly by being plunged directly into liquid nitrogen. They are also killed, irrespective of the cooling rate, if a cryoprotectant is not added to the medium. Dimethylsulphoxide, particularly at concentrations as high as 15 per cent, has been the most successful of the cryoprotectants that have been tested. Although it is slightly toxic to the cells, up to 50 per cent of those cells which are alive before freezing are still alive after being cooled slowly (1 to 3 degC per minute) to −70 °C and up to 20 per cent after being cooled to −196 °C and then warmed rapidly to room temperature.

Conclusions

These results, together with those of both Quatrano (1968) and Latta (1971) indicate that it should be possible to store plant tissue cultures successfully at the temperature of liquid nitrogen. It is not yet possible

355

to say whether plant cells would retain their viability during long-term storage in liquid nitrogen nor whether they would be undamaged genetically and retain their morphogenetic potential. Experience with micro-organisms and with animal cells under these conditions, however, is sufficiently promising to suggest that it should be possible to preserve at least some plant species by storing them as tissue cultures in liquid nitrogen. The value of the technique for the conservation of genetic resources would then depend upon the ease with which tissue cultures possessing high propagation potential could be isolated from those species which are not readily preserved by orthodox methods.

In those countries in which liquid nitrogen is readily available and relatively cheap, the actual storage at very low temperatures presents few technical problems because large-scale programmable cooling apparatus is available commercially, as well as liquid nitrogen refrigerators which only need replenishing every 60 days.

References

Ashwood-Smith, M. J. (1965). On the genetic stability of bacteria to freezing and thawing. *Cryobiology*, **2**, 39–45.

Ashwood-Smith, M. J., Trevino, J. & Warby, C. (1972). Effect of freezing on the molecular weight of bacterial DNA. *Cryobiology*, **9**, 141–3.

Cutter, E. G. (1969). *Plant Anatomy: Experiment and Interpretation. Part I. Cells and Tissues*. Edward Arnold Ltd., London.

D'Amato, F. (1965). Endopolyploidy as a factor in plant tissue development. In *Proceedings of an International Conference on Plant Tissue Cultures* (eds P. R. White and A. R. Grove), pp. 449–62. McCutchan Publ. Corp., Berkeley.

Gaff, D. & Okong'O-ogola, O. (1971). The use of non-permeating pigments for testing the survival of cells. *J. exp. Bot.*, **22**, 756–8.

Heslop-Harrison, J. & Heslop-Harrison, Y. (1970). Evaluation of pollen viability by enzymatically induced fluorescence; intracellular hydrolysis of fluorescein diacetate. *Stain Technol.*, **45**, 115–20.

Krasavstev, O. A. (1967). Frost hardening of woody plants at temperatures below zero. In *Cellular Injury and Resistance in Freezing Organisms*. Proc. 2 International Conference for Low Temperature Science (ed. E. Asahina), pp. 131–8. Hokkaido University Press, Japan.

Latta, R. (1971). Preservation of suspension cultures of plant cells by freezing. *Can. J. Bot.*, **49**, 1253–4.

Lovelock, J. E. (1953). The haemolysis of human red blood-cells by freezing and thawing. *Biochim. biophys. Acta*, **10**, 414–26.

MacDougal, D. T. (1926). Growth and permeability of century old cells. *Am. Natur.*, **60**, 393–415.

Mazur, P. (1965). Causes of injury in frozen and thawed cells. *Fed. Proc.*, *Suppl.* no. **15**, S-175–S-182.

Mazur, P. (1969). Freezing injury in plants. *Ann. Rev. Pl. Physiol.*, **20**, 419–48.

Mazur, P. (1970). Cryobiology: the freezing of biological systems. *Science*, **168**, 939–49.

Murashige, T. & Skoog, F. (1962). A revised medium for rapid growth and bioassays with tobacco tissue cultures. *Physiologia Pl.*, **15**, 473–97.

Popham, R. A. (1958). Some causes underlying cellular differentiation. *Ohio J. Sci.*, **58**, 347–53.

Quatrano, R. S. (1968). Freeze preservation of cultured flax cells utilising dimethylsulphoxide. *Pl. Physiol.*, *Lancaster*, **43**, 2057–61.

Reinert, J., Backs-Husemann, D. & Zerban, H. (1971). Determination of embryo and root formation in tissue cultures from *Daucus carota*. *Les Cultures de Tissus de Plantes*, Colloques Internationaux du C.N.R.S., No. **193**, 261–8.

Sakai, A. (1960). Survival of the twig of woody plants at −196 °C. *Nature, Lond.*, **185**, 393–4.

Sakai, A. (1965). Survival of plant tissue at super-low temperatures. III. Relation between effective prefreezing temperatures and the degree of frost hardiness. *Pl. Physiol.*, *Lancaster*, **40**, 882–7.

Sakai, A. (1971). Some factors contributing to the survival of rapidly cooled plant cells. *Cryobiology*, **8**, 225–34.

Sakai, A. & Otsuka, K. (1972). A method for maintaining the viability of less hardy plant cells after immersion in liquid nitrogen. *Pl. Cell Physiol.*, **13**, 1129–33.

Sakai, A. & Yoshida, S. (1967). Survival of plant tissue at super-low temperature. VI. Effects of cooling and rewarming rates on survival. *Pl. Physiol.*, *Lancaster*, **42**, 1695–1701.

Schenk, R. V. & Hildebrandt, A. C. (1972). Medium and techniques for induction and growth of monocotyledonous and dicotyledonous plant cell cultures. *Can. J. Bot.*, **50**, 199–204.

Shikama, K. (1964). Thermal and freezing denaturation of deoxyribonucleic acid. *Science Reports Tohoka University, 4th Series, Biology*, **30**, 133–41.

Steward, F. C., Kent, A. E. & Mapes, M. O. (1966). Growth and organisation in cultured cells: sequential and synergistic effects of growth regulating substances. *Ann. N.Y. Acad. Sci.*, **144**, 326–34.

Torrey, J. G. (1967). Morphogenesis in relation to chromosomal constitution in long-term plant tissue cultures. *Physiologia Pl.*, **20**, 265–75.

Widholm, J. M. (1972). The use of fluorescein diacetate and phenosafranine for determining viability of cultured plant cells. *Stain Technol.*, **47**, 189–94.

28. Possible long-term cold storage of woody plant material

B. H. HOWARD

Functions of cold storage

Plants are currently held in cold storage to reduce their rate of development and to extend their period of dormancy. In this way they are maintained in a physiologically optimum condition for regrowth, principally by avoiding desiccation, either from exposure to drying winds, or from premature foliation before root establishment has occurred in the new planting site.

Maintaining plants in a fully dormant condition aids the management of tree nurseries by facilitating post-harvest operations, shipment and replanting. The opportunity to delay replanting until spring or early summer in temperate climates is considered to be one of the most attractive features of cold storage because it allows pre-planting operations to be carried out under ideal soil conditions.

This technique could be developed to give storage of extended duration with relevance to genetic conservation.

Storage conditions and equipment

A temperature between $-2\,°C$ and $2\,°C$ is usually employed, the precise temperature being dependent on the time of year and hence state of dormancy of the plant material, the intended duration of storage, and the construction of the cold store in relation to potential desiccation.

Relatively higher temperatures are acceptable for storage periods of a few weeks, whereas at temperatures above $0\,°C$, bud development will occur during prolonged storage (Groven, 1965) and even at $-1\,°C$ callus and root development may take place (Howard & Garner, 1966). If plant material is taken into store as growth is commencing it is desirable not to subject it to temperatures at the lower end of the range, especially if it is liable to desiccate.

Relative humidity is a vital factor in the control of desiccation (De Haas & Wennemuth, 1962a). A relative humidity of 96–98 per cent is desirable which imposes limitations on storage procedure (De Haas &

359

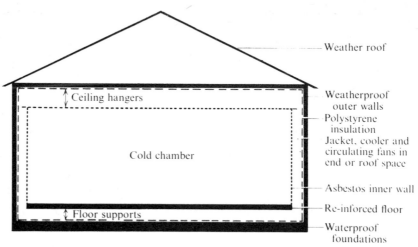

Fig. 28.1. Diagrammatic view of an indirectly cooled (jacketed) store.

Wennemuth, 1962*b*). Directly cooled stores, in which the cooling unit is erected inside the storage chamber are less suitable for maintaining adequate humidity because the relatively small cooling surface is necessarily maintained at a much lower temperature than the air. This leads to freezing of moisture in the air which is replaced by moisture removed from the plant material, leading to a continuous process of desiccation. It is usually found necessary in this type of store either to wrap the plants in polythene film, or to install humidifiers if plants are stored unwrapped. Excessive introduction of water may, however, lead to the development of moulds.

To overcome this limitation indirect cooling is used in the majority of installations for storing nursery material. The cold chamber is surrounded by a cavity or jacket space in which the cooler is installed, providing a large surface area (walls, floor and roof) for cooling (Fig. 28.1). This allows a reduction of the temperature differential between cold-chamber air and cooling surface from between 2.0 and 10.0 degC in a directly cooled store (depending upon specifications) to about 1.5 degC in an indirectly cooled store. Thus in the latter, desiccation of the plant material is minimised.

Relative humidity is indirectly influenced by leakage of heat into the cold chamber because it is this which determines the intensity and frequency of cooling operations. Whereas this may become a problem in directly cooled stores where air leaks directly into the plant storage chamber, leakage into an indirect system occurs into the outer air-space

360

and not into the cold chamber, and its effect upon humidity is therefore negligible.

Within the jacketed cold store, plant material is usually stored 'bare root' or in small bundles in the case of graftwood or unrooted cuttings. This should be done on racks to allow adequate circulation of air, which is essential for the quick removal of field heat in the absence of rapid air circulation.

The development of moulds and rots is rarely troublesome in a properly managed jacketed store if surface water is allowed to dry off before storage. In less favourable storage conditions pre-treatment with fungicides has been shown to be beneficial (Flint & McGuire, 1962; Howard & Garner, 1966) and in practice the outer surfaces of plant stacks are usually sprayed with a suitable fungicide once the racks in a jacketed store have been filled.

Rosaceous material is subject to damage from small traces of ethylene gas in the store (Curtis & Rodney, 1952) and similar damage has been obtained by storing ripening fruit and plants together (Howard & Banwell, 1970; Howard, 1973b), presumably caused by ethylene in the respiratory gases from the fruit. Care should be taken, therefore, not to introduce fruit collections even temporarily into plant cold stores.

Long-term storage as a means of genetic conservation

Terms such as 'gene bank' and 'clonal archive' imply the collective storage of well-documented material which retains its genetic integrity and is readily recoverable. Long-term cold storage of woody plant material has potential value in these respects.

The land area involved in genetic reservoirs in the form of living arboreta would be reduced by the addition of long-term cold-storage facilities. Land would then be required only for those subjects not amenable to storage and for periodic regeneration of stored material.

Because storage conditions are constant and controlled, plant material would not be exposed to extreme or unusual weather conditions such as drought or severe freezing. Exposure to such conditions is not uncommon in important conservation areas such as Turkey, and similar hazards may occur even in temperate insular climates. In south-east England, dormant cuttings of *Cydonia oblonga* cv. Malling C, failed to regenerate when collected from the field after a severe frost of -17.7 °C, whereas those held in cold store at 2 °C during this period subsequently rooted and grew normally (Howard, 1973a).

Plants could also be maintained by cold storage in areas with unsuitable climatic or soil conditions. Independence from a permanent living collection would provide an opportunity for the regenerative period to take place in artificial growing media or protected cultivation, as the case demanded.

The storage of vegetative plant material requires that it is amenable to asexual multiplication. A constant hazard to vegetatively propagated plants is the risk of virus infection, and its dissemination in the plant material. Maintenance in cold storage, with a concomitant reduction in plant exposure to insect and nematode virus vectors, and reduced exposure to wind- or insect-borne pollen, would reduce the risk of infection where these agents were involved. Because of the possible dependence upon grafting during the regenerative stage of some subjects, however, great care will be needed to select nurse plants which are as far as possible known to be virus-free.

From many aspects cold storage of vegetative material is similar in its objectives to seed and tissue culture storage, which are considered in Chapters 22–7.

Special attention may be necessary to ensure satisfactory seed germination in atypical conditions, and the ensuing juvenile period through which the seedlings of large woody trees pass may seriously delay their use in breeding programmes. The delay in flowering of vegetatively reproduced plants depends upon the size and type of material employed, but it is likely to be less than that for seed-derived plants. Relative to cold storage of vegetative material, practical tissue culture has the disadvantage that the technical requirements for maintenance and regeneration are considerable (Marston, 1969). In poorly maintained systems, selection pressures are likely to operate, and there is a possibility of mutation. Two of the four gooseberry plants raised from meristem culture (Jones & Vine, 1968) were found not to be true to type (Adams, personal communication) and polyploid plants are known to result from callus tissue cultures (Murashige & Nakano, 1966).

Storage of a specific genotype in the form of vegetative material is particularly valuable after an initial survey in which useful characters have been identified, although the particular disadvantage of cold storing vegetative organs is that material is not readily available for study and use, in which respect it is similar to seed storage, but less suitable than living arboreta.

Should it become possible to develop national cold storage facilities close to collecting areas, it would reduce the frequency of collection from the field and facilitate the maintenance of more diverse material in ideal conditions for shipment to other centres.

Future potential

Information is available to ensure the successful storage of deciduous and coniferous species in the form of graftwood, unrooted or rooted cuttings, or complete plants as appropriate. To date, this technique has been developed to provide a few months' storage as a tool to aid the management of tree nurseries during the busy lifting and dispatch season. Nevertheless, potential for longer-term storage has been demonstrated (Table 28.1). Unrooted and rooted woody cuttings of rosaceous fruit plants were held at − 1 °C for one year (thus missing out one growing season) despite a jacketed store not being available. Subsequent field establishment in excess of 50 per cent and frequently better than 80 per cent was achieved (Howard & Garner, 1966). In these experiments there was no evidence that material had deteriorated in cold store and it is reasonable to suppose that losses were partly attributable to conditions during establishment, as also occurs with non-stored material. It is likely that even higher survival levels would have been achieved had the cuttings been used as graftwood. Two-year seedlings of *Picea abies* were successfully stored at 0 °C from December 1968 to May 1970, 24 of a sample of 25 surviving when potted (Jackson, 1971). Plants of *Spiraea vanhouttei*, *Philadelphus coronarius* and *Potentilla fruticosa* were stored for a year at 0 °C and Forsythia bushes for 20 months at − 2 °C without damage (Green, 1963).

Conclusions

Results to date suggest that it would be worthwhile exploring the opportunities for greatly extended periods of cold storage for woody plants which at present are extremely brief compared with the storage of seeds.

To assess the value of this technique for genetic conservation the following information will be required:

The form in which material is most reliably stored

It is likely that with increasing duration of storage, plant mass will become a limiting factor for successful regeneration, since a decrease in dry matter with time has been observed (De Haas & Wennemuth,

Table 28.1. *Examples of species successfully cold stored as plants or vegetative organs*

Subject	Material	Duration	Reference
Buxus sempervirens	Rooted cuttings	6 months	Flint & McGuire (1962)
Cornus alba argenteo-marginata			
Chamaecyparis pisifera plumosa			
Euonymus alatus compactus			
Forsythia intermedia, Lynwood Gold			
Hydrangea petiolaris			
Pachistima canbyi			
Picea glauca conica			
Syringa, James Macfarlane			
Thuja occidentalis nigra			
Thuja orientalis aurea nana			
Rosa multiflora	Plants	4 months	De Haas & Bunemann (1962)
Prunus avium			
Ribes floridum			
Forsythia intermedia			
Prunus cistena	Rooted cuttings	5 months	Bailey (1963)
Prunus triloba			
Vaccinium spp.	Unrooted cuttings	1 to 2 months	De Haas & Hildebrant (1963)
Spiraea vanhouttei	Plants	12 months	Green (1963)
Philadelphus coronarius			
Potentilla fruticosa			
Forsythia spp.	Plants	20 months	Green (1963)
Rosa spp.	Budwood	'over-winter'	Mackay (1963)
Euonymus kiautschovica	Rooted cuttings	8 months over summer	Schneider (1965)
Ilex crenata nigra			
Thuja occidentalis nigra			
Cydonia oblonga cv. Malling A	Rooted or unrooted cuttings	12½ months	Howard & Garner (1966)
Malus pumila cv. Malling 26			
Malling 7			
Malling Crab C			
3430			
Prunus cerasifera cv. Myrobalan B			
Picea abies	Nursery plants	18 months	Jackson (1971)

1962*a*). The large woody 'cuttings' commonly used in Spain for the propagation of olive cultivars (Pansiot & Rabour, 1961) give a higher percentage of usable plants if the original cutting weight is greater rather than less than 1 kg (Humanes, personal communication). On the other hand, if regeneration is carried out more simply by grafting on to a suitable nurse plant, the depth of bud dormancy becomes an important consideration in ensuring successful regrowth, and it is unlikely that shoots older than two to three years would be suitable.

Where specimens are not unique, whole plant storage is likely to be most reliable in association with subsequent vegetative propagation to re-obtain a suitably sized plant for a subsequent storage cycle.

It is unlikely that plants other than those in the deepest rest or dormancy condition will be suitable for prolonged storage.

Storage conditions

Current evidence suggests that temperatures below freezing will be necessary for prolonged storage to arrest shoot and root development. It is unlikely that conditions of relative humidity suitable for long-term storage will differ from those preferred at present, and a standard fungicide treatment should be adopted.

The possibility exists of employing controlled atmospheres as successfully used in the post-harvest storage of some pome fruits, involving the substitution of inert gases or the accumulation of carbon dioxide at the expense of oxygen within the store. No information with respect to plant storage is available.

Regenerative procedure

Method and frequency of regeneration will be closely linked to the type of material stored. In the case of complete plants, regeneration involves their establishment and vegetative propagation to obtain plants of optimal age and size for re-storage. For such special purposes a much wider range of propagation methods than those normally employed commercially could be tried, including techniques such as marcotting and nurse rooting, separately and combined.

Where parts of plants are stored, regenerative methods apply similarly to parts of shoots or root systems. For subjects such as *Populus*, *Salix* and *Ribes* which are easily propagated from shoot cuttings, unrooted cuttings may be suitable, although pre-storage rooting may aid post-storage establishment by avoiding stresses from rapid shoot growth in the absence of roots, which often follows periods of chilling. The observed long-term regenerative capacity of roots remaining after fruit plantations have been uprooted suggests that root cuttings may be suitable for the storage of some subjects also.

For less easily propagated species recycling should take place by grafting on to suitable nurse plants. The identification of species with wide compatibility for this purpose would considerably aid management of the regenerative nursery.

Thanks are recorded to D. I. Bartlett, Storage Specialist, Agricultural Development and Advisory Service (ADAS), and M. G. Banwell, former ADAS Assistant Regional Fruit Adviser, East Malling, for access to unpublished data, and to Dr E. Antognozzi, University of Perugia, on study leave at East Malling, for assistance in the initial search of the literature.

References

(HA – refers to the volume and entry number in *Horticultural Abstracts* where references occur in inaccessible journals or languages other than English.)

Bailey, V. K. (1963). Propagation of *Prunus cistena* and *triloba* by softwood cuttings. *Comb. Proc. 12th ann. Mtg east. Reg. and 3rd ann. Mtg west. Reg. Plant Prop. Soc., 1962*, 115 (HA, **34**, 1178).

Curtis, O. F. Jr. & Rodney, D. R. (1952). Ethylene injury to nursery trees in
• cold storage. *Proc. Am. Soc. hort. Sci.*, **60**, 104–8.

De Haas, P. G. & Bunemann, G. (1962). Cold storage of young shrubs. IV. Jacketed cooling at −2 °C compared with direct cooling at −1.5 °C, and with the traditional heeling-in method. *Gartenbauwissenschaft*, **27**, 243–6 (HA, **33**, 3298).

De Haas, P. G. & Hildebrant, W. (1963). Experimental results on the vegetative propagation of blueberries. *ErwObstb.*, **5**, 162–6 (HA, **34**, 4389).

De Haas, P. G. & Wennemuth, G. (1962a). Cold storage of young shrubs. II. Cultural and physiological problems. *Gartenbauwissenschaft*, **27**, 214–30 (HA, **33**, 1107).

De Haas, P. G. & Wennemuth, G. (1962b). Cool storage of young shrubs. I. Climatological and technical problems in storing shrubs. *Gartenbauwissenschaft*, **27**, 199–213 (HA, **33**, 1106).

Flint, H. L. & McGuire, J. J. (1962). Response of rooted cuttings of several woody ornamental species to overwinter storage. *Proc. Am. Soc. hort. Sci.*, **80**, 625–9.

Green, S. (1963). Experiments on cool storage of nursery plants. *Viola-Tradgardsvarlden*, **69** (39) 10, 14 (HA, **34**, 3091).

Groven, I. (1965). Storage of nursery crops. *Tidsskr. Planteavl.*, **69**, 334–52 (HA, **36**, 6920).

Howard, B. H. (1973*a*). Winter cold injury to fruit nursery stock. *Rep. E. Malling Res. Stn for 1972*, pp. 193–8.

Howard, B. H. (1973*b*). Damage to the roots·of maiden trees held in cold stores with ripening apples. *Rep. E. Malling Res. Stn for 1972*, pp. 191–2.

Howard, B. H. & Banwell, M. G. (1970). Avoid holding maiden trees in cold stores containing ripening fruit. *Rep. E. Malling Res. Stn for 1969*, pp. 183–4.

Howard, B. H. & Garner, R. J. (1966). Prolonged cold storage of rooted and unrooted hardwood cuttings of apple, plum and quince rootstocks. *Rep. E. Malling Res. Stn for 1965*, pp. 80–2.

Jackson, H. (1971). Recent developments in cold storage at Tilhill. *Comb. Proc. Int. Plant Prop. Soc., 1970*, **20**, 351–2.

Jones, O. P. & Vine, S. J. (1968). The culture of gooseberry shoot tips for eliminating virus. *J. hort. Sci.*, **43**, 289–92.

Mackay, I. (1963). The collection, storage and use of budwood. *Comb. Proc. 12th ann. Mtg east. Reg. and 3rd ann. Mtg west. Reg. Plant Prop. Soc., 1962*, pp. 142–4 (HA, **34**, 1138).

Marston, M. E. (1969). *In vitro* culture – a technique for plant propagators. *Gdnrs' Chron.*, **166** (8), 15–17, and Technology of *in vitro* culture. *Gdnrs' Chron.*, **166** (9), 14–15.

Murashige, T. & Nakano, R. (1966). Tissue culture as a potential tool in obtaining polyploid plants. *J. Hered.*, **57** (4), 115–18.

Pansiot, F. P. & Rabour, H. (1961). *Improvements in olive cultivation. FAO Agricultural Studies*, no. **50**, 249 pp.

Schneider, E. F. (1965). Survival of rooted cuttings of three woody plant species after low temperature storage. *Proc. Am. Soc. hort. Sci.*, **87**, 557–62.

29. The storage of germplasm of tropical roots and tubers in the vegetative form

F. W. MARTIN

Tropical roots and tubers are staple foods and important sources of calories in many parts of the tropics. Their annual production is estimated at about 140 million tons, enough to feed 400 million people.

These crops have much in common, since they are all perennial plants grown as annuals and maintained as vegetatively propagated clones. As foods they are quite comparable in usage but differ in nutritive value.

They have been introduced throughout the tropics in a haphazard fashion and thus the most appropriate varieties have not always been developed for particular locations. Yields could undoubtedly be improved by plant breeding and better agronomic practices, and indeed the small amount of selection and breeding that has been carried out up to now has proved very successful.

Some few attempts have been made to gather together and evaluate the germplasm but the collections have often been discarded for want of continuing funds. There is a pressing need with these crops for the collection, evaluation and utilization of a full range of germplasm. Nevertheless, since the cultivation of the collections vegetatively year after year is costly and brings with it dangers of disease infection and many other hazards, new methods of long-term germplasm storage of vegetative organs are urgently required.

In the following sections the important roots and tubers of the tropics are discussed with respect to the storage potential of their propagating material.

Yams

The true yams are members of the family Dioscoreaceae and thus are monocotyledons. The principal genus, *Dioscorea*, consists of up to 600 species that are widely distributed throughout the tropics, and are sometimes found also in the temperate zones. At least 60 species bear edible tubers, of which about 20 are commonly cultivated and at least five must be considered important crop plants. The perennial nature of the species is made possible by a fleshy underground stem, in a few cases clearly a rhizome, but in the majority of the species an anodal, tuber-like structure.

Two kinds of propagating materials are commonly used in the culture of yams. Whole, small tubers are desirable because they resist rot, and germinate regularly. Such whole yams are produced as the second harvest in the case of some varieties of *D. rotundata*, or are the small tubers left over from the principal harvest. In species such as *D. esculenta* and *D. trifida* the production of large numbers of such small tubers is normal. In such species only small tubers are used for new plantings.

Yams are also propagated from tuber pieces. One large tuber is cut into pieces that may be as small as 50 or as large as 500 grams. The pieces, once cut, are dried slightly or are treated with fungicides to reduce rotting, and then planted.

Yam tubers are best stored at a low temperature. Sufficient work has not been done to characterize the optimum temperature, but it is probably between 15 and 20 °C for most varieties. If tubers are held at temperatures that are too low, internal decomposition occurs resulting in death of the tissue, and in the formation of bad colors and odors. However, if temperatures are too high, then other kinds of deterioration occur.

The deterioration of yams is of four principal types: fungal infection, which is the chief source of storage loss in most instances; water loss, which is due to evaporation transpiration; and sprouting, which is a result of maturation of the tuber. In addition, more subtle changes occur in flavor, no doubt associated with partial breakdown of stored substances.

Loss from rotting can be minimized by care, cleanliness, and appropriate curing and storage treatment. Most rots begin just after harvest and can thus be detected during the first weeks of storage. Water loss is reduced by storage at medium to high humidities and low temperatures.

In Puerto Rico storage at 16 °C after curing at high temperature and high humidity has effectively maintained yams in an edible condition for a complete year. Stored tubers are somewhat harder than fresh tubers but retain their flavor. Tubers develop small buds after about four months of storage, but these do not continue to grow. A more thorough investigation is needed to ascertain the effects of storage on the viability of tubers, the seasonal response of plants, and the fate of virus diseases maintained in the tuber.

370

The nature of the dormancy of yams is not well understood. Ethylene chlorohydrin has been shown to be useful in accelerating the rate of germination of tuber pieces, but it is toxic to plant tissues, and must be used with extreme care. Unpublished studies have shown that ethrel, an ethylene-producing substance, can also stimulate germination.

The problem of reducing the rate of germination by chemical treatment has received some attention. Trials have been made with tetrachloronitrobenzene, pentachloronitrobenzene, *iso*-propylphenyl carbamate, and maleic hydrazide, but have not yielded satisfactory results. One application of gibberellic acid can delay sprouting for several weeks. More treatments might possibly be useful in extending storage life still further. Methyl-α-naphthyl acetate sprayed on packing materials has inhibited sprouting for up to eight months.

The possibilities for long-term storage of yams in their vegetative form do not seem to be very favorable. The vegetative parts are annually renewed, and not easily modified. Therefore, short-term compromises such as the improvement of current systems of cold storage, which should permit the maintenance of yam tubers for one year or more in viable condition, should be used. Aerial tubers could be stored for two or even more years. Storage practices might be refined with carbon dioxide atmospheres or low pressure treatments. Chemical treatments might be combined with cold-storage treatments to give maximum storage life. The hormonal control of dormancy also requires study.

Tubers can be stored in the ground. If insect and rodent pests are controlled, it should be possible to grow plants for five or more years without transplanting. We have done this in the case of the sapogenin-bearing yams.

It may be feasible to maintain germplasm in the form of juvenile plants. These are obtained by rooting single node stem cuttings. Such plants can be grown in spaces as small as 10×10 cm and should be able to serve as sources for the rapid development of normal sized plants. Some yam species can be serially multiplied rapidly from rosettes or from young cuttings.

Thus, the techniques mentioned, while not perfected, should make possible a reduction in the maintenance aspects of a germplasm collection.

Cassava

Commercially, cassava is always propagated from stem cuttings, since neither the fibrous nor the tuberous roots possess the capacity to develop

371

new buds. When the above-ground parts of the plant are cut away, resprouting of the plant normally occurs, but such sprouts originate from stem tissue left in the ground, and not from the roots. Furthermore, when roots are removed from the soil they deteriorate quickly and after a week are not even fit for food purposes.

Often new cassava plantings are made directly from freshly cut sticks of old plantings. However, practical needs make some storage of propagating material necessary. Thus, sticks are normally tied in bundles until they can be used and are often left neglected in corners in the market place, or in the fields exposed to full sun. The maltreatment that such cuttings receive is ample evidence of their resistance.

In some areas cassava is planted at a definite season and harvested at about 10 months of age. In such cases cassava branches are bundled and stored, sometimes upside down, in shady, ventilated places for up to eight weeks. The limit of this common storage is about 10 weeks, but exact trials have not been made to estimate the maximum storage life. Branches that are harvested during very rainy seasons are said to have a much reduced storage life.

In areas bordering the temperate zone where cassava is grown, including Northern Argentina, propagating stakes are maintained for even longer periods of time during the winter season. Controlled experiments suggest that the most appropriate storage is in the field, with stakes placed in a horizontal position, completely covered with soil, and protected from excess moisture by a straw shelter.

Tropical plants usually respond unfavorably to the low temperatures necessary for such storage. Nevertheless, given that cassava sticks can already be stored for two months under uncontrolled conditions, it should not be difficult to improve the techniques to permit storage for one year or more. The variables requiring testing include humidity and moisture control, methods of avoiding fungi, and chemical inhibition of sprouting.

A thorough search of the literature has failed to reveal special chemical treatments for increasing the storage life of cassava.

In place of more adequate storage techniques, large collections of germplasm can be maintained in relatively small spaces by confining the plants to pots. Even normal planting sticks can be grown in this fashion. Periodic prunings would always be necessary, but the collection need not be transplanted more than once each year.

Sweet potato

A post-harvest curing period is highly desirable in the sweet potato. In the curing chamber the roots are exposed to temperatures of about 31 °C and humidities between 80 and 90 per cent for about five days. During the curing period high humidity reduces weight loss and drying of cut surfaces. High temperatures hasten the physiological processes of curing. In several of the outermost layers of cells near wounds the process of suberization or cork formation occurs. In addition, certain of the parenchymatous cells somewhat deeper than the suberized layer are modified to form a new meristem. Through the multiplication of these cells a new cork layer is formed, and thus the wound can be considered cured.

After curing, the tuberous roots are cooled rapidly to about 14 °C. Lower temperatures lead to increased decay while higher temperatures permit too much sprouting. The optimum humidity for long-term storage is high, about 85–90 per cent. In the temperate zone sweet potatoes are ordinarily not stored for more than about six months. The changes that occur during storage are well known. They include loss of moisture, loss of dry weight, and expression of cryptic virus diseases. Starch is lost, which is converted to sugar, and used up by respiration. In addition protopectin increases and soluble pectin decreases during storage.

No attempts have been made to test for the maximum storage life of the cured roots. Present storage systems have been studied in detail and are quite adequate for the purposes for which they were designed. However in order to maintain a germplasm collection of sweet potato, storage times of less than one year would not be of much value.

Studies have been made of the influence of growth-regulating substances such as thiourea and various hormonal treatments on the production of more uniform sprouting. The respiratory inhibitor maleic hydrazide applied to the foliage of the plants in the field, to the harvested tuberous roots, or by direct injection into the roots can also influence sprout production. Recently, dimethyl sulphoxide has been found to stimulate sprout production.

Thus, the regulatory effects of certain chemicals, while studied chiefly for their potentialities of increasing sprout production on over-wintered sweet potatoes in the temperate zone, suggest also the possibility of modifying storage life and subsequent sprouting. In this respect further studies are necessary.

373

Conservation and storage

A combination of known storage techniques and chemical treatments may very well permit the storage of sweet potato roots for a full year without loss of viability. Studies have not been designed to test for the possibility of longer storage, but new chemicals and new techniques suggest that much remains to be done in increasing the maximum possible storage life.

Sweet potatoes cannot be stored in the soil for long periods of time since they are quite susceptible to insect and disease attack. On the other hand, the lack of seasonal dormancy, the rapid germination of cuttings, and their facility for growth in small areas points to the possibility of maintaining large collections of sweet potatoes in a state of immaturity in which cuttings would be grown in small containers (minimum 10×10 cm spacing) and re-propagated periodically from small cuttings. A system such as this might make possible the preservation of a collection of 1000 varieties in a space of 50 to 60 m^2.

Aroids

Taro is the paddy form and dasheen is the dryland form of *Colocasia esculenta* (L.) Schott. The species of the closely related genus *Xanthosoma* are known as tanniers. The former first came from the Old World tropics, particularly China and South-East Asia, whereas the latter are species of South American origin.

A variety of different plant parts are used for propagation of aroids. With taros the preferred material is a cutting from the top of the corm, which is planted within a few days after harvest. Thus, harvesting and replanting are closely co-ordinated activities. Taros may also be re-planted from small corms, and although these are not so preferable they have the advantage that they can be stored for several months until ready for use.

The dasheens are propagated from small corms saved from the previous harvest.

The tanniers are normally multiplied from cuttings from the large central corm, which is usually tougher and more fibrous than the side corms which develop from it, and is seldom used as food. The section of the corm used must have an axillary bud, or 'eye'. Small corms, however, are equally useful as planting materials.

The literature contains conflicting reports on storage that reflect local practices based on climatic requirements. In strictly tropical areas corms are harvested when they are needed, and are often believed

374

to have little storage potential. This may be true of some varieties, but has not been rigorously tested. In most regions some type of short-term storage is desirable. Under conditions of common storage at ambient tropical temperatures resprouting occurs very readily either from the original shoot (top of corm) or, if this is destroyed or removed, from the axillary buds.

In cooler climates it is necessary to store propagating material until needed for new planting. Very little study has been made of storage conditions, but low temperature storage (10 °C) seems to be indicated for dasheens. Ventilation of the stored material is very important to reduce storage rots. Small corms for seed purposes have been stored successfully for up to six months.

In the few cases where germplasm collections of aroids have been brought together the labor of maintenance has not proved excessive, and no attempt has been made to simplify holding procedures. In Hawaii plants are grown in plastic mulch which reduces the weed control necessary. In the tropics replanting of aroids is not necessary more than once a year and a living collection does not therefore need much attention.

On the other hand, the possibilities of storage of corms have never been thoroughly investigated. Some problems might be anticipated with the taros that do not store well. Although this is a field that is not being worked on at present, it is quite probable that rapid progress could be made.

Potatoes

The tubers of potatoes are modified stem structures which make the plant perennial in nature, although herbaceous in habit. The bulk of the world's potatoes are grown in the temperate zone and the plant originated in high, temperate altitudes in the tropics, where it is still of great importance. In addition, potatoes are gradually finding their place in the lowland tropics where they should some day play an important role.

Potatoes may be harvested at several different times. Early-harvested potatoes are not mature, and cannot be stored for more than three months at about 5 °C with 90 per cent relative humidity. Tubers harvested near the end of the season are physiologically distinct, and once cured can be stored routinely for five to seven months.

The temperature of storage depends in part on planned usage. For table use potatoes are stored at 4–5 °C at 90 per cent relative humidity,

375

and with air movement to avoid carbon dioxide accumulation. Darkness is also necessary in order to avoid greening. Sprout growth is retarded at such temperatures, and little decay or shrinkage occurs. However, if potatoes are to be used for chips, they are stored at a higher temperature at which sugars do not accumulate. Storage of potatoes at temperatures as low as 1–2 °C results in cold injury, and internal discoloration called mahogany browning.

In place of cold storage other treatments are often used in the storage of potatoes. Maleic hydrazide is sprayed on the foliage of the plants several weeks before harvest. Subsequently, during storage at common or ambient temperatures, sprouting is reduced. Similarly, *iso*-propyl *N*-(3-chlorophenyl)carbamate is applied as a vapor during storage of tubers, or as a solution when the tubers are removed from storage, in order to reduce sprouting. Gamma irradiation at appropriate dosages is also useful as a sprout inhibitor.

Modified-atmosphere storage does not appear to be practical in the case of potato. Reduced oxygen increases the sprouting of potatoes, inhibits periderm formation, and thus wound healing, and at very low concentrations increases rotting and internal decomposition. High carbon dioxide concentrations increase rotting and susceptibility to internal black spot. Prolonged high concentrations inhibit sprouting and can prevent subsequent germination.

The germplasm of potatoes and related *Solanum* species is maintained in the form of seed collections whenever possible in European and American *Solanum* collections. Maintenance in this form is less costly, more flexible, and reduces the problems of virus diseases. Some special treatments are necessary to produce seed, including hand pollination among siblings and stimulation of flowering by short day-lengths.

Only a small proportion of the germplasm of potato is maintained in the vegetative form. Such materials include only the most elite available, including superior clones, outstanding breeding stocks, and disease differentials. The materials that are conserved in this fashion are propagated both from cuttings and from tubers, as necessary, and little is done in search of long-term storage methods.

Conclusions

In the case of all of the roots and tubers discussed, storage practices are concerned with extending the season of the crop or with maintaining the stored material between planting seasons. The possibilities of longer storage have hardly been investigated at all. Nevertheless, the little that is known suggests that storage life could be increased to what must be considered the useful minimum, that is, storage through a full growth cycle. In most cases this means storage for a year or more. Longer periods of storage would be much more difficult, and less rewarding in terms of the reduction in effort of maintaining a collection. Much study must be done to develop trustworthy methods for longer storage periods in all these crops.

In considering the banking of germplasm in the form of vegetative storage organs, it must be realized that such structures are actively metabolic. Their life spans follow closely controlled cycles associated chiefly with the annual cycle which are very difficult to break. Methods of inhibiting the natural processes, if too stringently applied, lead to undesirable side effects. Low temperatures, modified atmospheres, chemical applications, irradiation, and pesticidal treatments may be useful in inhibiting normal processes, and thus extending storage life, but they cannot be expected to extend the storage period indefinitely. Research should aim at what appears to be reasonable, that is to say, the extension of storage life of vegetative storage organs to one or two years, or perhaps a little more in some instances. In the opinion of the present author, however, methods of long-term storage of vegetative organs of tropical root and tuber crops can never compete with the really long-term storage of seeds or even of tissue cultures, as described elsewhere in this book (Chapters 22–7).

30. Genetic reserves

S. K. JAIN

The importance of wild relatives of many domesticated species as a germplasm resource is generally recognized; yet, as noted by Frankel (1970a; 1974), relatively little attention has been paid to the social and biological issues involved in their conservation *in situ*, i.e. the establishment of genetic reserves. In contrast to gene banks comprising seed collections or tissue cultures, genetic wildlife conservation must recognize the long-term needs – the need for continued evolution within natural environments, and the need for the preservation of wild biota as potentially useful resources for the future. Continued evolution in environments now changing at an accelerated rate (review by Detwyler, 1971, among others) will require genetic variation for adaptive changes (Frankel, 1970b), and certain basic ecological conditions for continued reproduction and persistence. Most nature conservation programs are aimed at the level of ecosystems, or multispecies communities; and they may give particular emphasis to some of the many threatened or near-extinct species which are recognized (e.g. *IUCN Red Data Book*). Genetic conservation goes further in recognizing the need for a wide genetic base, and should, therefore, be seriously considered as a part of, as well as in parallel with, nature conservation. Basic issues of survival (i.e. avoidance of extinction), adaptive evolution, maintenance of variation in populations, the long-term stability of population numbers and of species abundance within communities, are all central problems in modern population biology and relevant to nature conservation. Most nature conservation programs, for example, involve the use of from one to several areas for the preservation of a species or community; these areas would largely resemble isolated 'islands' with some natural or man-aided biotic exchange. The theory of island diversity and of the causes of extinction is relevant to the strategy of site selection, size, and management of reserves, and to the planning of long-term goals. The main objectives of this chapter are to examine briefly some of the relevant advances in population biology, and to outline a few basic principles underlying conservation programs for wild biota. Finally I shall argue that crop scientists need to get involved in making wide-ranging efforts in this area, collaborating closely with nature conservation programs.

379

Distribution and abundance of a species

A species may be widespread or endemic in a restricted area and, within its distribution range, it may be sparse throughout, or locally abundant in some areas. Periodic or seasonal cycles of abundance must be borne in mind in describing such features. The categorization of rare, threatened, or near-extinct species in any conservation program refers to the successive losses in population numbers, or changes in the distribution range over time. Rarity is a major but not the sole criterion of being listed as an endangered species. The California Native Plant Society, for instance, has undertaken an extensive survey through matching the present-day distribution of species with the floristic reports of the past. Two key-notes in such studies are (1) distinction between local and global extinction, and (2) uncertainty about the taxonomic boundaries themselves. A species may have several recognizable races or subspecies whose abundance may change at unequal rates. Evolutionary theory, in fact, predicts the presence of a dynamic picture of species identity, distribution, etc. through time. In many crop–weed–wild species complexes, the evolutionary dynamics make it even more difficult to observe the loss of genetic variation relative to the loss of habitats. Most recorded cases of local extinctions are anecdotal and, at best, based on broad habitat descriptions.

On a short-term basis, the most dramatic changes have occurred in the recent decades through man's increasing role in the destruction of habitats of the native vegetation. Often a continuously distributed species may only remain in scattered stands that resemble 'islands'. Some of the best-documented examples of recent extinction (e.g. passenger pigeon, peregrine falcon) indeed illustrate this process (Allen, 1942; Schorger, 1955; Hickey, 1969). Clearly the issues of survival and extinction need to be examined with the help of a theory of populations and communities living in 'islands'. Reduced numbers and modes of inter-island migration are primary features. In the case of certain endemics or species with only a few populations, there are now available detailed population ecology studies on numbers and reproductive processes. Often local population numbers as low as 100 cause serious concern to conservation programs in large birds and mammals; with numbers of 1000 or above, these species are considered safe. In the case of a rare orchid or forest tree species, the size of these 'island areas' may be of greater concern than the actual numbers due to perhaps low density, specialized niches, and/or longevity causing uncertainties about life tables and seed dynamics. For the wild relatives of crops, such as wild *Gossypium*, *Coffea*, or *Pistacia*, it might be desirable to ensure that both numbers and areas are large since the emphasis would be on retaining most of the genetic variation. Thus, many variables of niche breadth, reproductive rates and generation length are equally relevant. The role of genetic variation in assured survival is intuitively considered to be important. In order to understand the relationships among these variables, I shall examine some models of island diversity.

Diversity in islands

The problems of species diversity (number of species living in an island) and genetic variation within individual species are closely inter-related and bear directly on conservation theory. MacArthur & Wilson (1967), in their book *The Theory of Island Biogeography*, provide an elegant summary of this field as follows: Species diversity increases proportionally to the island size and the rates of immigration but the equilibrium composition depends on the relative rates of successful colonization versus extinction, the outcome of competitive niche evolution, and the physical environmental conditions. Rates of colonization in turn depend on the relative rates of birth (b) and death (d) (life tables), dispersal ability, and evolutionary changes in these parameters. For example, b/d ratio versus $[(b-d)/b]$ and the role of carrying capacity (K) determine the relative rates of extinction. Weedy species may rely largely on wide dispersal powers, or short life cycle, and high reproductive rates to take advantage of the high b/d ratios in opportunistic ways. Large K may result from lowered mortality and plastic responses to density. Many wild species are much less opportunistic and lack weedy characteristics; their survival in 'island' areas depends on the stability of habitat, and small, but non-negative r ($= b-d$). For constant population size, $r = 0$ (zero population growth), the time of extinction (T) depends on carrying capacity K and b/d ratio such that a small increase in b/d ratio, when d is density-dependent, raises the value of T dramatically. Hence, relatively small populations can persist with high positive r ($b > d$), or given b/d close to 1, a high K is required. Small changes in r and K can therefore determine the relative success of a conservation program.

The effect of dispersal among islands was also analyzed in detail. The stepping-stone model of dispersal is of great significance in the biotic exchange between the neighboring islands. This is relevant to the possible occurrence of many 'island' habitats within the confines of a large reserve, or many small reserves in a region. As pointed out by MacArthur & Wilson (1967), the emigration–immigration curves are known only for a few species (see Wright, 1962, for some examples of forest-tree species). Without this knowledge the models of island colonization and species diversity remain hypothetical.

In his book *Stability and Complexity in Model Ecosystems*, May (1973) re-examined the current dogmas on the conditions for high diversity and high stability. Stability was defined as continued existence of all member populations within limited size ranges. A surprise result was that increased total number of species interactions need not lead to greater stability. Random fluctuations in environments, if small, did not always change the patterns of diversity, but, if significantly large, tended to reduce the species diversity. The usual problem with such results is their empirical but uncertain nature due to the concept of 'randomness' itself in environmental changes.

A final point should be made about environmental changes and species turnover rates. A comparative study of tropical and temperate islands showed a higher rate of turnover for the latter group (Diamond, 1972). Here, the extinction rates were the same but the temperate island had lower immigration. On the other hand, a greater extinction rate needs to be balanced by allowing for greater immigration rates. In other words, reserves in temperate regions would have to be more numerous and closer together.

The genetic variation within a species subdivided into 'islands' (= isolates, demes) was extensively investigated by Wright (1969, and earlier) in relation to local versus global evolutionary changes. The role of subdivision can be briefly illustrated by two properties of genetic variation: the number of alleles in the total population (k_e, a direct measure of the degree of genetic polymorphism) and the amount of differentiation among subpopulations (measured by V_p, the variance of allelic frequencies p_i at a diallelic locus). Maruyama (1970) analyzed the effect of subdivision on k_e with a model of n colonies distributed in a circle, each of effective size N_e, having no selection, and exchanging genes through pollen and/or seed only with the two neighbors at a rate M_1. Table 30.1 gives a few of his results. Note that a total of 100 individuals ($N_e \times n$) can maintain a varying number of alleles at equilibrium depending upon subdivision pattern; high mutation rate, but *lower* M_1 (less gene flow) allow greater genetic variation in the total population where subdivision provides for greater allelic diversity.

382

Table 30.1. *Effective number of alleles* (k_e) *(after Maruyama, 1970)*

n	$2N_e$	M_1	u	k_e
10	20	0.1	0.01	6.2
50	4	0.1	0.0005	2.8
100	2	0.1	0.0005	5.1
40	5	0.05	0.01	15.8

n = number of subpopulations.
N_e = effective population number.
M_1 = migration rate between neighborhoods (stepping-stone model).
u = mutation rate.

Table 30.2. *Differentiation among subpopulations as measured by* V_p *(= Variance of allelic frequencies) under stepping-stone* (M_1) *and island* (M_∞) *models of migration*

M_∞	$M_1 (N_e = 10)$				$M_1 (N_e = 100)$			
	0.01	0.05	0.10	0.50	0.01	0.05	0.10	0.50
$p = 0.10$								
0.01	0.054	0.040	0.033	0.020	0.012	0.006	0.005	0.002
0.05	0.028	0.022	0.018	0.011	0.004	0.003	0.002	0.001
0.10	0.018	0.015	0.013	0.008	0.002	0.002	0.001	0.001
0.50	0.005	0.006	0.006	0.005	0.001	0.001	0.001	0.000
$p = 0.50$								
0.01	0.151	0.112	0.093	0.055	0.032	0.018	0.013	0.006
0.05	0.079	0.061	0.051	0.030	0.010	0.007	0.006	0.003
0.10	0.051	0.042	0.037	0.023	0.006	0.005	0.004	0.002
0.50	0.016	0.016	0.016	0.014	0.002	0.002	0.002	0.001

However, local selection may minimize this through the fixation of different alleles within subpopulations, in which case migration would counteract to produce greater allelic diversity. It is the *balance* of various forces of evolution that determines the pattern of diversity, and not any individual factor alone. With no selection, drift tends to produce greater genetic differentiation (V_p), whereas migration counteracts drift. Further, consider M_1 and M_∞ as the relative rates of migration under the stepping-stone model (only neighboring colonies exchange genes) and the island model (all colonies exchange genes at the same rate regardless of relative distances) respectively. These represent short-range and long-range gene flow. The values of V_p (Table 30.2) for some numerical cases show that a small increase in M_∞, as expected, has proportionally larger effect than changes in M_1 in reducing genetic differentiation.

Conservation and storage

In closing this section, we should note that island models have provided a great deal of insight into the dynamics of communities and the evolutionary origin of races and species on islands. Main & Yadav (1971) pointed out that islands could serve as outdoor laboratories for studies on population numbers and variation. From their own studies on marsupials in Western Australia's off-shore islands, they demonstrated the use of the concept of minimum viable size for the persistence of a species on an island. This is indirectly related to MacArthur & Wilson's discussion of K and the life table properties as noted above. Next, let us consider the causes of extinction in terms of size and variation in populations.

Causes and avoidance of extinction

The two basic causes of extinction are widely recognized: (1) failure to adapt to changing environments, and (2) overspecialization, resulting, again, in a failure to re-adapt quickly enough to the deterioration of resources due to overexploitation. The proximate result in all cases is negative r (birth rate smaller than death rate) for a prolonged period of time. Adaptive failures could result from one or more of the following reasons: (1) lack of genetic variation to cope with the changing environments, especially when a species faces rather rapid changes, (2) time response delay such as that due to life historical factors and (3) replacement by a related newly-evolved competitor species (pseudo-extinction). Lidicker (1966) studied in detail a declining population of the house mouse; several predator–prey systems have been analyzed in laboratory studies, and detailed studies have been published for classical examples of recent extinctions (e.g. peregrine falcon, passenger pigeon). In almost all cases, not any one factor but several in combination caused extinction. The 'last straw that broke the camel's back' in many cases might have been man's predatory or polluter role. The *IUCN Red Data Book* suggests (Fisher, 1971) that man has increased extinction rates at least four-fold, essentially through the large reduction in the number of habitats, in population numbers within habitats, and presumably through the loss of genetic variation. The three basic concepts are: *minimum viable size*, the number within a local population that allows reproduction with $b \geqslant d$ (or $r \geqslant 0$); *minimum genetic variation*, the allelic and genotypic diversity to allow for adaptation through natural selection, and *minimum time delay* in response (or conversely, slower change in environment since environmental descriptions are organism-

384

specific). In a bisexual, successful colonizer species, two or more individuals may rapidly build up a large population, but eventually this might fail due to the lack of genetic variation (founder effect). Autogamous colonizers are postulated to be successful through even single isolated founders. Genetic systems such as those involving mating-barriers or incompatibility require a larger group of founders. Main & Yadav (1971) showed for certain marsupial species that 200 to 300 might be the minimum viable size (MVS) in Western Australia's off-shore islands. In many rare, slowly reproducing species, such as the California condor, large habitat requirements resulting in low population numbers can barely keep up with the mortality rates. Under widely varying rates of survival in different age groups, the problem of defining MVS is somewhat more difficult. Under randomly varying conditions r must be somewhat greater than zero. Mountford (1971) showed that the non-reproductive reserve of mature individuals for optimal survival depends on both maximum r and minimum probability of extinction.

The strategies of survival are a focal issue in ecological theory. Genetic aspects have not been explored in any detail. Environmental changes may involve rapid climatic shifts; slow, secular changes in climate; man-related disturbances; ecological changes in biotic and abiotic niche components; and random (unexplained) changes. Most dramatic adaptive changes are recorded in the studies of stress environments, e.g. increase in melanic genes in Lepidoptera of smoke-polluted forests, disease resistance, pesticide resistance, or in controlled physiological experiments (temperature extremes, drought, toxic substances). Adaptive changes may involve individual homeostasis or populational buffering such as to elicit no genetic change, or spread of favorable major mutants, and subsequent adjustments in the co-adaptive properties. Levins (1968) has discussed in detail the outcome of such responses in terms of the genetic versus non-genetic components of fitnesses and various characteristics of environmental heterogeneity in space and time. The resulting adaptive changes may lead to monomorphism, localized or widespread polymorphism, or mixed strategies, with the result that *a priori* predictions of the *role of present-day genetic variation* are difficult. It is uncertain whether few individuals (all-purpose genotypes) can replace the strategy of heterotic and highly polymorphic genetic systems. In general, it is intuitively safe to consider a certain minimum variation to be essential for most adaptive changes. Chance and rapidly deteriorating environments may exceed tolerance limits, but generally speaking, more variable populations have a better chance for

385

survival. At the molecular level, Hochachka & Somero (1973) have provided elegant arguments for the following: major adaptive changes involve physiological, morphological and/or biochemical changes; coarse adaptations through qualitative or quantitative changes are readily acquired in the presence of genetic variation followed by some fine level regulatory changes. Most species retain flexibility (coarse tuning; near-best and not 'the best of all possible worlds') in line with the optimal evolutionary and ecological strategy arguments.

Among the basic conditions for maintaining genetic variation in populations are mutation, migration, heterotic selection, frequency dependency, and minimum drift losses in particular. Population genetic theory provides a large array of models to evaluate these postulates. Effective population size, defined in terms of the sampling variance of gene frequencies or inbreeding due to finite size, measures some very basic relationships between population size and genetic variation. Minimum viable size should therefore also take into account the maintenance of variation. Population fitness, evolving under selection, may in turn relate to r and K. In recent years, discoveries of a high proportion of multi-allelic polymorphic loci in natural populations have renewed discussions of these models. Ironically, finding large amounts of variation has raised questions on the widespread role of polymorphisms in adaptive responses. However, note that Kimura, Crow and others (labeled by some as 'neutralists', non-Darwinists, etc.) do not claim that *all* this variation is neutral, particularly under changing environments. All serious students of evolution assume the role of variation in evolutionary change, but how much variation is desired for avoiding extinction is not known and perhaps varies greatly among different species, genera, or higher groups.

A brief review of some protein variation studies brings home precisely this point. Table 30.3 gives a few of the summary results in terms of mean heterozygosis and allelic diversity, as measures of variation per population, summed over loci and populations. Three points to note are: (1) Wide range of values, as some species are monomorphic in large regions and highly polymorphic in others. The horseshoe crab is a 'living fossil' but showed a significant amount of variation. Another ecological analog of a fossil species (killer-clam, *Tridacna*) also shows variation (Ayala *et al.*, 1973). An inbreeding colonizer land snail (*Rumina* sp.) showed no variation, whereas another land-snail species (*Helix* sp.) shows a great amount of variation (Selander & Kaufman, 1973). Allozymic variation in populations of large mobile animals

386

Table 30.3. *Summary of some recent estimates of electrophoretic protein variation*

	Percent heterozygosity		Mean number of alleles per locus	Source
	Mean	Range		
Astayanax, fish				
Cave population	3.6	0–7.7	1.22	Avise & Selander (1972)
Surface populations	11.2	7.7–13.8	2.13	—
Uta, lizard	4.8	0–10.0	—	Selander & Kaufman (1973)
Rumina, snail	0	—	1.00*	Selander & Kaufman (1973)
Limulus, horseshoe crab	9.7	—	—	Selander & Kaufman (1973)
Drosophila willistoni				
Continental	18.4	17.2–21.0	2.40	Ayala, Powell & Dobzhansky (1971)
Island	16.2	14.2–18.4	2.50	—
Oenothera biennis	26.0	—	1.37	Levin, Howland & Steiner (1972)
Clarkia lingulata	8.5	6.0–11.0	2.63	Gottlieb (1974)
C. biloba	13.5	13.0–14.0	2.38	Gottlieb (1974)
Stephnomaria exigua				
subsp. *coronoria*	—	—	2.4	Gottlieb (1973a)
S. 'malheurensis'	—	—	1.2	Gottlieb (1973a)

* All ten populations monomorphic.

(vertebrates) was found to be higher than in those of small, relatively immobile invertebrates (Selander & Kaufman, 1973). Gottlieb (1974) attempted a test to compare variation in plant versus animal populations. However, there seem to be too few data available for such comparisons. (2) *Stephanomaria 'malheurensis'* is postulated to have originated from *S. exigua* ssp. *coronaria* through sympatric speciation (Gottlieb, 1973*a*) and *Clarkia franciscana* from *C. rubicunda* (Gottlieb, 1973*b*) and *C. lingulata* from *C. biloba*. Do we expect the derived species to have lost much of their variation but to recover new genetic diversity subsequently? The answers are not clear. Neither founder effects, nor any other historical aspects can be too well documented. Thus, neither rapidly evolving groups, nor the so-called living fossils seem to give any consistent pattern in the amount of genetic variation. (3) Furthermore, population size and polymorphism are not correlated in any simple way. Island and continental populations of *Drosophila willistoni* show about the same amount of allozyme variation, but reduced chromosomal polymorphism in islands. Avise & Selander (1972) found cave-dwelling fishes to have less genetic variation than surface populations and explained this on the basis of small population size and the cave environment. Rates of interpopulation gene flow are usually not known, but the genic heterozygosity may increase under gene flow whereas allelic diversity may decrease (see Maruyama's model above). Wheeler & Selander (1972) analyzed protein and morphological variation in *Mus musculus* populations of Hawaiian Islands in which they found, in spite of presumed founder effects, an even greater amount of polymorphism than in the Danish or North American populations. Gottlieb's (1974) data on *Stephanomaria exigua* ssp. *carotifera* show no correlation between genetic variation and as much as 40-fold difference in population size in the range of 250 to 10000. On the other hand, Jain (unpublished data) observed polymorphism in rose clover colonies to be related to size, in the range of one to 75; populations larger than 75 are all polymorphic within the same area.

The role of genetic variation in terms of population regulation is perhaps the most direct clue to certain stable population cycles. D. Chitty, C. Krebs and others have provided examples of qualitative changes related to variation in population density. Now it is more generally recognized that genotypic fitness values are likely to be frequency and density dependent (it follows from niche analysis and co-evolutionary properties). In this connection, a major area of interest is the study of life history patterns, variation in ecological parameters,

388

Table 30.4. *Fluctuations in population size as measured by LR ($= \ln N_{max}/ N_{min}$) and var R ($=$ variance of net reproductive rates). (After Reddingius & den Boer, 1970)*

Model	*LR*	*var R*
NM, SP	5.32	0.42
DIM, SP	2.37	0.06
DIM, DP	3.61	0.07
DDM, DP	1.76	0.08
NM, HP	2.11	0.11
DIM, HP	1.52	0.05
DDM, HP	1.77	0.04

NM, no migration among subpopulations; DIM, density-independent migration; DDM, density-dependent migration; SP, subpopulations with same parameters; DP, subpopulations with different parameters; HP, heterogeneity among groups of subpopulations.

and game theoretic arguments on the evolution of optimal adaptive strategies. Are polymorphic populations more successful? more likely to stabilize numbers? more productive? or better in colonizing properties? A large body of literature, mainly laboratory experiments, suggests that answers in most cases are definitely yes, but not invariably so. With better hindsight now, we can see that the theory does not predict any distinct pattern of polymorphisms to be directly related only to population size or breeding system alone, etc.

A recent study by Reddingius & den Boer (1970) may be cited as an excellent demonstration of strategy-building. Using computer simulation they showed that a species can stabilize population numbers by the so-called process of 'spreading-the-risk'. Heterogeneity of age structure, distribution in habitat, dispersal, all led to stable numbers, a criterion of long-term survival. Table 30.4 gives a few of their numerical results. Both *LR* and *var R* show the stabilizing effects convincingly. (See Birch, 1970, for evidence on the role of heterogeneity in survival of *Cactoblastis*.) Levins (1970) has also examined the problem of extinction in terms of interactions of selection among individuals, among population (groups), selective migration, and a balance between the processes of local extinction and recolonization. Here, again, all the best evidence comes from laboratory experiments (e.g., predator–prey or host–parasite systems). Local extinction rates in nature are only indirectly known or at best in a few cases only. We shall need data on population sizes and variation over space and time in order to get better evaluation of the kinds of balances and checks involved.

Finally, the regulation of population numbers and evolution in a multispecies community might involve a genetic feedback mechanism so as to shift selective responses between intra- and interspecies competition. Both numbers and genetic variation come into close interaction in determining the oscillatory patterns and long-term survival of such systems.

From this review of extinction theory, at the risk of some over-generalization, we may conclude that (1) adaptive responses to changing environments are largely based on genetic variation and natural selection; (2) such evolution often provides for continued evolution by retaining flexibility; (3) hence, the 'best' strategies are not necessarily 'perfect' in terms of the best possible monotypic populations; (4) numbers and variation jointly influence all features of the survival strategies; (5) spatial heterogeneity and population subdivision allow optimum strategies to evolve in terms of population differences; (6) temporal variations in the environment within certain limits may be coped with by individual as well as by genetic homeostasis. In summary, (7) the chances of extinction are higher in species with fewer sub-populations, lower dispersal rate, low N_e or high variance in N_e, low genetic variation, long but simple life cycles, and habitat specialization; however, (8) theoretical complexity of interacting parameters, and general lack of wide-ranging evidence from nature, at the moment allow only an intuitive and general view of the problem of survival (or extinction). Since each of these conclusions is derived primarily from theoretical work, but only meager empirical evidence from the real world, population biology can now look forward to polemic arguments (hopefully based on testable hypotheses) and to rapid developments in research.

Strategy of genetic reserves

It is clear, as emphasized by Frankel (1974), that 'wild species, increasingly endangered by loss of habitats, will depend on organized protection for their survival. On a long-term basis this is feasible only within natural communities in a state of continuing evolution.' The goals of genetic conservation encompass a wide range of wild biota, including many of direct economic importance in forage or forest production, for land reclamation, or as sources of food; the wild relatives of domesticated plants which are important gene pools for plant improvement; and many species of no immediate economic value yet of great social concern – not only as potential economic resources for the future, but for their ecological, social and aesthetic value as highly significant parts of the human environment. Strategies must be developed which encompass the essential elements of genetic conservation. They must consider (1) the appropriate level of biological organization – whether the species, the community, or the ecosystem; (2) the size and structure of the communities and populations to be protected; (3) the manner in which information on the dynamics of the system can be monitored. These considerations are essential for the planning and management of reserves if they are to be maintained as long-term resources for the welfare of future generations.

Many nature conservation programs now under way are aimed at individual endangered species of special aesthetic or scientific interest (e.g., Save-the-Redwoods League, Save-the-Poppies Foundation, etc. in California). Economically valuable species have also drawn wide attention (e.g. of the International Whaling Commission). The latest Conservation Directory of the United States National Wildlife Federation lists hundreds of organizations which generate publicity and collect funds for small nature reserves designed to protect individual species. There are 'success stories' (e.g. about the trumpeter swan, the Isle Royale wolf) that tell us about the re-establishment of sufficiently large populations by breeding and recruitment, by habitat management and the use of life table information. In fact, such individual case histories presently provide the best available ecological data on population survival rates. But to the best of my knowledge, the genetic and adaptational aspects have not been studied in any of these programs. Hence we have virtually no data on the role of genetic variation in these conservation efforts.

There is, however, a good deal of information on the effects of isolation and selection which have a bearing on the biology of species and populations under conditions as they may occur in reserves (see Chapter 3). The well-known examples of industrial melanism in various moth species in smoke-polluted forests, or the evolution of metal-tolerance in grasses in mine areas, suggest the rapid rates of micro-evolution under new environments. Likewise, the observations of occasional survivors during disease epidemics (e.g. Bermuda cider) are attributed to the presence of rare favorable genotypes. Recent surveys of genetic variation showed that most species have stores of allozyme variation at loci ranging from as low as 5–10 per cent to 60–70 per cent of the total number scored. Quantitative genetic experiments likewise reveal large capacities for response to selection. Molecular evolution will probably soon provide better clues to the adaptive role of variation in response to specific environmental challenges. We now have the theoretical insights into the mechanisms that generate and maintain variation in populations. While predictions about survival are subject to future environments a species may have to face, research on the genetics of endangered species and descriptions of environmental heterogeneity should be initiated for a good 'hindsight'. Relatives of domestic species should be emphasized. Galinat (1972) recently proposed that the Guatemalan teosinte is more primitive than the Mexican teosinte populations and, therefore, needs to be preserved, and Wilkes (1972) suggested that continued introgression between Mexican maize and teosinte should be facilitated, for which perhaps 100–200 strips (each 12×3 miles) should be set aside from modern agricultural developments. Such experiments are of value if adequately monitored throughout their currency. Otherwise one may well question what advantages they have to offer over the deliberately conducted breeding processes normally conducted by plant breeders, often with the inclusion of wild species.

Conservation programs aimed at the community or ecosystem level have gained wide support in recent years. 'The preservation of representative and adequate samples of all significant ecosystems' (UNESCO Conference on the Use and Conservation of the Biosphere, Paris, 1968) has been advocated on numerous occasions. The role of national parks, nature reserves, and other kinds of protected areas have been reviewed (see e.g. Stamp, 1969; Day, 1971; Usher, 1973; and numerous articles in the journal *Biological Conservation*). Many countries have plans to allocate as much as 1–2 per cent of their total land area for conservation, although sizes of individual parks or reserves may vary from very small (a few acres) to several thousands of square miles. Of course, area alone is not as important as community biology, surroundings, physical factors, or control of man's activities. Both faunal and floristic species diversity show a great variety of dynamic patterns in relation to environmental stability and gradient, successional stage, and management (see e.g. Duffey & Watt, 1971). In view of socio-economic as well as biological arguments in favor of ecosystem conservation, it would appear mandatory to integrate the conservation of selected species groups into the programs of ecosystem reserves. Special attention may still be given to particular species for aesthetic, scientific, or other reasons.

It is relevant to consider the ecological background in the past evolution of domesticated plants and animals. Our primary crops and their ancestors were not parts of climax vegetations but were pre-adapted to disturbed open habitats and both seed culture and vege-culture species probably evolved in areas where large food reserves facilitated survival under strongly seasonal environments (Hawkes, 1969). Thus, although many present-day crops may have originated in only a few localized regions of the world, there is ample evidence from the ethnobotanical literature that many other species have potential agricultural value. Moreover, the patterns set by 'weediness' and 'seasonality requirements' do not apply any longer under conditions of modern agriculture. Agricultural scientists, therefore, must be concerned not only with the known areas of crop diversity, but with the components of all ecosystems, as sources of potential economic plants.

The number, size and spatial distribution of genetic reserves constitute the basic and most difficult decisions in strategy-building. On theoretical grounds, one might be tempted to conclude that 20 to 50 populations, each of a size in the range of 100 to 1000 individuals, could provide a minimum viable size, minimum adaptive variation, and spatial multi-niche flexibility. On the other hand, many rare endemics, which have appeared on various Endangered Species lists, perhaps cannot thrive on such a large scale. There are no clear guidelines from real world observations about the role of population subdivision.

Management needs may be minimal for most species but relatively large in others. Policies need to be open-ended, so as to be responsive to the yearly or periodic observations, and to social factors. Nature conservation reports and a recent book by Usher (1973), among others, provide more details along these lines. The recent *British Ecological Society Symposium* (Duffey & Watt, 1971) provided a summary of views and approaches to the scientific management of communities. However, genetic reserves and evolutionary arguments were not adequately considered. For example, Berry (1971) assumes that most populations rapidly adapt to new environments, variation is rapidly acquired in small, recently founded colonies, and environmental risks are rarely great, so that one gets the impression that on the whole chances of extinction are small. His arguments may be more justified for rapidly reproducing small mammals, but they certainly warrant re-examination. Hooper (1971) considered the question of genetic deterioration due to small population size and inbreeding. He argued that, given agreement on the time scale of concern and on the tolerance limits for losses due to random drift and extinction, minimum population numbers could be specified for an array of small reserves. However, this will be largely academic unless some data are gathered for deriving probability distributions of such losses.

I hope that population biology models and arguments presented here will at least provide a basis for further discussions of alternative strategies for the genetic conservation of wild biota. We must urgently get involved both as crop scientists and as naturalists in defining the issues clearly, in developing inventories of endangered genetic resources (species, communities and areas to be protected from extinction), and in making every possible scientific contribution to the nature conservation programs. The crop genetic resources are ignored by most conservationists (e.g. see Curry-Landahl, 1972; Usher, 1973). The only article in *Biological Conservation* on this topic was that of Frankel (1970*a*). It can be argued that an economic argument can further re-inforce the public view on the need for nature conservation and it is my own experience in connection with my work on *Avena* and *Limnanthes* that various agricultural and educational agencies in California have shown great interest in collaboration. Surveys of germplasm resources of the wild relatives of crop plants should be discussed in the broader context of environmental issues. The recent efforts of Gómez-Campo (1972) for the preservation of many west Mediterranean crucifers could provide a good model. My colleague Professor C. M. Rick has found a large series of valuable genes in the wild Galápagos tomato species that need protection and have attracted the attention of the local conservationists. We must get involved in these issues for the sake of our own survival.

I am very grateful to Sir Otto Frankel for critically reading the manuscript and for many useful discussions.

References

Allen, G. M. (1942). Extinct and vanishing mammals of the western hemisphere. *Am. Comm. Int. Wildlife Protection, Special Publ.*, no. **11**, 1–620.

Avise, J. C. & Selander, R. K. (1972). Evolutionary genetics of cave-dwelling fishes of the genus *Astayanax*. *Evolution*, **26**, 1–19.

Ayala, F. J., Hedgecock, D., Zumwalt, G. S. & Valentine, J. W. (1973). Genetic variation in *Tridacna maxima*, an ecological analog of some unsuccessful evolutionary lineages. *Evolution*, **27**, 177–91.

Ayala, F. J., Powell, J. R. & Dobzhansky, Th. (1971). Polymorphisms in continental and island populations of *Drosophila willistoni*. *Proc. Nat. Acad. Sci. USA*, **68**, 2480–3.

Berry, R. J. (1971). Conservation and the genetical constitution of populations. *Brit. Ecol. Soc. Symp.*, **11**, 177–206.

Birch, L. C. (1970). The role of environmental heterogeneity and genetical heterogeneity in determining distribution and abundance. *Proc. Adv. Study Inst. Dynamics Numbers Popul.*, pp. 109–28. Wageningen.

Curry-Landahl, K. (1972). *Conservation for Survival.* Morrow, New York.

Day, M. F. (1971). The role of national parks and reserves. In *Conservation* (eds A. B. Costin and H. J. Frith), pp. 190–213. Penguin Books, Australia.

Detwyler, T. R. (ed.) (1971). *Man's Impact on Environment.* McGraw-Hill, New York.

Diamond, J. M. (1972). Comparison of faunal equilibrium turnover rates on a tropical and a temperate island. *Proc. Nat. Acad. Sci. USA*, **68**, 2742–5.

Duffey, E. & Watt, A. S. (eds) (1971). The Scientific Management of Animal and Plant Communities for Conservation. *Brit. Ecol. Soc. Symp.*, **11**.

Fisher, J. (1971). Wildlife in danger. In *Man's Impact on Environment* (ed. T. R. Detwyler), pp. 625–39. McGraw-Hill, New York.

Frankel, O. H. (1970*a*). Genetic conservation of plants useful to man. *Biol. Conserv.*, **2**, 162–9.

Frankel, O. H. (1970*b*). Variation the essence of life. *Proc. Linn. Soc., NSW*, **95**, 158–69.

Frankel, O. H. (1974). Genetic conservation: our evolutionary responsibility. *Proc. XIII Int. Genet. Cong., Genetics*, **78**, 53–65.

Galinat, W. C. (1972). Preserve Guatemalan teosinte, a relict link in corn's evolution. *Science*, **178**, 323.

Gómez-Campo, C. (1972). Preservation of west Mediterranean members of the cruciferous tribe Brassiceae. *Biol. Conserv.*, **4**, 355–60.

Gottlieb, L. D. (1973*a*). Genetic differentiation, sympatric speciation, and the origin of a diploid species of *Stephanomaria. Am. J. Bot.*, **60**, 545–53.

Gottlieb, L. D. (1973*b*). Enzyme differentiation and phylogeny in *Clarkia franciscana, C. rubicunda* and *C. amoena. Evolution*, **27**, 205–14.

Gottlieb, L. D. (1974). Genetic confirmation of the origin of *Clarkia lingulata. Evolution*, **28**, 244–50.

Hawkes, J. G. (1969). The ecological background of plant domestication. In *The Domestication and Exploitation of Plants and Animals* (eds P. J. Ucko and G. W. Dimbleby), pp. 17–30. Aldine, Atherton, Chicago.

Hickey, J. J. (ed.) (1969). *Peregrine Falcon Populations. Their Biology and Decline.* Univ. of Wisconsin Press, Madison.

Hochachka, P. W. & Somero, G. N. (1973). *Strategies of Biochemical Adaptation.* Saunders, Philadelphia.

Hooper, M. D. (1971). The size and surroundings of nature reserves. *Brit. Ecol. Soc. Symp.*, **11**, 555–62.

Levin, D. A., Howland, G. P. & Steiner, E. (1972). Protein polymorphism and genic heterozygosity in a population of the permanent translocation heterozygote, *Oenothera biennis. Proc. Nat. Acad. Sci. USA*, **69**, 1475–7.

Levins, R. (1968). *Evolution in Changing Environments. Some Theoretical Explorations.* Princeton Univ. Press, Princeton.

Levins, R. (1970). Extinction. In *Some Mathematical Problems in Biology* (ed. M. Gerstenhaber). Amer. Math. Soc., Providence, Rhode Island.

Lidicker, W. Z. (1966). Ecological observations on a feral house mouse population declining to extinction. *Ecol. Monogr.* **36**, 27–50.

MacArthur, R. H. & Wilson, E. O. (1967). *The Theory of Island Biogeography.* Princeton Univ. Press, Princeton.

Main, A. R. & Yadav, M. (1971). Conservation of macropods in reserves in Western Australia. *Biol. Conserv.,* **3**, 123–33.

Maruyama, T. (1970). Effective number of alleles in a subdivided population. *Theor. Pop. Biol.,* **1**, 273–306.

May, R. M. (1973). *Stability and Complexity in Model Ecosystems.* Princeton Univ. Press, Princeton.

Mountford, M. D. (1971). Population survival in a variable environment. *J. Theor. Biol.,* **32**, 75–9.

Reddingius, J. & den Boer, P. J. (1970). Simulation experiments illustrating stabilization of animal numbers by spreading of risk. *Oecologia,* **5**, 240–84.

Schorger, A. W. (1955). *The Passenger Pigeon: Its Natural History and Extinction.* Univ. of Wisconsin Press, Madison.

Selander, R. K. & Kaufman, D. W. (1973). Genic variability and strategies of adaptation in animals. *Proc. natn. Acad. Sci., USA,* **70**, 1875–7.

Stamp, L. D. (1969). *Nature Conservation in Britain.* Collins, London.

Usher, M. B. (1973). *Biological Management and Conservation.* Chapman and Hall, London.

Wheeler, L. L. & Selander, R. K. (1972). I. Genetic variation in populations of the house mouse, *Mus musculus,* in the Hawaiian Islands. *Studies in Genetics VII,* Univ. Texas Publ. no. **7213**, 269–96.

Wilkes, H. G. (1972). Genetic erosion in teosinte. *Plant Genetic Resources Newsletter,* **28**, 3–10.

Wright, J. W. (1962). *Genetics of Forest Tree Improvement.* FAO, Rome.

Wright, S. (1969). *Evolution and Genetics of Populations,* vol. II, *The Theory of Gene Frequencies.* Univ. of Chicago Press, Chicago.

PART V

Documentation and information management

31. Documentation for genetic resources centers

D. J. ROGERS, B. SNOAD & L. SEIDEWITZ

Introduction – definition of documentation

Documentation is the most active function in any Genetic Resources Center because we use information about the plant material which is the basic material of a genetic resources center (called hereafter GRC), rather than the material itself when we wish to communicate. Documentation may be considered the thread from which the whole fabric of GRCs is woven. Whether we know it or not, scientists and other workers spend at least 30 per cent of their time taking care of the data associated with the collections. Since this is true, it becomes imperative that we give this function, documentation, the attention which it deserves. From each data point, we must derive the most benefit for the cost of time and effort expended in that effort. It should also be pointed out that the most expensive part of the whole documentation function is the actual gathering of the data in the first place and also the preparation of that data for any function. By contrast, the computer manipulation of the data is very inexpensive. Documentation of the data in any GRC will have much wider application than data in the normal setting in which scientists have worked in the past, because the data must serve many different functions beyond that which the individual scientist may have conceived himself.

Data may serve primarily for the research and development in the GRC, but the same data may be used in slightly different ways by the administration of the GRC for purposes of reports, planning, and fiscal policy. They may be used for publicity purposes. They may be used by agronomists, plant breeders, geneticists, and biologists in general; by economists, sociologists, anthropologists, and many others whom we have not identified. But since the data in documentation do have these multiple applications, we must be certain that the best possible treatment is given to this fundamental aspect of GRC function.

Documentation has, of course, been done for a very long time, using whatever techniques were extant at a particular time. Clay tablets of the ancient Sumerians were documentation; papyrus was used by the Egyptians, and then sheep skin (parchment), then beaten bark, then paper.

When the printing press was invented, documentation began in earnest. With the invention of the typewriter, more documentation was accomplished. Finally, when the computing machine became available for use, documentation reached a stage where we could accomplish tasks of storage, rapid retrieval and communication of vast quantities of data, never before achieved. However, the use of computing machines by themselves does not accomplish the task – it requires considerable knowledge and skill to make a machine do the bidding of the scientist in the GRC.

Documentation specialists do *not* determine the information to be included in the work of the GRC – that task is the responsibility of the scientists who work in the centers. The documentalist can, however, aid the scientist in determining the best means to handle the data to guarantee that the data can be used to their best advantage. Properly done, documentation can prevent loss of material and can trace the source of accessions which are exchanged.

Usually, we associate (at least in the minds of most computer-oriented people) computer-based information retrieval systems with documentation. We have come to think about information retrieval as a primary function of documentation, and although the retrieval of information is indeed one of the major functions, as stated above, documentation is a much larger function. It includes, in the best sense of the word, management.

We must ask the question then, what is documentation in a GRC? Every item that goes into the GRC, whether it is a seed collection, a living plant, a dried herbarium specimen, a photograph, or a book, requires that some record be kept. That record is documentation. We can, by efficient management techniques, assure ourselves that we will derive the greatest benefit from that record.

The functions of documentation in a GRC

The major functions of a GRC include: (1) exploration, (2) quarantine, (3) conservation, (4) utilization, and (5) documentation. A flow-chart which shows the inter-relation of each of these, and documentation is given in the accompanying Fig. 31.1. To these we should add communication and exchange between all GRCs. After the work or functions listed above are completed within each GRC, it is then possible to accomplish the tasks of communication and exchange through the documentation function.

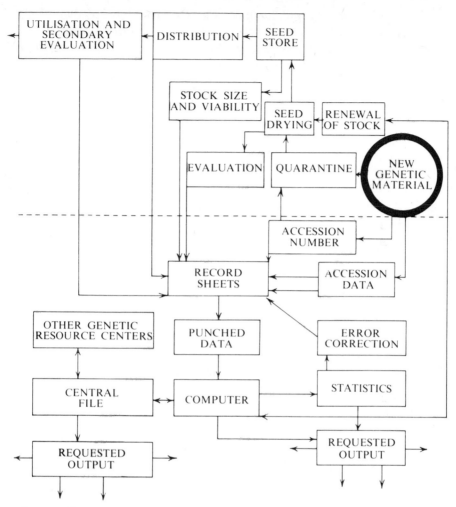

Fig. 31.1. The major functions of Genetic Resources Centers are illustrated. Functions other than documentation are shown above the dotted line, those of documentation are given below. Interconnections are indicated by double-ended arrows. Each of the boxes gives a general indication of function, but each box could have a separate flow-chart of internal functions.

We must remember that there are already a number of GRCs operating in various parts of the world, financed by states, nations, or internationally. These clearly already possess valuable materials and many of them have some type of documentation. Since this is true, we must try to share material and information from these with any new center that will be developed. We may be able to prevent useless repetition if we can have an accurate means of communication and documentation.

It is difficult for the individual scientist, interested in his own crops, to visualize the enormous quantity of data which must be handled in a large GRC. Needless to say, the quantity is enormous. Because of this fact, there is no question but that we use electronic computing machines. Since applications of computers in our endeavors is very recent, we do not yet have all the necessary systems to do everything that we may wish for, although we do have a system available which will accomplish most of the tasks. However, to get the desired results from documentation using computing machines, it requires that there be close co-operation between the computer personnel, the documentation specialist, and the scientific staff. We must establish some minimum standards which we can all accept.

Following is a preliminary list of information types. We will designate the information under each type as *descriptors*. We have not by any means exhausted the many possible descriptors under any of the classificatory types.

(1) *'Finding' information*
   Accession numbers
   Collection information (collector, collector number, date)
   Nomenclature information (scientific and common)
   Origin (country, state or province, and precise locality)
   Storage information (location, date placed in storage, etc.)
   Distribution information

(2) *Environmental information*
   Altitude
   Climatic
   Ecological

(3) *Organismic information*
   Morphological
   Genetic
   Physiological
   Biochemical
   Environmental damage (cold, drought, wind or other susceptibility)
   Pest and disease information (bacterial, fungal, viral, insect, nematode)
   Rejuvenation (viability of seeds, time in storage)

(4) *Use information* (including breeding, genetics, performance tests, etc.)

(5) *Ethnobotanical information*

(6) *Food characteristics*

(7) *Industrial characteristics*

(8) *Bibliographic*

Using the above types of descriptors, almost any data to be found in any GRC can be placed appropriately. Different functions within the GRC will use some different combination of these descriptors, but at least they will be one (or a set) of the above types and this improves the chance that the documentation will be more general and more useful.

For example, for that function listed as exploration, the following is probably to be generally found:

Classification	Descriptor
'Finding' information	Date of collection Collector, collector number Country, state, exact locality Scientific and common names Photographic information Herbarium specimen
Environmental information	Altitude Ecology or plant association Special growing conditions
Organismic information	Morphology Disease resistance, or insect damage Environmental damage
Use information	Medicinal, fiber, food, etc.
Ethnobotanical information	Local knowledge about the collection

The other functions within the GRC will clearly not employ exactly the same set of descriptors used by the exploration function. There is no need here to list what each of these are, but clearly, for quarantine, storage, rejuvenation, exchange, utilization, etc., there will be a set of descriptors which particularly appertain to that function.

Standards for data presentation

In order to facilitate description of accessions, and results of tests of accessions, and also to improve communication between individual scientists in different localities, there has been a continuing effort in Germany and elsewhere to provide documentation standards acceptable to all interested individuals. Attempts to achieve these standards have enlisted the aid of a large number of experts, and the consensus of their work is presented below.

403

(1) Scalar information – intensity of expressions, where scale '1' is lowest, or least, or smallest; scale '3' is next lowest; scale '5' is intermediate; scale '7' next higher intensity, and scale '9' is the greatest possible expression. Where variation of expression is found, additional information may be provided by combinations of the above, or by inserting even-numbered scales between any of the above odd-numbered scales. If any descriptor is variable for a particular accession, this may be denoted by the convention 'X'. If information is lacking altogether a blank should be left.

Where the expression of a characteristic is stated to be absent, the convention 'O' is chosen to denote this fact. If, on the other hand, there is no gradation to the descriptor, but one wishes to note merely the presence of the characteristic, the convention ' + ' has been chosen. (See Chapter 32, pp. 407–43, for fuller details.)

Variability of expression of descriptors dealing with reaction to pests and disease may also be indicated as follows: 'H' for hypersensitive; 'I' for immune; 'R' for resistant; 'S' for susceptible, and 'T' for tolerant.

(2) Thesaurus. In order to provide a set of terms, a list of descriptive terminology for data usually associated with accessions of various crops has been developed in Germany. The thesaurus of terms is provided with exact (or nearly exact) equivalence for each term in some of the major languages, German, English, Spanish, etc. By use of these terms, workers in various crops and in various parts of the world can accurately describe their collections and the results of their tests, and be certain that the terminology will be understood by others.

The computer functions for a GRC

At this moment, there is no standard computer program which has been designated for all genetic resources centers. Several different systems, each written for different types of computing machines, are in operation. For example, at the John Innes Institute, a system was developed to handle the documentation of *Pisum* collections. This system has been working for several years, and for most of their purposes, is satisfactory. At the Institut für Pflanzenbau und Saatgutforschung der FAL, at Braunschweig, Federal Republic of Germany, the GOLEM system, written for the Siemens computing machine, has been applied. In the United States, many different systems have been developed. One of these, TAXIR, has been applied to GRC work with such crops as *Manihot*,

Phaseolus, Zea, Solanum (tuber-bearing species), and for general GRC data at the National Seed Storage Laboratory at Fort Collins.

While each of these systems has some merit, it should be recognized that each has some problems still associated with it, and that no central decision has yet been made on which of these should be used to the exclusion (if ever) of any of the others. The important aspect is that it is possible for each of these systems to use data prepared for any other system, provided the data are placed in the format required for each of the specific systems. However, it is not necessary for all GRCs to adopt any one or the other of the systems, provided whatever system may be in operation at that station is known to others working in other GRCs. A discussion of the general applicability of TAXIR is given in Chapter 32 by Hersh and Rogers.

For efficient operation of any of these systems, there must be close co-operation between the computer experts, the documentalist, and scientists working in the GRC and in user institutions. Without any one of these, there is little likelihood that there will be successful operations of the GRC.

Exchange between GRCs

One of the most important needs or requirements for exchange between GRCs is a communication system which will permit the determination of the types of accessions, some of their qualities, their origins, and other important data that any one GRC maintains. With this information, the exchange of the accessions becomes more efficient and rapid. Eventually, some type of information and documentation network for GRCs anywhere in the world must be established. Although there have been some preliminary efforts towards establishment of such a network, to date, no plan has been agreed upon. Until such agreement has been reached, little discussion can be given in this paper. Close co-operation of many organizations will be required in order for a network to succeed.

32. Documentation and information requirements for genetic resources application

G. N. HERSH & D. J. ROGERS

The starting point for any meaningful systems analysis is learning about the extant system to be studied and changed. In order to do this the analyst must gain his knowledge from the personnel at work in the genetic resources community. He must understand fully their problems, their tools, their budgets, their data, their institutions and their hopes.

The following preliminary analysis is based on such information, but from a limited, although reliable, group. All the planning will fail if not based on this type of analysis.

Often the collective group of Genetic Resources Centers (GRC) personnel have excellent ideas of how their systems can be changed to become more effective. 'If only we could do this . . .' is a familiar cry. The analyst must attempt to solve the problem of 'how to do this' effectively and within realistic cost constraints. In this paper we attempt to demonstrate one of these 'how to do techniques' that may provide real, asked-for assistance to genetic resources personnel.

In turn, the organization of the systems to deliver these techniques must be capable of providing every worker with opportunity to use these systems regardless of the institutional size in which he works.

The advantage of this philosophy is that one may begin without major shifts in philosophy, or operational procedures now extant. At the same time, systematic analysis of the problems gives a much more real knowledge base for determining costs, user demand, etc.

It can be demonstrated that there is already a genetic resources community, with an operating 'system' even though it is rudimentary. It can, by actual process, also be demonstrated that by development of an integrated system (or network) the productivity of Genetic Resources Centers can be markedly improved. We trust that the methods here discussed will be viewed as one set of steps in this direction.

Introduction

The purpose of this paper is to establish the systems approach for documentation for a network of Genetic Resources Centers (see Appendix 1, *Definitions*). The global network is to provide an efficient, flexible and continuous mechanism through which information and physical material maintained at any set of Genetic Resources Centers can be exchanged, shared and used by any other set of authorized users.

The following must be done to realize a global network: (*a*) identify and rigorously understand the implicit genetic resource work functions; (*b*) design and implement an operating system to fulfill each functional need.

An interdisciplinary team has undertaken two specific tasks: to analyze the implicit communication–information–documentation function for all aspects of genetic resource work; and to prepare and propose a design and implementation strategy for a global network Communication, Information and Documentation System (CIDS) (see Appendix 1, *Definitions*).

These tasks were partially undertaken in a Genetic Resources Centers (GRC) information system pilot project initiated as a result of the recommendations of the Birmingham Workshop.* The pilot project was designed to test the complex hypothesis that TAXIR† could effectively meet information system requirements for genetic resource work, and could be used as the core computer-assisted information storage and retrieval module for the global network, CIDS.

The preliminary results of the pilot project indicate that the requirements of an information system for Genetic Resources Centers can be met by the TAXIR System. Insights gained while doing the pilot project provide the basis for the design of the CIDS, an implementation strategy for CIDS, and a preliminary set of evaluation criteria for the CIDS.

The first activity in the pilot project was to develop an initial set of test criteria to judge TAXIR and CIDS performance.

---

* The Birmingham Workshop was an informal meeting of a group of genetic resources experts held at the Department of Botany, University of Birmingham, July 1972. The group met to consider common problems on computerized documentation systems for genetic resources work.

† *TAXIR* is the acronym for *TAX*onomic *I*nformation *R*etrieval, a general purpose, computer assisted information system developed at the Taximetrics Laboratory, Department of EPO Biology, University of Colorado, David J. Rogers, Director. The development of the initial system was supported by a grant from the National Science Foundation, Office of Scientific Information Services GN656.

The test criteria

The TAXIR System (the CIDS) must: (*a*) meet basic user requirements and demands (see Appendix 1, *Definitions*); (*b*) have sufficient flexibility to handle considerable changes in these requirements and demands; (*c*) have sufficient flexibility in design to be changed as more effective computer assisted CID techniques are developed and become available; (*d*) have, as a component of the CIDS, a means to anticipate changes in user requirements/demands and to locate or develop more effective CID techniques; (*e*) have the capacity to be implemented for operations within three to six months after a decision is made to adopt the system; (*f*) minimize the costs associated with development, implementation and continued use of the system, relative to the system itself (continuous effort to make the system more efficient), or compared to other systems meeting the same effectiveness criteria.

Each criterion is general and requires further definition, and the second activity in the pilot project was the selection/development of a method which would further define these criteria and, in so doing, would provide the necessary information for designing the CIDS as well as evaluating the TAXIR System.

The systems approach

The team chose the 'systems approach' drawn from the inter-discipline of management science. The systems approach (often referred to as systems analysis, which in reality is only one step in the systems approach) is relatively simple to understand, but has been cloaked in an aura of confusion and complexity. It rests on basic assumptions.

(1) There is a system currently in use for the function under study, in this case the Genetic Resources Documentation function.*

(2) There are persons who are using the current system and they (*a*) to some extent understand the system and its effectiveness, (*b*) are willing to work with an analyst to make this understanding explicit.

(3) The current system can be changed, if necessary, to become more

---

* There is a special case for which this assumption may not hold – where the center under study is new and has no current system. In fact this is two cases: (*a*) the 'new' center will employ personnel and consultants who will install a system based on their past experience and will be similar to other systems in use or (*b*) the center will use a new system which is *not* similar to other systems currently in use. If (*a*) pertains, then the assumptions are valid since the system that will be used can be predicted. This has been the case recently with at least one new GRC.

effective at some attendant change in cost (or less costly with some attendant change in effectiveness).

(4) As a corollary, there are analysts who are willing to work on this problem and who will work closely and continuously with knowledgeable persons in the field.

From these assumptions several operating rules and steps emerge. The initial step is to 'understand' the system, thus the step of systems analysis. At the outset the analyst attempts to get a set of users to tell him what they know. The analyst (1) is given a list of pertinent literature in the user's discipline; (2) converses with the user about the user's work; (3) asks the user to provide a written description of his information system; and (4) works through an information problem with the user.

The analyst prepares an initial descriptive model of the system and improves it as he has more and better information. He presents the model to the user who 'corrects' it. In effect, a set of users 'teaches' the analyst about the system. The analyst learns, and should develop some insights which then should be rigorously defined in the descriptive model. In the process of description, he may define systems characteristics that will allow for the measurement of the system's effectiveness in meeting users' needs and the cost associated with this effectiveness.

The model allows the analyst to determine what types of change in the system are desired by the users collectively. He can further identify points at which change can be made effectively. He also reviews the state of the technology to determine what operating techniques can be introduced at those points. If there are none available, he determines the costs of developing suitable techniques. When the systems model provides sufficient understanding of the system and changes desired, an initial design change is suggested by the analyst to the various users. This design change emphasizes one or two basic changes in the system. The analyst should have a good feel for the available techniques needed to implement these changes (i.e., computer systems, programs, etc. that are available) and at what approximate cost. Note that, for the GRC pilot project, one of the conditions was the use of an available computer-assisted technique. If the available techniques were not satisfactory, the analyst would suggest a design that could be developed, carefully estimating costs.

The users working with the analyst would react to and modify the suggested design change. The procedure of design-change-reaction/ modification would continue until a design mutually acceptable and feasible emerged. During this process, the analyst would continuously point up the cost/effectiveness trade-offs, essentially working from the rule that the design change will cost some amount and that this cost varies depending on the extent of the change. If the user group wanted the change to meet their utmost desires, the cost would be higher for this level in change in effectiveness than if the minimum desires were to be met. There would be a range of costs associated with the degree of effectiveness desired. (When the users decide on a cost/effectiveness level, it is said that this is their level of demand, no longer that of desire.)

The analyst usually will present a minimum change in effectiveness for a minimum change in cost. This minimum is the level at which any change is noticeable to the user. The analyst will suggest further changes, showing probable attendant changes in effectiveness and cost.

After some decisions are made the analyst may suggest a pilot project to test and evaluate the design change to determine whether the design change will meet the predicted necessary requirements in effectiveness and costs. The pilot project will allow a re-evaluation of the analysis to date (because when the pilot experiment is done, most of the errors will appear, and all faulty understanding of the existing system will be exposed). The results of the pilot project give a concrete basis upon which to redesign the system (if needed), and can provide important information in the implementation process. (During the course of the pilot project, users are involved at all steps. Watching their reactions provides insights into the design of orientation/training programs for future implementation.) If redesign is required, changes are made and pilot project 2 is run. This process continues to improve the system's design, testing the capacity of computer techniques as required.

Included in this series of comparative testing, there is a dynamic 'dialog' between the system design under consideration and the user. Often the user and all too often the user's computer consultants are unaware of the latest developments in information science. A well-versed analyst will assist in designing a system which, when tested, may 'reorient' the user. The reorientation should demonstrate how the most current computer technology may be applied to his requirements. The user may increase his requirements on the system having learned of new opportunities available to him through a new design. Thus, as the user sees new opportunities arising from orientation to the system's capacity, he may add new requirements. This dynamic process is constrained by the imagination of the user/analyst *and* the cost for each extended application.

At some point, a decision is made to implement the system. There are three major steps: installation of systems at some centers, either in whole or in part (see later discussion); preparation of current and back-log data to enter into the system at these centers; and connection of systems at these centers to the network of global CIDS, on a priority basis. The priority is established by users. Each of these steps requires instruction of personnel in proper management and use of the system.

Thus, the pilot system is expanded to an operating system on a priority schedule while analysis continues to determine that the expanded system will meet requirements as each data bank and center is added, inserting appropriate changes in the system as implementation continues. To avoid confusion, the system is designed in modules (self-contained techniques) which are integrated into a complete system. A modular approach allows considerable flexibility in modification while maintaining the effectiveness of the system and enhances the continuous process of upgrading. A good initial system is thus improved during operation. The process from analysis through design and implementation to continuous operation is iterative. The several steps are repeated as often as necessary.

Fig. 32.1 summarizes the process of the systems approach.

The systems approach has some drawbacks: it is highly sensitive to the competence of the systems team chosen to do the work, and the willingness of the users to participate in the process; it costs more than a static approach but the costs are usually well justified. (Most static approaches try to become iterative after the initial implementation. When this occurs, costs are exceptionally high and all too often the users are 'locked in' to a less-than-desired system.)

Analysis of CIDS requirements for a global network

Background

Given the framework of the systems approach above, members of the Taximetrics Laboratory assembled the information at hand to conduct a pilot project (recommended by the Birmingham Workshop) to test the TAXIR System as the *core module* for the CIDS.

An initial understanding of the CIDS problem was drawn from team members who had been involved in work related to genetic resources centers for some time. The previous experience of the team included

412

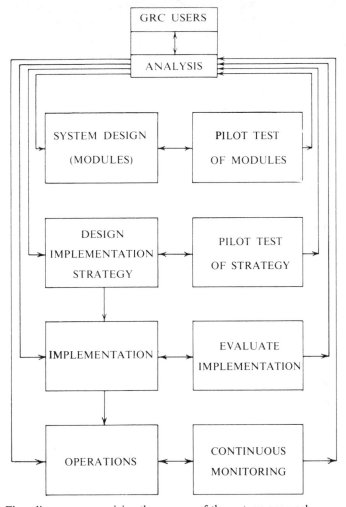

Fig. 32.1. Flow diagram summarizing the process of the systems approach.

documentation of genetic resource data at the international, national, and local levels and development of the computerized TAXIR systems. These experiences further include management science methods which guided the processes.

Aside from a review of the literature, the team communicated with substantive scientists with varying experiences in genetic resources centers. A site visit was made to the National Seed Storage Laboratory at Fort Collins, and to the Agricultural Research and Introduction Centre, Izmir, Turkey.

The analysis and design test (pilot project) is based on a limited data sample, but one which provided most of the problems which will be found in any GRC. The initial CIDS design based on the pilot project is an adequate starting point for implementation of a global network.

The following sections present:

(1) An initial description of users' requirements and demands.

(2) A brief description of the TAXIR System and a detailed description of the results of the pilot project. This will give a fairly good picture of the capacity of the TAXIR System and its uses.

(3) A suggested design, program and strategy for the implementation of the TAXIR based initial global network Communication–Information–Documentation System (CIDS).

There are two types of CIDS users, the primary user and the secondary user. The distinction is relatively simple:

The primary user *both* enters and retrieves information; the secondary user *only* retrieves information.

The information input by primary users derives from the operation of GRCs and affiliated institutions such as germplasm collection, general evaluation and breeding; studies related to plant protection, evolution, genetics, agronomy, taxonomy, etc.; GRC management, germplasm conservation; etc.

The primary user may be an individual worker, a Genetic Resources Center or a set of Genetic Resources Centers.

Any control over who is a 'recognized' user must be exercised by a users' organization and is outside the scope of this analysis. For the purpose of design, however, we have assumed the broadest definition of the position.

Basic user requirements

User requirements flow from the process of genetic resources work. There appear to be four basic operational processes during which information is 'produced' and/or required.

(1) *Basic Operations*
   (*a*) collection of germplasm
   (*b*) distribution of accessions of germplasm (hereafter = accession)
   (*c*) continuous evaluation of accessions
   (*d*) breeding
   (*e*) new accessions from breeders

(2) *Storage* of accessions

  (*a*) active collection

  (*b*) base collection

  (*c*) working collections*

(3) *Planning* for research and other activities

(4) *Reporting*

  (*a*) on the accessions in the system or inferences made from them (areas in which collections have been made)

  (*b*) on activities underway in the GRC global network

The collector's (or breeder's) descriptive information for each accession is the first information set to enter the CIDS. Additional information on any accession must enter the CIDS as work is done on that accession. This additional information may be new, in that it was not known previously, or it may correct or update previous information already in the CIDS.

The core module requirements are:

(1) Any qualified user can enter initial or subsequent information or any accession to the CIDS. The user is constrained by certain standard operating procedures to control information changes in the CIDS, but the burden of this constraint is on the CIDS and not the user.

It follows from the above that the CIDS must be able to handle any type or any number of updatings or corrections that may be needed, including addition of more descriptive information about an accession and changes in measurements or observations about any descriptive characteristic for any accession. Any qualified user will have access to the CIDS for the purpose of obtaining the most current information on any set of accessions in the CIDS.

(2) Each collector, evaluator, in effect each user, may have different methods of working and probably has different ideas about what information should be used to define, describe, annotate, etc. any accession. Some users are exceptionally rigorous and wish to include vast quantities of information while others may provide information that is too scant. Several systems problems emerge which place requirements on the CIDS:

The CIDS must be able to:

(*a*) accept any amount of accessions as determined by an individual user;

---

* This term applies to the more ephemeral collections in use by breeders, geneticists, etc. at any given time. Although these are not parts of the formal GRC collection system, the user may wish to use the CIDS for his work facilitating information gained through the use of these collections.

(*b*) accept any amount of information for each accession as determined by an individual user;

(*c*) accept information in any form that is required by an individual user (e.g., codes, numerals, and alpha-numeric combinations – names, dates, locations, botanic binomials, in any language for which there is a computer typeface);

(*d*) accept information in any recorded format from an individual user as long as the user carefully, clearly, and completely defines that format.

It is imperative to add that the core module of the CIDS be able to meet these requirements efficiently and with a minimum of human dependence.

(3) Clearly, the above are maximum requirements on the CIDS. It is important to note that although the CIDS should meet these efficiently, there will be a high attendant cost for this high level of effectiveness.

These high costs are of three types.

(1) Rapid use of the resources (personnel, equipment, money, etc.) of the CIDS.

(2) Much user time needed to understand and utilize the variety of information sets in the system. This leads to a potentially higher cost, and subsequently lower user-dependence on the system for sharing information.

(3) Within a structure of limited resources, the loss of opportunity to do other work within the CIDS.

On the other hand, the high effectiveness in being able to meet these requirements is shown in four ways.

(1) Any user can conduct research as rigorously as he desires and will have a flexible CIDS to assist him.

(2) Potentially valuable information will not be lost if it is recorded and included within the CIDS.

(3) General user resistance to the CIDS is minimized because he will not have to conform to some standards forced upon him.

(4) Any users can view the 'thinking' of others through examination of the information they collect and the processes used.

At this point, the difference between requirements placed on the core module and the entire CIDS become clear. The core module must be able to meet these requirements, but on the other hand, there is a matter of policy which must be formalized: what are the cost/effectiveness trade-offs concerning the size, complexity, and format of the information set for any accession? Complete individual freedom of choice has value in working collections, but not in the documentation files. However, the exclusion of information from the document files may mean permanent loss of potentially important information.

The policy issue on information standards is not the concern of this paper, although it is an exceptionally important matter. Standards are means by which CIDS effectiveness can be increased – better sharing of information, rules on what must be in a documentation file, etc. Standards can also lower CIDS costs per accession, but they can also constrain a researcher's freedom of choice and may inadvertently exclude important information.

From the above, the following specific requirements emerge: the CIDS should assist in providing information to a group of users studying and setting standards. This information must include:

(*a*) The variety of information sets in use for any accession at any time, including synonymy, measurement techniques used, etc.;

(*b*) the relative value of any component 'piece' of information in that set;

(*c*) the marginal costs associated with adding another piece of information for any accession;

(*d*) user reaction to a set of standards in terms of comparative utilization of the CIDS;

(*e*) the capacity to synonymize, reformat or change information entered in the CIDS to meet standards – and at any time using the core module's internal capacity and minimizing the dependence on personnel.

It must be made very clear that the standards that emerge need not be constrained by misconceptions of what the computer can do, but based on what the CIDS is doing, with useful, probable, cost estimates. The CIDS should be operational before decisions are made on standards.

There will be a great number of accessions in the CIDS and much data associated with each. At any time a single user may not be concerned with all the information in the entire GRC data bank. The user will require that the CIDS assist in structuring working data banks, subsets of the entire data bank, for his specific interests. These subsets must be flexible and can be expanded or contracted as necessary.

The CIDS must provide the following capabilities and assistance without requiring specialized programming or user reliance on programmers.

(1) Assistance to the user in determining the subsets of information he wishes to extract from the system.

417

(2) A query system that will then extract exact subsets of information as needed by the user, with no limit to the complexity of the query.

(3) Capacity to format this information in any manner specified to present information to others and to use data as input to other programs for further processing (statistical, clustering, mapping, graph drawing, simulation, etc.).

(4) Access to additional computer program modules to further process various information subsets. The user should be able to get access to these modules easily.

(5) Write standard reports (lists or inventories of accession records) on a periodic basis easily, quickly and cheaply, using the query system.

Overall, the CIDS must be designed to place maximum burden on the system, making efficient use of the computer. The burden on the user should be restricted to understanding how to use the system effectively. This burden should not require more than a few days of the user's time to learn the fundamentals of the system. The user should be free of the need for a programmer or for specialized programming to use the system, but should be confident that he can get assistance should he need more training. The 'language' needed to communicate with the system should be simple and flexible.

In summary, using a sophisticated core module that makes efficient use of computer hardware, the CIDS must be able to meet individual user requirements for his own work as well as for the sharing of information. In order to learn to use the system there must be continuous orientation programs, including written and oral presentations, and briefing instructions. Time taken to learn to use the system must be minimal. The system must be easy to use, without specialized programming nor dependence on a programmer for continued use. Furthermore, it must be easy to update and correct information in the system. The system should be able to accept data in almost any well-defined format and should be able to deliver to the user data in a pre-specified format. The system should have other programs (modules) to further process the data and to which the user should have easy access.

Restrictions placed on the user in terms of standards of procedure or of terminology should not be inherent in the system, but specified by a policy group of users. The CIDS should be able to provide information to such a group based on actual data examples. Once standards have been set, the CIDS should be able to make data conversions to bring information in the system into use without losing the initial records. Further, the CIDS should be able to change standards and make the necessary conversions at the will of the standards committee.

Analysis of user demand

At the present time it is difficult to estimate user demand for the CIDS because the data needed for such estimates are either sketchy or not available.

The following text briefly specifies the major determinants for demand analysis and should provide some idea of the magnitude of demand. It is important to note that demand is highly cost dependent and cost estimates are as difficult to make as any other component of demand. Interaction with actual users is essential in order to move from a demand–cost guess to successively better estimates. As stated earlier, this is the basis of this paper – to provide the structure for such inter- active analysis and to demonstrate that there are current core techniques in rudimentary forms which provide a worthwhile service basis for users. From this service basis, user behavior can be measured as an interactive basis.

The following analysis of demand is in three parts: (1) basic determi- nants of demand; (2) specific expected patterns of demand based on a GR user classification; (3) planning, organization and implementation problems.

Basic determinants of user demand. Demand is defined as the amount of service a user will purchase at any given price per unit of service. Clearly the major questions raised are: what is the value of the proposed service to the user and what is the attendant cost.

In the case of GR work the basic criterion for estimating the expected value of the service is what may be the expected change in a user's productivity or effectiveness with the use of this service. This expected change in productivity must then be compared to expected changes in cost.

After a certain transition period of experimentation and adjustment then 'expected changes' become actual changes and decisions on effec- tiveness–cost relationship are somewhat easier to make.

From some experience (although very limited), the proposed changes may substantially affect expectation of changes in productivity which have to be more carefully tested. If in fact productivity is changed

419

substantially final demand will be very hard to estimate for some time. Further, the initial desire on behalf of certain users to enter the experimental phase is fairly high – some users are faced with problems in the information area they have not been able to solve – 'a bottleneck' – and the new system techniques seem to offer a 'breakthrough'.

Two broad cost categories must be considered: the cost with CIDS development, implementation, training, maintenance and research; the cost associated with the data in operations of the CIDS.

The first category includes the cost to obtain the system and make it operational for any single user (or, in the case of a larger GRC, any group of users). These include: type of equipment needed and its cost (or if equipment is already available is it practical for this purpose; if equipment is not practical, what are the expected costs to obtain or rent other equipment?); type of essential staff specialists and the cost to train them in the use of the system; the costs of improvements in the system (research and development personnel); cost of changes in the system as well as staff reorientation time, and actual money outlay.

The second category of costs is possibly even more important: the cost to get the necessary data into the system and the cost in equipment time and ancillary supplies once the data are prepared and the system is in use.

These variable cost aspects are the easiest to control and at the same time are the hardest to predict. Arbitrary cost control may actually lessen the change in productivity brought about by the use of the system, while lack of all cost controls can be budgetarily disastrous (see Appendix 2, *Costs*).

Specific demands based on a user classification. The general determinants discussed in the preceding section can be made more specific (and useful) given the presently understood nature of GR work. A simple classification based on the type of user activity or concern improves understanding. The categories of this classification are:

(*a*) the entire global set of genetic resources for a given crop;

(*b*) a locally available subset of the entire global set of genetic resources (i.e. a user's own active or working collection) for one crop;

(*c*) locally available subsets of the global set of genetic resources of many crops (i.e., a local or national gene bank);

(*d*) the entire global genetic resources.

Virtually no one, except planners and organizational personnel, is expected to be in class (*d*). Personnel involved in gene bank maintenance should be interested in (*c*). These are basically the gene bank curators, another small group. Categories (*a*) and (*b*) are not mutually exclusive, and almost all workers will fall into (*b*) on a continuing basis.

Category (*b*) is probably of primary concern to breeders, evaluators, geneticists, taxonomists, etc. The major concern to users of type (*b*) is the relationship between their own subset collections and the global collection for any given crop. These users may attempt to be cognizant of changes in the global collection of interest to them, but they probably encounter great difficulties because the information is not easily available or because the data do not exist in a form that is easily or effectively communicable. Because of the difficulty in communication, the group in (*b*) who have very specific problems with their own work, tend to concentrate on their own materials and disregard the global GR collection. In their own work, the systems they have evolved/developed seem to meet their needs. They may realize the limitations of their own systems and ignore, or work around, the limitations, but little time is given to improvement of the system.

Group (*a*) is theoretically composed of all those who are involved in GR work. The term 'theoretically' follows from the above discussion. The individual user has determined not to be dependent on the global GR collection because he cannot use it with any assurance. Many informal networks have been developed which allow individual users to have access to subsets of the global GR collection, but these are informal and incomplete.

There may be substantial user demand on a CIDS for the information contained in the global GR collections if the information about the collections were complete (all competent, valid collections were referenced), and there is an effective CIDS available to them.

This analysis suggests that the scope of the CIDS problem is to provide information that references the entire global GR collection for any given crop and that the CIDS will be easily accessible and available to any user, if the user perceives that the CIDS core techniques may increase his productivity. Availability may mean that the user can either install them on his own equipment or equipment he has access to, or can have access to a service bureau that is available through an organized equipment network.

Planning and organization. It is clear then, that the global GR CIDS should have information that references every competent collection for every crop (or plant) deemed important. It is difficult to estimate the number of accessions that would be referenced, or the complexity of the information maintained about each accession or the expected rate of growth of these collections. A very open-ended guess is that there may be from 5.1 million to 9.5 million accessions for all crops.

The estimate for the expected number of accessions for the global resources collection for all crops is calculated from the following equation: total accessions $= \sum_{i=1}^{n} C_i$, where C_1 is the number of accessions for any crop indexed by i and n is the total number of crops. The staff of the National Seed Storage Laboratory at Fort Collins, Colorado, estimate that there are 200 crops. Of these about 30 currently have 'large collections', 50 have 'medium' size collections and 120 have 'small' collections.

Collection size for each crop (C_1) is estimated on the basis of three crop accession sizes:

(1) The *Zea mays* collections from Mexico and South America which number approximately 40000 accessions. The maize experts estimate that this may be 40 per cent of the global collection of 100000 accessions. This is seen as a representative 'large collection'.

(2) The *Solanum* (potato) collections from three competent sources number about 25000 accessions and may be 50 per cent of the global collections, or about 50000 accessions. This represents a 'medium collection'.

(3) The *Vigna* (cowpea) collection at the International Institute of Tropical Agriculture, Ibadan, Nigeria, is currently about 6000 accessions and probably represents 50 per cent of the world collection of about 12000–15000. This may represent a 'small collection'.

Using the above estimates for the distribution of collection size over n crops the following total estimate is obtained:

Collection size	Number of such collections	Number of accessions/collections	Total
Large	30	100000	3000000
Medium	50	50000	2500000
Small	120	15000	1800000
Total	200	—	7300000

The range of accuracy is ± 30 per cent for a range of 5.1 to 9.5 million accessions.

The CIDS planning/organization problem has two major aspects about which decisions must be made.

(1) Given the distinct aspect of user demands, what portion of a CIDS team's activities and resources should be expended on the following.

(*a*) Establishing the global genetic resources collection information system.

(*b*) Assisting users on a local basis (or network basis by installation/training) to use CIDS core techniques on their active-working collections. (It is expected that priority must be given to (1*a*) even though to date the most striking requests from users is (1*b*).)

(*c*) Determining the crops to be given priority for input to the CIDS in establishing the global GR collections information system, and for each crop, establishing priorities of entry of collections from each existing collection.

(*d*) Determining priorities for installation of CIDS core techniques (or service bureau availability) for various GRC users requesting assistance.

(2) Given the above criteria,

(*a*) what type of organization is needed, both to set these methods of determining priorities and setting the priorities themselves; and

(*b*) what are the expected sources of income to execute these programs.

The level of income will determine how fast the overall program is carried out (see Appendix 2, *Costs*).

Immediate CIDS implementation

Since there is a degree of urgency in GR work, there is great need for the immediate implementation of the CIDS for the global network. The design of the global network CIDS is based on a core module (information storage and retrieval system), and it is necessary to have one that is immediately available.

Further, the CIDS core module must be a generalized system, one which is not based on specific data structures, data types or data content, but a system that can handle inventories of automobiles, furniture, bibliographic material, soil types, pesticide registration, or accessions on germplasm. This generalized system must be able to meet the changes that may occur in GR work.

There are immediately available generalized systems. TAXIR is one such system. With an immediately available generalized system, an operating CIDS can be available within three to six months of a decision to proceed and fund the project.

Minimization of CIDS costs

The concept of cost–effectiveness trade-offs was introduced earlier. In this section the potential effectiveness of the system is seen as being constant, and perceived by the users as being exceptionally high, i.e. they want to use the system. The requirements discussed here, then, are of minimizing those costs which, if too high, will disallow any user access to the system. The term 'too high' is relative, and will have different value levels for each user. The concept 'minimize' here is some means to reduce costs so that they fall below the 'too high' for most, if not all, users. This should be possible. Of course, once the CIDS is in operation and a user has access to it, the costs of continued use may become 'too high'. These costs are of a different type from those faced in initial implementation/installation. These will be discussed separately.

The most difficult costs to minimize are those of resistance to the use of the CIDS by any user of his GRC. The user, as an individual, must be convinced of the effectiveness of the system for his needs and for the purpose of the global network with emphasis on the latter. Also, the costs associated with the format conversion of his data for input must be met. Generally, the user will be willing to overcome these costs – with CIDS assistance. However, non-user internal resistance in any GRC presents another set of problems.

(1) Funds may not be budgeted for the user's participation in the CIDS, should there be a cost-sharing burden on the GRC.

(2) If there is an internal computer group (or computer) there may be

(*a*) professional 'jealousy' towards the CIDS by computer personnel;

(*b*) restrictions placed on expending funds for use of external systems because the internal computer/systems group as a whole is budgeted to provide all necessary services.

(3) The CIDS is external to the control of any GRC and an administrator is not sure that the CIDS can meet his GRC's needs nor that CIDS's costs can be controlled. (The argument goes: the current price of entry and use may be low enough, but how can the administration know that these costs will not accelerate greatly in the near future? Should we be dependent on the CIDS at that point, what can we do?)

These 'costs' can be minimized in three ways.

(1) Adequate orientation for the users and GRC personnel and their involvement at all phases of CIDS development.

(2) Emphasis on the global network aspect of the CIDS: it is basically for common information-sharing and for those GRCs that do not have a satisfactory documentation capacity.

(3) The initial implementation costs and all data conversion costs will be covered at the CIDS level.

The continuing costs should be shared in some manner among GRCs, but should always be at a predetermined level. This level should be known sufficiently in advance to allow for proper GRC budgeting. Cost excesses (over the pro-rated GRC shares) should be made up by an international agency.

The TAXIR pilot project

As stated above, the formal work in studying Genetic Resources Centers documentation function began with the TAXIR pilot project. The project was undertaken to test the hypothesis that TAXIR could effectively meet all requirements placed on the core module of the CIDS; and to assist further in the understanding of the documentation function for any individual GRC – independent of size – or any set of GRCs working jointly.

This section presents in detail the method, results, and conclusions drawn from the TAXIR pilot project. There are two reasons for a detailed presentation: (*a*) to explain fully and demonstrate the capacity of TAXIR with respect to the Genetic Resource Centers documentation function; and (*b*) to indicate the specific steps necessary to implement the CIDS for any user.

Information systems and TAXIR

It is useful to review, briefly, the concept and general design of information systems and to describe TAXIR specifically.

Informations systems in general. An information system, in general, is any integrated set of procedures to extract, record, store and retrieve, and process information. These procedures are usually based on a concept of information content or structure for a given area of work. The procedures may use any combination of the following.

(1) Manual techniques.

(2) Some simple machine assistance (typewriters, sorting folders, filing cabinets).

(3) Some specialized data equipment (tabular cards, tabulator card sorter, tab card interpreter, etc. – sometimes called unit record equipment).

(4) General equipment with memory such as computers with various peripheral devices (tape drives, card readers, drum or disc storage units, high-speed printers, etc.).

Effective information systems can be designed using any combination of (1), (2), and (3), depending on the requirements placed on the system. There are some very effective manual, simple machine systems that cannot be greatly improved by the use of computers; and there are some computer-based systems which would be better as manual systems!

The mass of data and its inherent complexity requires a computer-based CIDS for the global network. Computer-based information systems have been in wide use for the past 10–15 years. Considerable strides have been made in the development of computers and computer equipment (hardware) for information processing; but, surprisingly, very limited improvement has been made in the design of information processing (software) that controls the computer.

Software must have two basic attributes: the internal process for information storage and retrieval, and the interface with the user (what the user must do to control the computer operations). Significant strides have been made in improving the user interface with the operation of the computer. These strides have been of two types: (1) ease of use, somewhat less dependence on highly trained programmers; and (2), meeting more of the user's requirements. However, the basic internal program design has not changed significantly. Software for information storage and retrieval programs generally stores information in a master file (data added sequentially into the system) and, usually, inversion of this master file into a series of special files. For example, the telephone directory is a special file taken from the chronological listing of all telephone subscribers. It is ordered according to last name, then other attendant data. The telephone company has another special file ordered by telephone number, and other attendant data. Most information systems software are variations on this theme.

Information retrieval is executed by specifying the type of information (search criteria) desired and then a comparative search is made of all records of a master file (or for greater efficiency all records in a special file) until the search criteria are satisfied. The process is not very differ-

426

Table 32.1

DESCRIPTORS	Genus name	Species name	Plant height	Plant habit	Country of collection	Collector's name
DESCRIPTOR	TRITICUM	aestivum	80 cm	2	Ethiopia	Harlan
STATE	SECALE	cereale	40 cm	4	Turkey	Sencer
	SOLANUM	tuberosum	55 cm	5	Bolivia	Hawkes
	MANIHOT	esculenta	150 cm	3	Mexico	Rogers

ent from that which would be performed manually, but, of course, computer execution is exceptionally fast by comparison. Several good general information systems that work in the above manner are commercially available.

A brief description of TAXIR. TAXIR is a set of programs, software, that can currently operate on virtually any larger scale (core $\geqslant$ 32k words; word size $\geqslant$ 36 bits) computer. It was designed to meet the interface requirements of biologists with vast quantities of complex data. The internal mechanism is entirely different from that described above. Essentially, the computer does not locate specific information by search–compare methods on actual data files, but by actual Boolean calculation on highly compressed data structures. This process enables greater flexibility for the user in querying the system, and exceptionally effective retrieval at very low costs. The computer is used, as initially intended, as a computing machine. For a rigorous and complete discussion of the theory behind TAXIR see Estabrook & Brill (1969).

TAXIR requires a simple data structure that the user must follow. The user determines the ITEM under study. The main ITEM for Genetic Resources Centers is an accession. The ITEM is defined and described by a set of DESCRIPTORS, e.g. genus name, species name, plant height, plant habit, country of collection, collector's name.

Usually input is in variable-field format with separators between DESCRIPTORS (usually a comma ',') but the system also accepts fixed-field format.

For each specific ITEM there is one mutually exclusive DESCRIPTOR STATE for each DESCRIPTOR (Table 32.1).

There can be as many ITEMS or DESCRIPTORS as needed. The user may have as many DESCRIPTOR STATES for each DESCRIPTOR as he wishes, and need not be bound by a predetermined dictionary. The dictionary can be produced from the data as they are described, or

Table 32.2

DESCRIPTOR	DESCRIPTOR STATE
Collector's name	Brooke-Smyth/Alfred J. Imbouyu/Jomo
Principal reagent	4,2-Dimethyl-8-alphagammadamadiobenzyl-chlorideamine
Flower color	Red or red-violet with white spots Rojo Rouge Ret Rubot

the user may use a pre-defined dictionary. (TAXIR will compile a dictionary on request.)

For each DESCRIPTOR one of four options may be used to record a DESCRIPTOR STATE. These are: (1) CODE, (2) ORDER, (3) NAME, (4) TEXT.

(1) The *CODE option* accepts data input in the form of numeric codes, which have been defined to the system.* Queries and output provide the full alphanumeric name represented by the input code. This option is useful to enter previously coded data, or to reduce input costs.

(2) The *ORDER option* accepts either numeric (including decimals) or alphanumeric data.* If the data are numeric, the inherent order in the number system is assumed. If the data are alphanumeric, then a specific order must be defined to the system, e.g. street names in a city, or location names on a map or places along a river.

An order DESCRIPTOR STATE can be labeled. PLANT HEIGHT is an example of an order. Uses for the order option are obvious, but there are two special cases:

(*a*) In querying the system, order ranges may be used, e.g. when looking for a plant height *from* 30 cm *to* 80 cm. The query scan will select the entire set meeting this range requirement. Another example is: All collections made at altitude *from* 500 m *to* 1000 m and latitude *from* 5° S *to* 5° N, and year *from* 1950 *to* 1973.

(*b*) When using a data code for input, for querying and for output, the order option is used.

(3) The *NAME option* accepts any alphanumeric combination (Table 32.2).

---

* TAXIR will accept 80 alphanumeric symbols for any one DESCRIPTOR STATE. This limit can be changed if needed.

(4) The *TEXT option* allows alphanumeric input of any length to record histories, textual material, etc. for recall upon request.

TAXIR produces a Dictionary for any DESCRIPTOR on request. The Dictionary contains all unique DESCRIPTOR STATES for the DESCRIPTOR in either alphabetic or numerical order.

The user prepares a QUERY to retrieve desired subsets of information from the data bank. From among all possible combinations of ITEMS and DESCRIPTOR STATES, the query selects the exact subset of information desired. The query has two parts:

(1) Boolean or logical expression that designates specific criteria for the identification of a subset of ITEMS in the data bank; and,

(2) a list of attendant information to be retrieved from those ITEMS. For example, PRINT: [collector name, year of collection] for ITEMS with genus, Manihot AND species, esculenta.

Part 1 is within the bracket, part 2 without.

The DESCRIPTOR/DESCRIPTOR STATES composing the Boolean expression are coupled by the *logical operator* 'AND'. Thus, the subset of ITEMS meeting this criterion are the intersect of both conditions – Manihot AND esculenta. Other *logical operators* are 'OR' and 'NOT'.

An example of a complex query is:

PRINT: STORAGE LABORATORY, STORE ROOM, STORE RACK, STORE TRAY,
    VARIETY NAME, DATE of collection for Items with Genus, oriza
    AND Species, sativa AND
    Year of collection, From 1960 to 1973 AND
    ALTITUDE, from 0 meters to 500 meters AND
    Country of collection, Japan OR Thailand OR Malaya AND NOT
    VARIETY, HR3 AND NOT
    VIRUS X REACTION, Hypersensitive.*

This query locates germplasm that may be needed in a breeding program.

A query can use any number of logical operators in any sequence and can retrieve any amount of information.

A simple interface module can format the output from a number of modules (mapping, graphing, statistical, etc.). TAXIR has a complete

---

* In this particular place and in similar places throughout the rest of this chapter, the asterisk is a part of the query, indicating to the computing machine 'END OF QUERY.'.

update and correction facility. The query mechanism also makes corrections. For example, suppose the Dictionary for country of collection shows:

Algeria
Belgium
France
Franxe
Spain
Espana

Obviously the spelling FRANXE is incorrect, and SPAIN and ESPANA refer to the same country. With two correction statements the user can change the Dictionary as follows:

1. CORRECTION COUNTRY of Collection, SPAIN for all ITEMS with country of Collection, Espana.*
2. CORRECTION COUNTRY of Collection, FRANCE for all ITEMS with country of collection, FRANXE.*

These corrections are *not* made on the original input material, but internally. By recompilation of the original input material, the mistake can be regenerated. This is useful to prevent permanent loss of information. Further, this facility can be used in translation from one language to another.

The basic characteristics of TAXIR have been explained above. The following section explains in some detail the pilot project recommended by the Birmingham Workshop.

The pilot project

To test TAXIR and to work through several typical GRC information problems, sample data banks from four centers were prepared. From 50 to 200 accession documents were received from the Commonwealth Potato Collection, the USDA Potato Collection, the International Potato Center, and the USDA National Seed Storage Laboratory.

The documentation for the accessions was received in the same format in which they are used at each respective GRC. A description of the accession record structure and content accompanied each set of documents except for those from the Seed Storage Laboratory. Descriptions of two accession record structures were developed during a visit to the National Seed Storage Laboratory.

The accession record formats were all different. The Commonwealth Collection consisted mostly of retyped field observations. The ITEM was obviously a single collection which became an accession; however, the DESCRIPTORS were not in a specific order. The DESCRIPTOR STATES were usually identified by a DESCRIPTOR notation. The published Commonwealth Potato Collection *Inventory of Seed Stocks 1968* provided a clear tabular format in which most of the information had not been reduced to numeric codes.

The accessions from the International Potato Center were also in tabular form, with an ITEM constituting a row, each with a descriptive characteristic in each column. The information was almost completely in numeric codes and a well prepared set of code lists was provided.

The data on accessions received from the USDA Potato Collection were in machine readable, fixed-field format on punched cards. Information was mostly numerically coded, with abbreviated country names.

Two types of accession records came from the National Seed Storage Laboratory:

(1) General collection, with data on several crops. These data were on edge-punched cards, and on the face of each card clearly typed data were also presented, although without precisely defined descriptors in many cases.

(2) A sample of 'computerized' (that is, punched on machine-readable cards) data for guar (*Cyamopsis tetragonolobus*). Data structured in fixed-field, numerical code, the codes standing for country of origin, collector, etc.

The pilot project began with these varied types of documents. The following outline-narrative describes the process, the same to be used for preparing any GRC collection for entry to the CIDS. There were certain specific procedures for each data bank that are emphasized – these procedures were 'experiments' to test TAXIR and to demonstrate its capacity. The analyst worked through these steps and called upon the user when necessary.

Initial data description

Step 1. The accuracy and consistency of the written descriptions about each sample data bank were checked against the actual records received. Clarification was requested for any problems that could not be understood and solved without assistance. There were relatively few such problems that could not be solved by the analyst.

Step 2

(*a*) Determine the nature of each ITEM. In all sample banks, the ITEM was a single, unique accession. (The serial, or identification number, assigned to each accession by each organization was maintained as one of the DESCRIPTORS.)

(*b*) Transform the user's set of attributes defining each accession to a set of formal TAXIR DESCRIPTORS and DESCRIPTOR STATES.

This process is usually straightforward; one attribute is one DESCRIPTOR; e.g., attribute 'Plant Height' becomes DESCRIPTOR 'Plant Height'. However, sometimes the transformation is complicated by having a single attribute that will contain more than one DESCRIPTOR. This complex attribute must be divided into simple attributes, and then transformed into DESCRIPTORS, because one may wish to ask questions about each DESCRIPTOR separately.

Most of the transformations were straightforward, but there were two good examples of complex transformations from the International Potato Center (IPC) and from the USDA Potato Collection (USPC). These two examples will be explained in detail and the TAXIR solution to this problem demonstrated below.

IPC/USPC example. One of the 70 attributes for each accession was 'Tuber shape'. This was a complex attribute consisting of three possible observations. These observations were: general shape of the tuber, the shape of the apical end (or crown) of the tuber, and if the tuber were of an odd shape.

Each observation carried a one digit code representing a condition carefully explained on a diagram sheet. A three digit code, then, was used for the attribute Tuber Shape. It was necessary for there to be three DESCRIPTORS.

TUBER SHAPE – GENERAL
TUBER SHAPE – END
TUBER SHAPE – ODD FORM

each represented by the one digit code.

The USPC example was the representation for 'other' virus reactions. There were three 'other' viruses for which a reaction could be recorded. These were: Virus S = code 1, Spindle Tuber Virus = code 2, Virus

432

F = code 3. An accession could be either resistant = code R or susceptible = code S to each virus. Other virus reaction for an accession, then, might be R13 or resistant to Virus S and resistant to Virus F.

The designation for the complex attribute 'other virus reactions' had to be broken into simple attributes, and the machine-readable code R13 had to be unpacked.

The simple attributes and DESCRIPTORS were

Virus S reaction	VIRUS S REACTION
Spindle tuber reaction	SPINDLE TUBER REACTION
Virus F reaction	VIRUS F REACTION

Step 3. The selection of DESCRIPTOR titles. The titles of DESCRIP-TORS should convey meaning, but should be as short as possible to lower costs when making queries. Usually the DESCRIPTOR title selected for the pilot project corresponded directly to the attribute title created by the user. A change was made only with user consent.

Step 4. Identify the form of the DESCRIPTOR STATES for each DESCRIPTOR from each data bank: Is a code used?, is a name used?, are the DESCRIPTOR STATES in some order? For each, determine the OPTION to be used. If a code, determine and list in sequential order the terms represented by the code, e.g.

TUBER SHAPE – END (CODE ROUNDISH, OVATE, ELLIP-TIC, FLATTENED, FLAT).

If an order option, indicate the order associated with the DESCRIPTOR e.g.

VIRUS F REACTION (ORDER 1, R, T, S, H,)
or,
Month of collection (order JAN, FEB, MAR, APR, MAY, JUN, JUL, AUG, SEP, OCT, NOV, DEC,)
or,
Plant Height (order from 1 to 1000 in cm)

If the NAME option is used, merely indicate NAME and estimate the number of DESCRIPTOR STATES for each NAME option DESCRIP-TOR. When the number of DESCRIPTOR STATES was difficult to estimate, it was indicated that the number of DESCRIPTOR STATES could not exceed the number of ITEMS in the data bank. This estimate was used as the first approximation, e.g.

COLLECTOR   (NAME 50)
FLOWER COLOR (NAME 15)

433

NAME was used to determine the range of terms used for any DESCRIPTOR. In this way it can assist in building the Dictionary. For example, the National Seed Storage Laboratory data included SOURCE as an attribute, indicating the group or person who sent the accession to the data bank.

Step 5. If a user wanted the flexibility of either code in, name printed out, or code in, code printed out, then both the CODE and ORDER OPTIONS were used, e.g.

Where 0 = not applicable, 1 = pineapple, 2 = uveous, 3 = concertina, 4 = digitate.

then TUBER SHAPE – ODD FORMS CODE (ORDER from 0 to 4)
    TUBER SHAPE – ODD FORMS (CODE not applicable, pine-
                                   apple, uveous, concertina,
                                   digitate).

Step 6. After completing steps 1 through 5 above, the next procedure is to instruct the computer as to the title of each DESCRIPTOR, the number of DESCRIPTOR STATES for each DESCRIPTOR, the serial order of DESCRIPTORS, and the OPTION, to be used in the data bank. The example below is illustrative:

ACCESSION NO.	(1 ORDER from 1 to 5000),
COLLECTOR	(2 NAME 200),
VIRUS S REACTION	(3 CODE immune, resistant, toler-ant, susceptible, hypersensitive),
VIRUS S REACTION CODE	(4 ORDER from 0 to 5),
COUNTRY OF COLLECTION	(5 NAME 100),
YEAR OF COLLECTION	(6 ORDER from 1950 to 1973)*

Data format conversion

Step 7. None of the data banks selected for the pilot project were in TAXIR input format. If the data were machine-readable, then a modular computer program was used for format conversion. If the data were not in machine-readable format, then a set of instructions was written to be followed by the key punch operator. From our continuing examples:

(1) IPC data were not machine-readable and the instructions for the key punch operator were to insert separators, where needed, to break the complex attribute into simple DESCRIPTOR STATES.

(2) The USPC data were machine-readable and, in general, had to be converted to TAXIR format. TAXIR could be used to make the format conversion of the complex attribute 'other virus reaction'.

(*a*) A DESCRIPTOR was selected to take in the pertinent information:

OTHER VIRUS REACTIONS (2 NAME)

(*b*) A set of DESCRIPTORS was selected to correspond to each VIRUS:

VIRUS S REACTION   (3 ORDER 1, R, T, S, H,)
SPINDLE TUBER REACTION (4 ORDER = 3)
VIRUS F REACTION   (5 ORDER = 3)*

Note the 'ORDER =' designation, which saved rewriting of a previously defined order. With long lists this procedure was exceptionally useful.

There were no recorded attributes in USDA Potato Collection data to correspond to these DESCRIPTORS, and TAXIR would note this and stop the compilation of the data bank. To handle this, the system was instructed to ignore these DESCRIPTORS through the use of an ADJUST DESCRIPTOR statement for the initial compilation of data.

(*c*) Once compiled, a Dictionary for DESCRIPTOR Other Virus Reactions was requested. This showed:

OTHER VIRUS REACTIONS
R1
R13
S3

This indicates that some accessions have these designations.

(*d*) A set of correction statements were then used to 'load' the information into the specific DESCRIPTORS:

CORRECTION VIRUS S REACTION, R for items with OTHER
    VIRUS REACTION, R1 OR R13*
CORRECTION VIRUS F REACTION, R for items with OTHER
    VIRUS REACTION, R13*
CORRECTION VIRUS F REACTION, S for items with OTHER
    VIRUS REACTION, S3*

(*e*) To test that this had been done, TAXIR indicated that the correction had been made. Also, a test is made for all items to see if the information had been properly 'loaded'. See below.

435

Compilation and correction

Step 8. After the data format conversions were completed, the data were compiled (converted to computer storage under the direction of the program). After compilation of ITEMS and their associated DESCRIPTOR STATES, if there were any ERRORS, the ITEM or INSTRUCTION was identified by the TAXIR program. A series of ERROR messages indicated the type of ERROR, and where it was located.

Mistakes in any part of the data bank were corrected, using a facility of the TAXIR program. An example of the use of this correction facility is useful. In the National Seed Storage Laboratory data there is a DESCRIPTOR = SOURCE, which means the person or agency sending the accession. The USDA Regional Introduction Stations throughout the country are sources for much of the material in the laboratory.

The formal designation for such a station is

United States Department of Agriculture
Regional Plant Introduction Station W-6
Pullman, Washington

However, the Dictionary for the DESCRIPTOR SOURCE shows that the following were recorded for the accessions, each different one referring to the same source:

P.I. W-6
USDA/PI/W6 Pullman, Wash.
W6 Pullman
Pullman W6

There were several designations for the other Regional Plant Introduction Station as well. It was necessary to synonymize (p. 438) the various designations for one station. Discussions with Ms Doris Clark of the National Seed Storage Laboratory suggested the use of the following type designation:

'P1 W-6 pullman/Wash,'

The synonymy was executed by the use of a correction statement:

CORRECTION SOURCE, P1 W-6 Pullman/Wash.
for all items with SOURCE, P.1. W-6 or
USDA/P1/W6 Pullman, Wash or W6 Pullman*

Thereafter there was a single designation for this station.

436

The original input data, however, were *not* altered. The only alteration was to the internal TAXIR compressed file, and if, for any reason, a user wanted to return to the original designations, it would be possible, and in fact simple, to do. Then, should the user wish to redetermine the appropriate synonymy and change the single designation, it could be done easily and cheaply.

There is still a different type of synonymy: use of different symbols for similar concepts by various users. An example was found in PEST REACTION as described in the IPC and USDA potato collection DATA. The pests are basically the same and there was basic agreement on the recording of the range of reaction; however, IPC use the following order and codes:

CODE		REACTION
1	=	susceptible
2	=	hypersensitive
3	=	tolerant
4	=	resistant
5	=	immune

There is an inherent range of reaction indicated in the numeric code range.

The USPC uses:

H = Hypersensitive
I = Immune
R = Resistant
S = Susceptible
T = Tolerant

There is no inherent range of reaction indicated in the alphabetic code. The code letter directly represents the reaction term, nothing more.

The information on accessions was synonymized as follows:

(1) The NAME option was used for DESCRIPTOR 'DISEASE REACTION VIRUS X', and a Dictionary term for this DESCRIPTOR was automatically produced by the computer, including both the IPC and the USPC data.

(2) A decision was made to use the code for input and the full expression for query and output. (This decision was arbitrary and for example only.) The OPTION to be used was CODE, and all future data descriptions would use the IPC code list.

437

(3) Current data changes were accomplished as follows:

(*a*) Select a new DESCRIPTOR

VIRUS X REACTION (CODE Susceptible, Hypersensitive, Tolerant, Resistant, Immune)

(*b*) Write a set of CORRECTION statements to load the proper information into this DESCRIPTOR:

CORRECTION VIRUS X REACTION, Susceptible for items with DISEASE REACTION VIRUS X, 1 or S*

CORRECTION VIRUS X REACTION, Hypersensitive for items with DISEASE REACTION VIRUS X, 2 or H*

CORRECTION VIRUS X REACTION, Tolerant for items with DISEASE REACTION VIRUS X, 3 or T*

CORRECTION VIRUS X REACTION, Resistant for items with DISEASE REACTION VIRUS X, 4 or R*

CORRECTION VIRUS X, Immune for items with DISEASE REACTION VIRUS X, 5 or I*

Note the facility to add (or delete) any number or type of DESCRIPTORS even after the initial data compilation.

Synonymy. The above two examples demonstrate how TAXIR can assist users in making decisions on synonymy, then, easily execute these decisions without destroying the original data. TAXIR provides for the development of a wide variety of specific techniques to handle the synonymy problem.

A brief discussion of synonymy and coding may be useful at this point. Synonymization is the process by which DESCRIPTOR STATES for a single DESCRIPTOR are divided into subsets based on their similarity, and a single term is synonymized with the entire subset. Coding is the process of selecting and using a set of symbols to represent exactly a corresponding set of terms, again for DESCRIPTOR STATES for a single DESCRIPTOR.

These processes are interrelated, and can be complex. There are several decisions associated with each process and these must be made by a user group. The role of the analyst is to develop and present the decision framework, i.e. what decisions must be made in what order, and to provide the information to make satisfactory decisions. The information should include an analysis of changes in costs and effectiveness associated with each alternative. The CIDS should assist by providing information for the decision makers, and the capacity to execute the decisions, once made.

The study of synonymy begins with data in the CIDS. For a specific set of items the DESCRIPTOR STATES come from several field scientists who have used various terms to record their observations and measurements. The user group places these DESCRIPTOR STATES into synonymous subsets and chooses a single DESCRIPTOR STATE to represent each subset. The CIDS then executes the synonymy without loss of the original data.

With a predetermined thesaurus, there may be loss of effectiveness for the individual field scientist, but the predetermined thesaurus may produce an attendant decrease in the cost to others using the CIDS. Of course, the synonymized DESCRIPTOR could be added to the system without removal of the original DESCRIPTOR, and this will maintain the level of effectiveness for an individual field scientist, but will decrease the cost in using the CIDS for others. However, there will be an attendant increase in the cost of operating the CIDS.

Again, changes could be made in the formal documentation file in one way, and the working file in another – with different cost changes. These cost variations must be taken into account in the decision process.

Codes can be exceptionally useful or exceptionally costly. The cost savings for using a code over an exact term for computer storage and retrieval may cause a severe decrease in effectiveness if the user must constantly refer to complex tables to decide upon the correct code. Mistakes made in coding are very costly to find, while mistakes made in writing longer terms are not difficult to locate. On the other hand, codes can provide lower costs for initial field data recording, and a useful structure for the field worker. In general, the analyst should assist the user group in making decisions about code use for each DESCRIPTOR.

Effectiveness. The pilot project has demonstrated the effectiveness of TAXIR in meeting stringent user requirements; however, the pilot project was too small to provide meaningful data on relative costs. Costs will be studied in detail when the CIDS accepts the first major data bank. Information from this study will provide the basis for accurate cost projection for CIDS installation and use. It will also provide the baseline data to compare any module in the system with other external programs and to judge internal development.

Costs. There is an inherent cost structure in GRC work which is presented in Table 32.3. This merely shows the steps in the GRC process

Table 32.3. *GRC process and functions*

Process step order (highest = 1)		Process cost order (highest = 1)
1	Plant collecting and evaluation program	8
2	Install CIDS and instruct in use	3
3	Set up standards	9
4	Collect germplasm	2
5	Store germplasm	4
6	Evaluate germplasm (all steps)	1
7	Record and submit information to CIDS	7
8	Convert DATA format for input to CIDS	5
9	DATA correction	10
10	DATA synonymization	12
11	Information storage and retrieval	13
12	CIDS evaluation and development	11
13	—	6

for any data bank. The ranking of steps according to costs is at the right. The costs associated with collection and full evaluation are probably 5 to 20 times the cost of installation and use of the CIDS (Process steps 2, 8, 9, 10, 11).

Once the information is in the system, the cost of storing and retrieving data is very small. As an example the computer charges and human costs to execute a complex query (10 Boolean operators) on a data bank with 50 DESCRIPTORS and 5000 ITEMS retrieving 8 DESCRIPTORS was less than $0.50.

The question of how much will it cost to develop and install the CIDS at some central computer center and the cost of continuous operation cannot be answered here; but the design of the CIDS allows for this information as the work begins and continues.

Summary of the pilot project

We have given a very short, step-wise description of the activities of the pilot project, to indicate the many considerations necessary in preparation for documentation of genetic resource materials. Indeed, the description given may be too short to provide a thorough understanding of the philosophy, the methods, or the desired results, but limiting factors of space dictate this abbreviated presentation.

We learned from the pilot project that a systematic, well-designed approach is necessary to accomplish the goals of documentation for a global network. A random application of a set of computer programs, no matter how elegant and sophisticated, will be ineffective unless they are understood by the users, and a set of guiding principles are employed under the direction of the scientists responsible for genetic resource conservation.

We learned that many of the problems associated with attempts to 'computerize' the data from genetic resources are unnecessary problems because of poor understanding of the capabilities of computing machines. Attempts at coding of data, or use of fixed-field format *on the basis of misunderstood computer requirements* have unfortunately forced the genetic resource scientist to lose much of his valuable data, and accept an unnecessary set of coding procedures with which he is not familiar.

By careful reading of the steps listed to use the TAXIR System, it is possible to see how the computing machine becomes a valuable tool in documentation, not only as a secure resting place for the data, but also as an active means by which the scientist is aided in his substantive work. A pragmatic problem is that we already have much valuable information, though incomplete, in coded or fixed-field format. We must use these data for some time to come, and the pilot project has demonstrated that they can be placed in the TAXIR System at relatively low cost. The correction facility, the updating (or deletion) and addition facility of the system make it possible to use the data with more precision and accuracy.

From our experience in the pilot project, we start the processes of documentation of Genetic Resources Center data with a core system which so far accomplishes the tasks required. We expect that there are areas requiring improvement, and we must give constant attention to the needs and methods.

Implementation strategy and program

The systems analysis through the pilot project has to date provided much of the information necessary to begin the implementation of global network CIDS. This information can be summarized in a simple strategy, and further defined in an initial work program.

(A) Strategy

(1) A CIDS design/implementation/operations team should be chosen and funded to begin the installation of the CIDS.

(2) TAXIR should be used as the initial core module for a global CIDS.

(3) A CIDS base installation center should be chosen. The base installation center should have sufficient computer hardware capacity for the installation of TAXIR and to handle the expected CIDS demand.

(4) Users from around the world should be solicited for co-operation with the CIDS team. These users should advise on every phase of analysis, design, implementation (including orientation and training) and evaluation.

(5) The initial costs to all users must be minimized for reasons indicated earlier. This means that all data format conversion should be done at the CIDS base, and most initial installation costs be met by the CIDS central organization (perhaps with financial assistance from established GRCs).

(6) The global network policy group should assist in setting up several working committees on standards and evaluation.

(7) Financing necessary to implement the system should be sought, and in sufficient amount to insure successful initial operation.

(B) Program

To follow through on the above strategy the following steps should be taken.

(1) Hire the chief of the CIDS team.

(2) Select CIDS base, negotiate cost–service aspects.

(3) (*a*) Continue analysis to determine initial user demand and factors that will cause changes in demand.

(*b*) Install TAXIR.

(4) Set implementation priorities.

(5) Visit GRCs within implementation priority list. Do a specific and detailed analysis of each. Begin to input data.

(6) (*a*) Meet with standards committee, begin to determine standards with information from priority GRC Banks.

(*b*) Test standards.

(7) Continue priority GRC Data input.

(8) (*a*) Work with Evaluation Committee on initial work to date. Determine CIDS changes to be made.

(*b*) Plan, develop and implement CIDS changes.

(9) Continue priority work.

(10) Policy Board reviews priorities. Enter new GRC Banks and continue. Specific time phasing and budgets will be further developed

as the work continues. All this should be under way in the very near future for the time is right to begin.

The Taximetrics Laboratory team is grateful to the following for presentation of sample data and for advice on the structure of the pilot project.

Dr L. Bass and Mr D. Clark, USDA National Seed Storage Laboratory (NSSL), Fort Collins, Colorado, USA; Dr J. Craddock, USDA Regional Plant Introduction Station, Beltsville, Maryland, USA; Dr S. Deitz, USDA Regional Plant Introduction Station, Pullman, Washington, USA; Sir Otto Frankel, CSIRO, Canberra City, ACT, Australia; Mr D. R. Glendinning, Scottish Plant Breeding Station, Commonwealth Potato Collection (CPC), Roslin, U.K.; Professor J. G. Hawkes, University of Birmingham, Birmingham, UK; Dr J. León, Crop Ecology and Genetic Resources Unit, FAO, Rome, Italy; Dr P. R. Rowe, formerly at Agricultural Research Service, Crops Research Division, University of Wisconsin, Madison, Wisconsin, now at IPC, La Molina, Peru; Dr R. L. Sawyer, International Potato Center (IPC), La Molina, Peru; Miss Ayla Sencer, Agriculture Research and Introduction Centre, Izmir, Turkey, and many others.

Selected references

Cohen, B. J. (1971). *Cost-Effective Information Systems*. American Management Association, Inc., New York.

Estabrook, G. F. & Brill, R. (1969). Theory of the TAXIR accessioner. *Mathematical Biosciences*, **5**, 327–40.

Fano, R. M. (1961). *Transmission of Information*. The M.I.T. Press, Cambridge, Mass.

Frankel, O. H. & Bennett, E. (eds) (1970). *Genetic Resources in Plants – their Exploration and Conservation, IBP Handbook*, no. **11**. Blackwell, Oxford and Edinburgh.

Gruenberger, F. (ed.) (1969). *Critical Factors in DATA Management*. Prentice Hall, Englewood Cliffs, New Jersey.

Langefors, B. (1966). *Theoretical Analysis of Information Systems*, vols. I and II. Studente Litteratur, Lund, Sweden.

McMillan, C. & Gonzales, R. F. (1968). *Systems Analysis*. Richard D. Irwin, Homewood, Ill.

Menzel, H. (1966). Information needs and uses in science and technology. In *A. Rev. Inform. Sci. Tech.* (ed. Carlos Cuadra), pp. 41–71. Wiley, New York.

Rogers, D. J. (1970). Theoretical and practical considerations on data structuring for a computerized information retrieval system. In *Archeologie et Calculateurs, Marseilles*, pp. 145–59 (ed. Centre National Recherche Scientifique, Paris VII).

Rogers, D. J. & Appan, S. G. (1969). Taximetric methods for delimiting biological species. *Taxon*, **18**, 609–24.

Appendix 1

Definitions

CIDS

In this paper the terms Documentation, Documentation Function, Documentation System are replaced with the concept of a Communication–Information–Documentation System (CIDS). Each term has a specific definition.

Information is either

(*a*) primary data, or

(*b*) secondary source material derived from the use of primary data, e.g. aggregate statistics, reports, papers, etc., or

(*c*) specific research plans, judgements, opinions, evaluations, programs, methods, etc. that have value to any global network user.

Communication is the systematic transfer of information among global network users,

(*a*) upon user demand and

(*b*) periodically to a prespecified group.

Documentation is the application of a general set of rules and procedures to the recording, preserving, deleting and communicating of information in the global network.

System is a set of integrated and interdependent processes and procedures for attaining a set of objectives.

The most important of these is a general computer based information storage and retrieval technique (IRT). This IRT is probably the single most important component in the design of the CID system.

Demand. In the technical language of economics a demand schedule or a demand curve demonstrates the functional relationship between the quantity of goods or services that will be purchased from a market at a given price for a unit of goods or services.

In this paper the concept of demand subsumes the above technical definition and goes beyond it encompassing the analysis of a series of demand schedules pertaining to the genetic resources documentation work.

Genetic Resources Center(s) – GRC(s). In this paper a Genetic Resources Center is given the broadest definition possible:

A facility at which a competent collection of genetic resources resides (or will reside). The collection may be single or multi-crop and may be designated as either a Base or Active Collection (see Chapter 37). It is expected that there are one or more users of the collection at the center, but many more in other institutions.

Appendix 2

Costs

A predominant question is what is the expected capital (resource) requirement for completely executing 1(*a*)–(*d*) (p. 420)? At this stage it is exceptionally difficult to make an estimation. One then is asked, and rightly so, what might be an expected level of magnitude. A hint can only be given at this time.

(*a*) The establishment of the CIDS for global genetic resources for all crops from all competent collections is a major task. If the data are in machine-readable format for most collections the costs are, in fact, a magnitude less than if the data are not in machine-readable format. At current prices, it may cost from $0.10 to $0.50 to transform the information on one accession to machine-readable format. (There are low-cost international services available for this type of bulk work.)

Setting up, maintaining and getting the data into the system is not a technical problem. The costs from planning to implementation of the basic system may run at from $250000 to $500000 depending on network complexity. Some data entry costs are covered in the above. But should the data exceed an expected two million accessions in the first two years the above cost may be greater.

The cost of the use of the system for updating and querying is the most difficult to estimate and could run at a low of $50000 per year to a high of $250000 per year – depending on the computing system used.

(*b*) The costs of installation at any given center are highly dependent on the equipment available – and the skill level of the personnel available.

Most often the facility to be used is large enough to have competent personnel. Given a relatively standard set of equipment with which to work (IBM, Siemens, CDC, UNIVAC, Honeywell, some 'minis') – the conversion, installation, and training costs might run from a low of $2000 per center (the center personnel doing most of the work) to about $7500 per center. This is clearly dependent on the continued existence of a competent CIDS team to do this work.

The actual use cost for any given center depends on the center's price/use relationship with the equipment leaser. This is so variable that cost estimates are almost impossible to give. The actual use a center will make of the system is also difficult to judge. It is expected that use may start slowly but, as the value of the system is judged, use rates will accelerate greatly.

By appropriate budgeting, each center can control its use costs. Realistically, a center with two or more users should expect minimum use costs of $50–$100.

Please remember these figures are dependent on current prices and at best are exceptionally tentative.

PART VI

Genetic resources centres

33. Crop germplasm diversity and resources in Ethiopia

M. H. MENGESHA

There are many cultivated crops in Ethiopia which show considerable genetic diversity (Vavilov, 1950; Zohary, 1970; Frankel, 1973; Harlan, 1969; 1973). The most important ones are the following, as listed in their approximate order of importance: coffee, teff, barley, sorghum, wheat, chick pea, horse bean, field pea, lentil, flax, neug (*Guizotia abyssinica*), ensete (*Ensete* sp.), millet, sesame, fenugreek, brassica, pepper, onion, ground nut, and other vegetables. Forage grasses, legumes and castor bean also show a high degree of diversity although they are generally not cultivated and are mainly treated as wild species. Among the diverse list of crops, the broad genetic diversity of the wheats, barleys, and coffee is probably under the greatest threat, in wheat due to new varietal influx, in coffee to clearing of forests and forest fires, and in barley due to the gradual change of cropping patterns.

An estimate of the diversity of some major Ethiopian crops is given in Table 33.1 which is compiled on the basis of personal observations and communications. More thorough data collection on the nature of genetic variability of crops in different ecological areas would be useful for future collection and conservation purposes.

The tremendous genetic diversity that exists in Ethiopia deserves much more attention than it has received so far. Scientists from many parts of the world have collected Ethiopian germplasm mainly from areas that are easily accessible. The less accessible areas are still largely untouched and serve as important natural reservoirs of germplasm which deserves urgent collection and conservation.

The major areas that need more concentrated efforts of genetic exploration are shown in Fig. 33.1. Most of these areas are not easily accessible and as such they have been more or less neglected by previous collectors. Some of the areas like Yerrer and Keryu are abundant in crop types and are historically important but have not been exhaustively explored. The only organized collection in this area was made in 1967 by the Kyoto University team which reported high variability in various crops.

Many of the high elevation areas of Ethiopia, ranging from 8000 to

449

Table 33.1. *Estimate of crop diversity in Ethiopia based on personal observations and communications*

Province	Teff[2]	Barley	Sorghum	Wheat	Coffee	Neug[3]	Millet	Horse bean	Pea	Chick pea	Lentil	Ensete[4]	Flax
											Estimate of crop diversity[1]		
Arusi	L	M	M	M	—	T	L	L	L	L	L	L	T
Bale	L	T	M	T	T	—	M	T	—	—	—	T	—
Beghemdir	H	M	M	L	—	M	M	L	L	M	L	T	T
Eritrea	L	M	H	H	—	T	M	L	M	M	L	—	T
Ghemugofa	T	T	H	T	T	T	L	L	L	—	—	M	—
Gojjam	H	H	M	M	T	L	L	T	M	M	M	—	H
Harerghe	L	L	H	M	H	—	L	L	T	T	—	L	—
Illubabor	L	L	M	T	M	T	L	T	L	T	—	L	—
Kefa	L	T	M	T	H	T	L	T	T	—	—	M	—
Shewa	H	H	M	H	T	M	—	H	H	H	M	M	L
Sidamo	L	L	L	T	M	—	T	T	—	—	—	H	—
Tigre	H	H	H	M	—	M	M	M	M	M	M	—	M
Wellega	M	L	M	L	M	M	M	T	T	L	—	M	T
Wello	H	M	H	M	T	M	L	M	M	M	L	L	T

[1] H, M and L represent high, average and low diversity; T represents trace appearance, and — represents no data.
[2] Teff = *Eragrostis tef* (Zucc.) Trot.
[3] Neug = *Guizotia abyssinica* L.
[4] Ensete = *Ensete* spp.

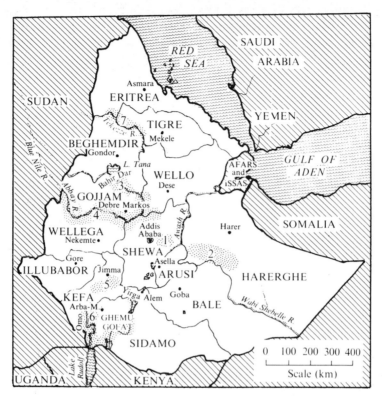

Fig. 33.1. The major areas of Ethiopia needing more concentrated efforts of genetic exploration.

12000 feet, also have little known germplasm that has been neglected in the past. The wheats, barleys, sorghums, pulses, grasses and legumes of the high elevations in central and north-central Ethiopia are renowned for their genetic diversity.

In the past many plant breeders have discovered some highly desirable genetic characteristics such as disease resistance (see Chapter 5 by Qualset), high lysine content in barley (Munck, Karlsson, Hagberg & Eggum, 1970; Munck, 1972), and rust resistance in wheat (Creech & Reitz, 1971) in relatively small Ethiopian germplasm collections. For example, Qualset reported that two per cent of the world collection of barley showed virus resistance and that all resistant strains were of Ethiopian origin. A more extensive collection of germplasm from the inaccessible areas of Ethiopia could therefore extend the range of desirable genetic characteristics for breeding purposes.

So far the traditional Ethiopian farmer has played a significant role

451

in germplasm conservation by knowingly or unknowingly maintaining a broad genetic base in his seed stocks. He did this by using his own seed stocks which often trace back for several generations. He used his ancestral seeds in spite of their very low yields. We are no longer certain that this situation will continue since the farmer has now started to abandon his traditional seeds and is looking for the newly bred, more homogeneous, high-yielding varieties. While such a step results in yield improvement, it threatens germplasm erosion. This situation presents a dilemma which can be resolved only by systematic efforts to conserve the ancient stocks.

Some germplasm collecting has been undertaken by the College of Agriculture of the Haile Selassie I University and the Institute of Agricultural Research. Both institutions have maintained their collections as 'breeder's' or 'active' collections. So far, however, there is no centre in Ethiopia sufficiently equipped to house a long-term or 'base collection' (see Chapter 37 for definitions).

Clearly there is an urgent need for a comprehensive centre in Ethiopia, equipped and staffed for an active, well-planned and co-ordinated programme for exploring, collecting and conserving the rich and now threatened germplasm resources of the country. There is also a need for conserving, *in situ*, germplasm of coffee where it can still be found in relatively undisturbed rainforest communities. Ecologically diverse habitats should be identified and managed as nature reserves in the interest of the nation and the world.

There is a strong case for the establishment of a regional genetic resources centre in Ethiopia, as recommended in the reports of the Beltsville group and the FAO Panel of Experts on Plant Exploration and Introduction (see Chapter 37). The centre is to serve in the first instance as a base station for the exploration, evaluation and conservation of germplasm in Ethiopia. It will form an important link in the proposed international network of genetic resources centres. It will also provide university-oriented teaching and research facilities in the whole field of genetic resources, and particularly in germplasm conservation, which will strengthen the cadre of trained workers in this field.

In years to come this centre should be in a position to guide a co-operative effort of genetic conservation in the East African region, including Sudan, Kenya, Somalia, Uganda, Tanzania, Zambia, Malawi and, across the Gulf of Aden, Yemen and Saudi Arabia.

The proposed genetic resources centre could be established in close association with the College of Agriculture, Haile Selassie I University, and with the support and co-operation of the Institute of Agricultural Research and the Ministry of Agriculture. The host institution should be selected on the basis of its location, availability of technical and supporting staff, possibilities for university teaching, an already developed and genetics-oriented research programme, interest in the development and operation of the genetic resources centre and its participation in the international network, and freedom to operate under relaxed quarantine regulations. The majority of these requirements seem to point to the Debre Zeit Agricultural Experiment Station of the College of Agriculture in the Haile Selassie I University, which is located only 45 kilometres east of Addis Ababa.

The College of Agriculture and the Institute of Agricultural Research will form the two most important national centres in the country which will support and strengthen the Ethiopian Genetic Resources Centre. Organized plant exploration will be handled by the two national institutes as a joint responsibility. The base collection of genetic material will be at the genetic resources centre. Active collections will continue to be maintained at the national centres which will undertake medium-term storage, regeneration, multiplication, distribution, evaluation and documentation.

References

Creech, J. L. & Reitz, L. P. (1971). Plant germ plasm now and for tomorrow. *Adv. Agron.*, **23**, 1–49.

Frankel, O. H. (ed.) (1973). *Survey of Crop Genetic Resources in their Centres of Diversity. First Report.* FAO/IBP, Rome.

Harlan, J. R. (1969). Ethiopia: a center of diversity. *Econ. Bot.*, **23**, 309–14.

Harlan, J. R. (1973). Genetic resources of some major field crops in Africa. In *Survey of Crop Genetic Resources in their Centres of Diversity. First Report* (ed. O. H. Frankel), pp. 45–64. FAO/IBP, Rome.

Munck, L. (1972). High lysine barley – a summary of the present research development in Sweden. *Barley Genetics Newsletter*, **2**, 54–9.

Munck, L., Karlsson, K. E., Hagberg, A. & Eggum, B. O. (1970). Gene for improved nutritional value in barley seed protein. *Science*, **168**, 985–7.

Vavilov, N. I. (1950). The origin, variation, immunity and breeding of cultivated plants. *Chronica Botanica*, **13**, 1–366.

Zohary, D. (1970). Centers of diversity and centers of origin. In *Genetic Resources in Plants – their Exploration and Conservation* (eds O. H. Frankel and E. Bennett), *IPB Handbook*, no. **11**, pp. 33–42. Blackwell, Oxford and Edinburgh.

34. A regional plan for collection, conservation and evaluation of genetic resources

E. KJELLQVIST

As one of the most ancient centres of cultivation, the Near East is the richest source of genetic diversity in the eastern hemisphere. The survey of genetic resources (Chapter 37) shows that it is also the most acute emergency area, with many or most of the germplasm reservoirs depleted or in jeopardy. The countries most concerned, Turkey, Syria, Iraq, Iran, Afghanistan and Pakistan, determined to salvage what is left of their genetic heritage, have agreed on a regional plan for collaboration in germplasm collection, conservation and evaluation. The plant genetic resources unit of the Agricultural Research and Introduction Centre at Izmir in Turkey is to act as the regional centre (see Chapter 37). The Izmir unit was established in 1964 under a joint project between the Turkish Government, the United Nations Development Program/Special Fund, and FAO. Since that time it has gained considerable experience in the exploration, conservation and documentation of the genetic resources of Turkey.

The aims of the regional plan are as follows:

(*a*) to carry out a systematic plant exploration programme in the participating countries on a co-operative basis;

(*b*) to provide safe long-term storage of base collections in order to preserve the genetic resources of the region both for its own use and that of the rest of the world;

(*c*) to provide a regional documentation centre;

(*d*) to provide facilities for short-term storage of active collections and introductions in each of the member countries;

(*e*) to provide regional co-operation in germplasm evaluation and utilization.

The overall direction and co-ordination of the plan will be the responsibility of a Board of Management, composed of representatives of each country's programme and an international co-ordinator. An International Scientific Advisory Council, which will also be represented on the Board, will advise on technical matters.

The countries will carry out a jointly agreed programme for plant exploration and maintenance of collections, which is to be reviewed annually. A co-ordinator will be appointed by an international agency who will organize meetings of the scientific staffs in one of the network centres or at the regional headquarters.

The Izmir Centre, already well equipped with storage facilities and auxiliary services, will maintain a basic collection for the region as a whole. It will receive a portion of all collected samples together with collection records and herbarium specimens, and carry out a final identification and documentation. Subsamples will be lodged for safe-keeping at one or more co-operating centres in other regions, and the Izmir Centre will offer similar facilities to other base collections.

All seed collections will be available to the members of the network and also on request to the scientific community elsewhere.

The Izmir Centre will be the documentation centre for the region and will seek co-operative data exchange with other centres.

Regeneration of seed stocks will be carried out with due regard for ecological requirements. For accessions from within the region this will as a rule require regeneration to be carried out in the country of origin.

It is proposed that evaluation of accessions be co-ordinated so that the greatest benefit can be derived from observations over a range of environments.

Training is an important aspect of the plan, with practical training courses within the region and professional training at a university.

35. IRRI's role as a genetic resources center

T. T. CHANG, R. L. VILLAREAL, GENOVEVA LORESTO
& A. T. PEREZ

When the International Rice Research Institute was started in 1960, it was to establish a varietal collection of rice as one of the most useful ways by which the Institute could be of service to rice breeders. With the assistance of many national rice research centers and the support of international organizations, the IRRI germplasm project has grown well beyond the scope and expectations of the initial plan.

Building the rice collection

Early contacts with national research leaders, the United Nations Food and Agriculture Organization and the US Department of Agriculture resulted in the first genetic stocks from diverse sources reaching IRRI. During 1961 and 1962 the geneticist, T. T. Chang, and the plant breeder, P. R. Jennings, approached many rice experiment stations in Africa, Asia, Australia, and Latin America, requesting seed samples for the IRRI germplasm collection, and half of them responded. The collection of rice cultivars was rapidly built up from the 256 accessions in 1961 to 6900 from 73 countries a year later. The bulk of the collection came from Ceylon, India, Indonesia, Japan, Malaysia, Pakistan, Philippines, Taiwan, Thailand, USA, and Vietnam.

Since then the collection has grown steadily in size and genetic coverage. Researchers in India, Japan, Taiwan, and the USA deposited with the bank their genetic testers, mutants, African rices (*O. glaberrima* Steud.), and wild species of the genus *Oryza*. Further exchanges of seeds were prompted by visits of IRRI staff members to rice experiment stations abroad, or by foreign researchers to Los Baños. Farmers in Ceylon, India, the Philippines, and Latin America, after reading about the work of IRRI, sent in varieties that they considered unusual.

During 1970, leading rice breeders reviewed the composition of the IRRI collection and its relationships to existing major national collections (Chang, 1972a). These breeders suggested that a co-ordinated action program be organized quickly to collect existing rice germplasm.

457

A meeting of representatives of national rice research programs, USDA, the Rockefeller Foundation, and IRRI at Hyderabad later in the year led to the formation of the International Rice Collection and Evaluation Project (IRCEP). It appointed a technical committee to organize a co-ordinated program by pooling available financial and technical resources. At the symposium on rice breeding held at IRRI in 1971, the 100 rice breeders present recommended that IRRI assume the leading role in organizing a comprehensive genetic conservation program (IRRI, 1972).

During 1971 and 1972 IRRI's staff participated in national collection projects of Ceylon, Indonesia, and Khmer and this resulted in the collection of 4000 samples, mostly from remote areas. IRRI also assisted projects funded by the US Agency for International Development (USAID) in Nepal and South Vietnam, which added 1133 samples to the IRRI bank. The IRRI collection now contains 23 560 cultivars and breeding lines, 316 genetic testers, 326 mutants, 375 African rices, and 1078 strains of wild taxa.

Composition of the rice collections

The samples in the IRRI germplasm bank came from rice scientists, botanists, university staff, managers of private farms, educated farmers, extension workers, anthropologists, Peace Corps members, missionary workers, and officers of commercial firms.

A comparison between national collections and the estimated total numbers of rice cultivars in each country shows that few national collections have a comprehensive coverage of indigenous germplasm (Chang, 1972*a*). Moreover, most national collections contain many leading commercial varieties, but are deficient in less important, obsolete, and primitive varieties. Most national collections contain many foreign introductions that are duplicated in several other collections. Thus rice collections have a redundancy of improved varieties and a deficiency of primitive ones.

Services provided by IRRI

A person requesting seed from IRRI may be a rice breeder, an agronomist, a plant pathologist, a soil chemist, a plant physiologist, an extension worker, a university professor, a biochemist, a private farm manager, a missionary worker, a Peace Corps member, or an educated rice farmer. Except when the requesting party explicitly indicates the desired cultivars or strains, we select the accessions that meet their specific needs under their ecological conditions. IRRI provides the necessary quantities of seed from the most readily available seed stocks. Often an exchange of letters is needed to clarify either the objectives of the experiment or the prevailing environment at the test site. The plant quarantine regulations in the particular country are consulted and the seeds are treated with the appropriate chemicals and later certified by the Bureau of Plant Industry of the Philippines. In cases where quarantine regulations restrict transfers of seed outside a quarantine zone, IRRI arranges where possible to meet requests by supplying seed from a source within that region. When the requested seed is not available at IRRI, the requesting party is advised to try an alternative source. Technical information related to the seed material is supplied, often with a word of caution about the applicability of crop data obtained at Los Baños to ecological conditions at a different site (Chang, 1971). In some cases IRRI supplies seed of new varieties or breeding strains that the original breeding station is unable to supply because of restrictions imposed by government policy.

To serve research needs on a regional basis, IRRI has supplied hundreds of selected accessions to three centers for further screening: the International Institute of Tropical Agriculture (Nigeria) – for drought and blast resistance; the Centro Internacional de Agricultura Tropical (Colombia) – for tolerance to acid sulphate soils; and the US Department of Agriculture in collaboration with the rice stations in California – for early maturity, plant type, and short stature from tropical and temperate areas.

All seed samples for foreign shipment are airmailed or airfreighted free of cost, except when large quantities are involved or when repeated requests are made within short periods of time. From 1963 to 1972, IRRI sent out about 34000 seed packages from the bank in response to 930 requests from more than 100 countries and territories. Moreover, more than 70000 seed packages from the Institute's breeding program were furnished to numerous requesting parties.

By providing the above services, IRRI has established an efficient system of seed increase, characterization and evaluation, and seed preservation and distribution. Its scientists use all available channels of communication to disseminate information on promising genetic material: publications in international scientific journals, IRRI publications, IRRI symposia and international research conferences, other

459

international scientific conferences, staff visits to experiment stations in other rice growing countries, and discussion with foreign visitors to Los Baños. The vast amount of information accumulated at IRRI on varietal resistance to insect pests and diseases (see Chang *et al.*, Chapter 16) and response to ecological factors (Chang, 1967; Chang & Vergara, 1971) has substantially assisted IRRI researchers in selecting the appropriate genetic material for experiments in many countries. Persistent efforts to grow and preserve poorly-adapted or unproductive accessions have made possible the perpetuation of such types.

The tropical climate at Los Baños makes possible year-round planting of rice. Well laid out experimental plots with independent irrigation and drainage facilities place no restriction on different dates of planting or different types of experiment. Screened nursery beds protect valuable materials from damage by birds or rats. Seeds can be dried by heat or dehydrating agents. IRRI is also fortunate to have competent young Filipinos to assist in the experimental operations. Finally, refrigerated seed storage at 8 to 9 per cent moisture content, 3 to 4 °C, and 25 to 40 per cent relative humidity, facilitates longevity of seed in spite of humid tropical conditions and reduces the need for repeated regeneration by field plantings. All these factors have enabled the IRRI bank not only to satisfy thousands of requests but also to restore hundreds of varieties which had been lost from national collections and to supply several countries with obsolete varieties that are no longer available in their national collections. The availability of ample field space made possible the initial supply to India of the one ton of seed of Taichung Native 1 in 1965 which started the large-scale adoption of semidwarfs in that country (Dalrymple, 1971; Pal, 1972). IRRI has supplied the bulk of the seeds required for international trials sponsored by FAO/IAEA and by IBP (Chang, 1970). On the other hand, the continuous planting of many potentially susceptible genotypes on the IRRI farm has raised the level of diseases and insects and necessitated intensive plant protection measures (Chang, 1971).

Recent activities in genetic conservation and evaluation

It is gratifying that IRRI's conservation and evaluation efforts have stimulated national activities in the same directions. Since evaluation is described in Chapter 16, discussion will be confined to recent field collection and preliminary evaluation which are co-ordinated under IRCEP.

Since IRCEP was organized in 1970, its technical committee has helped several national agencies to organize and implement field collection programs. The committee consists of representatives from the All-India Coordinated Rice Improvement Project, USDA, the Rockefeller Foundation, and IRRI. It helps in reviewing the status of conservation and evaluation in the country concerned, in planning further operations, mobilizing the required financial resources, and in providing the necessary technical assistance (IRRI, 1972). The national contribution generally consists of technical personnel, transport facilities, travelling expenses, equipment and supplies. Outside assistance may come from USAID, USDA, IRRI, the Ford Foundation or the Rockefeller Foundation. The support from USAID or USDA may consist of funds for local travelling, transportation facilities, or domestic project personnel employed for the collection and evaluation period. IRRI may partly fund an exploration project (Oka & Chang, 1964) or contribute a staff member who organizes the collection operations, trains the field staff and participates in collecting activities. IRRI also supplies the necessary materials and a manual for field collectors (Chang, Sharma, Adair & Perez, 1972) and assists the collaborating national center in processing the collected samples. In return, the national center provides IRRI with a duplicate set of collected samples for evaluation and storage.

While the collection of indigenous rice germplasm in South and South-East Asia is progressing well, IRRI is directing its effort to the next phase of evaluation and maintenance by national centers. The lack of refrigerated facilities for seed storage in most tropical countries, the accompanying need to renew seeds every year, and the lack of trained personnel have placed such a heavy burden on most national centers that valuable accessions continue to disappear from the annual plantings or become badly mixed (Chang, 1972*b*). While seed storage facilities are gradually becoming available to some tropical countries, improvements need to be made in operations related to (1) the registration of accessions and the removal of duplicate samples; (2) the morpho-agronomic description of accessions for cataloging; (3) systematic screening of indigenous germplasm and foreign introductions for useful traits; and (4) the storage and distribution of seed. IRRI has helped by providing short-term training at Los Baños and training programs at national centers. We are preparing a manual on evaluation, characterization, genetic maintenance, storage, and distribution. IRRI is collaborating with FAO on the handbook for plant introduction and on the bulletin on rice terminology (FAO, 1972).

461

Recording and documenting crop data is another area where IRRI has collaborated with other international organizations in standardizing nomenclature and methods. The recommendations of the International Rice Commission and the Symposium on Rice Genetics and Cytogenetics (IRRI, 1964) led to the IRRI bulletin on the morphology and varietal characteristics of the rice plant (Chang & Bardenas, 1964) and the monitoring of rice gene symbols (Chang & Jodon, 1963) and of genome symbols (IRRI, 1964). The coding of data in the IRRI variety catalog (IRRI, 1970) followed the principles worked out by the FAO/IAEA Working Group on International Standardization in Crop Research Data Recording, of which IRRI was a member. We shall continue to collaborate with the EUCARPIA working group on the standardization of records on plant genetic resources.

IRRI's role in a global network of genetic resources centers for rice

There are limits to the execution of genetic conservation within any one institution, which render both international collaboration and technical improvements indispensable.

The IRRI collection has 25000 accessions and is growing by several thousand a year. To facilitate efficient maintenance, duplicate accessions will need to be discarded. This will call for critical comparative studies at IRRI as well as the collation of information from national centers to facilitate the work of consolidation. At present such information is largely lacking. Refined biometrical techniques and biochemical analyses may assist in identifying true duplicates.

The regeneration of seed at a single center poses serious problems. We have experienced difficulty in growing plants and gathering seeds from (1) cultivars from Europe, Japan, and Korea because of their thermo-sensitivity or photoperiod-sensitivity and their susceptibility to tropical virus diseases and to the insect vectors (Chang & Vergara, 1971); (2) the African rices because of their extreme susceptibility to bacterial leaf blight and virus diseases (Bardenas & Chang, 1966); (3) the Ifugao rices from high elevations in the Philippines due to their erratic growth behavior at Los Baños; and (4) directly-seeded Australian, Surinam, and US types that are extremely low tillering and generally susceptible to the virus diseases and to stem borers. Moreover, many tropical cultivars show discernible variability in ecological features or disease reactions. Most wild species are susceptible to the virus diseases

462

of the tropics. Therefore, it appears most desirable to maintain different ecological groups at centers where the environmental conditions are closest to those of their native habitats, to facilitate good seed production and minimize changes in genetic composition (Frankel, 1970).

These problems indicate the need for a network of co-operating centers. It should consist of one or more institutes which maintain major collections and assume the over-all task of preservation and exchange of material and information; regional centers which maintain sizeable working collections, each center serving an ecologically-based region; and national centers which maintain selected indigenous germ-plasm and breeding lines.

At present cultivars in the IRRI collection are handled as a 'library collection', i.e. each accession, after judicial purification or splitting, is maintained as an individual accession in evaluation, utilization, and conservation. The 25000 accessions pose a heavy burden. Although duplicate samples can be eliminated by critical comparisons, there is no foreseeable means of further discarding duplicates among the less important cultivars. One way to reduce the number is to preserve all accessions in long-term storage, while only a selected segment is stored in large quantities and distributed for experimental use. This would require a genetic cataloging of the sources of such useful traits as short plant stature (Chang & Vergara, 1972) and pest resistance (Chang *et al.*, Chapter 16) so that the set would contain all of the useful and non-allelic genes.

The long-term storage of rice seed is not as difficult as the storage of oil seed crops, but existing facilities at IRRI can only ensure from 10 to 15 years of seed longevity between regeneration cycles. New storage facilities similar to those at the National Seed Storage Laboratory of Japan (Ito & Kumagai, 1969) are needed. We have deposited a duplicate set of 10450 seed samples with the US National Seed Storage Laboratory at Fort Collins. We look forward to the co-operation of other gene banks, such as the Institut für Pflanzenbau at Braunschweig, to assist IRRI with seed storage. Multiple sites for seed preservation not only guard against possible losses due to human error and change of control, but also provide insurance against typhoons and earthquakes which occur frequently at Los Baños and cause electric blackouts. High temperature and humidity in the tropics greatly reduce the efficiency and life span of refrigeration equipment.

Conclusions: the role of a crop-specific institute

The research program of a crop-specific institute focuses on the primary needs of the crop. Its multi-disciplinary staff should include workers trained in exploration and conservation as well as specialists in the various disciplines involved in evaluation and utilization. Adequate facilities for field, laboratory, and greenhouse experiments, coupled with a comprehensive library, permit the staff to execute basic studies that are vital to conservation and evaluation. Moreover, when the primary function of the institute is oriented toward the solution of basic problems that limit crop yield, the researchers are encouraged to make immediate use of research findings and disseminate the information and material to other research organizations, thus stimulating collaborative research that leads to further advances.

Through research at the center, co-operative research abroad, frequent reviews of current research, counsel to foreign research programs, and resident training at the center, the staff can develop wide experience in their own fields of specialization. The free supply of research output from the center and the staff's professional experience in other countries should be especially helpful in their relations with national research centers. Collaboration with other centers is strengthened by their young technicians who had received training at the institute. We believe that it is the international and autonomous character of such an institution which makes it possible to establish with modest financial resources a meaningful and dynamic program in collaboration with national and international organizations. Indeed it is through these close links that an institute like IRRI can overcome the disadvantage of not being located in the center of maximum genetic diversity for its crop.

References

Bardenas, E. A. & Chang, T. T. (1966). Morpho-taxonomic studies of *Oryza glaberrima* Steud. and its related wild taxa, *O. breviligulata* A. Chev. et Roehr. and *O. stapfii* Roschev. *Bot. Mag., Tokyo*, **79**, 791–8.

Chang, T. T. (1967). The genetic basis of wide adaptability and yielding ability of rice varieties in the tropics. *Int. Rice Comm. Newslett.*, **16** (4), 4–12.

Chang, T. T. (1970). The description and preservation of the world's rice germplasm. *SABRAO Newslett.*, **2** (1), 59–64.

Chang, T. T. (1971). Field experiments with rice in the tropics with emphasis on variety testing. *SABRAO Newslett.*, **3** (1), 59–69.

Chang, T. T. (1972*a*). IRRI rice germ plasm project and its relation to national varietal collections. *Plant Genetic Resources Newslett.*, **27**, 9–15.

Chang, T. T. (1972*b*). International cooperation in conserving and evaluating rice germ plasm resources. In *Rice Breeding*, pp. 177–84. International Rice Research Institute, Los Baños, Philippines.

Chang, T. T. & Bardenas, E. A. (1964). The morphology and varietal characteristics of the rice plant. *IRRI Tech. Bull.*, **4**, 1–40.

Chang, T. T. & Jodon, N. E. (1963). Monitoring gene symbols in rice. *Int. Rice Comm. Newslett.*, **12** (4), 18–29.

Chang, T. T., Sharma, S. D., Adair, C. R. & Perez, A. T. (1972). *Manual for field collectors of rice*. 32 pp. International Rice Research Institute, Los Baños, Philippines.

Chang, T. T. & Vergara, B. S. (1971). Ecological and genetic aspects of photoperiod-sensitivity and thermo-sensitivity in relation to the regional adaptability of rice varieties. *Int. Rice Comm. Newslett.*, **20** (2), 1–10.

Chang, T. T. & Vergara, B. S. (1972). Ecological and genetic information on adaptability and yielding ability in tropical rice varieties. In *Rice Breeding*, pp. 431–53. International Rice Research Institute, Los Baños, Philippines.

Dalrymple, D. G. (1971). Imports and plantings of high yielding varieties of wheat and rice in the less developed nations. *U.S. Dep. Agr. Foreign Econ. Dev. Rept.*, **8**, 1–43.

FAO (1972). Bulletin on rice terminology. 25 pp. *Int. Rice Comm. Working Parties Sessions, Oct. 30–Nov. 10, 1972*, Bangkok, IRC/PP–72/VII/29.

Frankel, O. H. (1970). Genetic conservation in perspective. In *Genetic Resources in Plants – their Exploration and Conservation* (eds O. H. Frankel and E. Bennett), *IBP Handbook*, no. **11**, pp. 469–89. Blackwell, Oxford and Edinburgh.

IRRI (1964). Recommendation of the committee on genome symbols for *Oryza* species. In *Rice Genetics and Cytogenetics*, pp. 253–4. Elsevier Publishing Co., Amsterdam, London and New York.

IRRI (1970). *Catalog of rice cultivars and breeding lines (Oryza sativa L.) in the world collection of the International Rice Research Institute*. 281 pp. International Rice Research Institute, Los Baños, Philippines.

IRRI (1972). Discussions of international cooperation. In *Rice Breeding*, pp. 707–712. International Rice Research Institute, Los Baños, Philippines.

Ito, H. & Kumagai, K. (1969). The National Seed Storage Laboratory for genetic resources in Japan. *Jap. Agr. Res. Quart.*, **4** (2), 32–8.

Oka, H. I. & Chang, W. T. (1964). *Observations of wild and cultivated rice species in Africa – report of trip from Sierra Leone to Tchad, 1963*. 73 pp. National Institute of Genetics, Misima, Japan (mimeographed).

Pal, B. P. (1972). Modern rice research in India and its impact. In *Rice, Science and Man*, pp. 89–104. International Rice Research Institute, Los Baños, Philippines.

36. Maize germplasm banks in the Western Hemisphere

W. L. BROWN

Maize (*Zea mays* L.) is a good example of a crop species whose indigenous germplasm has been collected, described and preserved on a fairly sound and consistent basis for almost forty years. A brief summary of maize conservation activities in the Western Hemisphere may, therefore, be of interest not only to maize specialists but also to others engaged in the various aspects of plant genetic resource management.

In most parts of the Western Hemisphere where maize is a crop of economic importance, native strains have been replaced by hybrids and improved varieties. Consequently, many of the original land races, the primary source of improved varieties, no longer exist in the countries concerned. Fortunately, however, most of the indigenous strains were salvaged prior to their disappearance and are still available in germplasm banks.

Some maize was included in the early plant collections made by the USDA which was established in 1862. Additional national emphasis was placed on collecting following the establishment of the Seed and Plant Introduction Section of the USDA in 1898. When it became obvious that hybrid corn was rapidly replacing the open pollinated varieties of the US Corn Belt, federal funds were made available to several Mid-western states for the purpose of collecting old varieties still existing in those states. Although some strains were lost, a large number were salvaged and increased, and are still maintained at one or more of the Regional Plant Introduction Stations of the USDA.

The collecting activities just described resulted in the preservation of some maize germplasm which otherwise would have been lost. Yet these efforts were fragmentary, sporadic, and inadequate. It was not until 1943 that an integrated and co-ordinated plan for the assembly of maize germplasm began to evolve. The Rockefeller Foundation, in co-operation with the Mexican Ministry of Agriculture, initiated in Mexico a program of practical maize improvement. To provide materials for breeding, varieties were collected from all parts of Mexico and from most countries of Central America. The primary purpose of the collection was for use in breeding programs, yet an effort was made to collect all

467

varieties and strains regardless of their agronomic value. These collections formed the basis of the first maize germplasm bank in Mexico.

Later, when the Rockefeller Foundation extended its agricultural program to Colombia, similar maize collecting activities were initiated in that and surrounding countries. Thus, a second Latin-American maize germplasm bank was established at Medellín, Colombia.

As the collections grew, the tremendous diversity of Latin-American maize became apparent and the need for a systematic classification of the multiplicity of varieties became obvious. Furthermore, it became clear that additional collecting to include all of the Western Hemisphere was essential to an understanding of the total variability of maize, and to the preservation of existing land races rapidly being replaced by improved varieties. The recognition of this need by a few interested scientists resulted in the formation of a Committee on the Preservation of Indigenous Strains of Maize within the National Academy of Sciences–National Research Council (NAS–NRC). This occurred in 1951. The Committee consisted of leading geneticists, maize breeders, botanists, and agricultural administrators. Financial support was obtained from the United States Department of State, and funds were administered through the Office of Foreign Agricultural Relations of the USDA and through the Institute of Inter-American Affairs. In co-operation with the Rockefeller Foundation and numerous Latin-American governments and institutes, the Committee planned and supported comprehensive maize collecting in all countries of South America, most of Central America, and the major islands of the Caribbean. It was also able to augment to some extent the collections of North American maize which had been made previously. Through its efforts, duplicate samples of most of the Latin-American collections were placed in storage at the National Seed Storage Laboratory at Fort Collins, Colorado. Funds were made available to expand seed storage facilities at Chapingo, Mexico and Medellín, Colombia, and a third Latin-American seed storage center was established at Piracicaba, Brazil.

The persistent efforts of the Committee for the Preservation of Indigenous Strains of Maize in stimulating the study and classification of the assembled Latin-American collections is especially worthy of note. These studies resulted in the publication by NAS–NRC of ten monographs which have become popularly known as the 'race bulletins'. These are recognized world-wide as an invaluable source of information for students of maize.

Following the completion of the work of the Committee for the

468

Preservation of Indigenous Strains of Maize, there ensued a period of several years of limited activity relative to maize germplasm preservation. The seed banks in Latin America and the USA continued to function, but lack of sufficient support limited additional collecting and prevented the development and functioning of the most efficient systems of germplasm maintenance.

As a result of its long-time interest in germplasm conservation and because of its concern about the status of germplasm of the major crop species upon which man depends for food, the Agricultural Division of the Rockefeller Foundation, in 1969, asked a small group of scientists to consider again germplasm needs of a number of crop species, including maize.

Subsequently, a small committee of maize specialists was organized and asked to investigate the current status of collections, to identify the geographic areas, if any, from which additional collections were needed, to estimate costs of obtaining these collections, and to make recommendations generally for the implementation of measures needed to insure the orderly preservation of maize germplasm.

The Maize Committee completed its study in 1972. It found that despite the loss of a few collections, most of the maize germplasm assembled under the sponsorship of NAS–NRC was still existing in viable form. Most of these accessions are stored in three banks, (1) CIMMYT (International Maize and Wheat Improvement Center, Mexico), (2) ICA (Instituto Colombiano Agropecuario, Colombia) and (3) INIA (Instituto Nacional de Investigaciones Agricolas, Mexico). The material from Brazil and other South American countries, formerly stored at Piracicaba, Brazil, has been transferred to the CIMMYT bank. In addition to these centers, some collections are maintained at the Universidad Agraria, La Molina, Lima, Peru and at the Estación Experimental Regional Agropecuaria, Pergamino, Argentina. These banks are working primarily with national accessions.

The total number of accessions in the various Latin-American banks are distributed approximately as follows:

Argentina	1 000
Brazil	90 (transferred to CIMMYT)
Colombia (ICA)	2 000
CIMMYT	12 000+
Mexico (INIA)	7 000+
Peru	1 600
Total	24 000+

The bank at CIMMYT (El Batán) has become the largest single storage facility for maize germplasm. Physical equipment is new, well designed, and large enough to permit considerable expansion.

Plans are to transfer to the CIMMYT bank the collections from Bolivia and Peru now in storage at ICA. As mentioned previously, the collections from eastern South America have already been transferred to CIMMYT.

Following transfer of the Bolivian and Peruvian collections from Colombia to Mexico, the bank in Colombia (ICA) will concentrate on accessions from Colombia, Venezuela, and Ecuador. This will reduce to approximately 1000 the number of accessions to be maintained at ICA.

The INIA bank in Mexico is the depository of samples of individual collections of most of the Mexican and Guatemalan accessions. These are essential for certain types of investigations requiring precise knowledge of geographic origin. Since many of the individual collections are not available elsewhere (due to bulking of samples), they represent a critical part of the indigenous maize germplasm of Mexico and Guatemala.

The geographic distribution of the maize germplasm now in storage indicates the existence of some gaps in the collections. By and large these are isolated areas, difficult to reach and with little contact with modern agriculture. Consequently, the indigenous maize of these areas is not likely soon to be replaced by modern varieties. Notwithstanding, plans are now under way to begin filling these gaps.

The two most critical areas from which collections are needed are the Amazonas of South America and the Himalayas. CIMMYT, in cooperation with the governments of countries within which the areas of interest lie, is moving forward with plans for expeditions to each of these areas.

When collecting is completed in the two critical areas, an effort will be made to fill small existing gaps in the collections from Peru, Ecuador, Colombia, and Bolivia.

Storage facilities at the major banks in Latin America are being upgraded. Yet, there is still need for additional, permanent, duplicate storage for insurance purposes. This problem is now being solved through an arrangement with the National Seed Storage Laboratory, whereby small samples of all individual and composite collections will be placed in the Laboratory at Fort Collins for permanent storage. These samples will be for insurance purposes only. Their use will be limited to cases of emergency.

470

The vast collection of maize germplasm now being maintained in the seed banks of Mexico, South America, and the USA represents a priceless and irreplaceable natural resource. There is now sufficient organized effort directed toward the preservation and maintenance of these materials to lead to the conclusion that the risk of serious loss is minimal. However, germplasm resources are of little value unless they are used and, unfortunately, they have not been utilized to the extent they should have been. A part of the neglect is due to a lack of knowledge of most maize breeders and geneticists as to what is available in the major seed banks. Each bank maintains its own inventory, yet a complete catalog of all accessions of banks has not been prepared. To alleviate this need, plans are now in progress to develop an organized and uniform system of documentation which will lead to the development and distribution of a complete catalog encompassing all accessions in each of the three major Latin-American maize seed banks. This, it is expected, will stimulate interest and additional utilization of one of the New World's most important resources.

Relevant references

Brieger, F. G., Gurgel, J. T. A., Paterniani, E., Blumenschein, A. & Alleoni, M. R. (1958). *The Races of Maize in Brazil and Other Eastern South American Countries*, 283 pp. Nat. Acad. Sci.–Nat. Res. Counc. Publ. **593**. Washington, DC.

Brown, W. L. (1960). *Races of Maize in the West Indies*, 60 pp. Nat. Acad. Sci.–Nat. Res. Counc. Publ. **792**. Washington, DC.

Grant, U. J., Hatheway, W. H., Timothy, D. H., Cassalett, D. C. & Roberts, L. M. (1963). *Races of Maize in Venezuela*, 92 pp. Nat. Acad. Sci.–Nat. Res. Counc. Publ. **1136**. Washington, DC.

Grobman, A., Salhuana, W. & Sevilla, R. (1961). *Races of Maize in Peru*, 374 pp. Nat. Acad. Sci.–Nat. Res. Counc. Publ. **915**. Washington, DC.

Hatheway, W. H. (1957). *Races of Maize in Cuba*, 75 pp. Nat. Acad. Sci.–Nat. Res. Counc. Publ. **453**. Washington, DC.

Ramírez, E. R., Timothy, D. H., Díaz, B. E., Grant, U. J., Nicholson, C. G. E., Anderson, E. & Brown, W. L. (1960). *Races of Maize in Bolivia*, 159 pp. Nat. Acad. Sci.–Nat. Res. Counc. Publ. **747**. Washington, DC.

Roberts, L. M., Grant, U. J., Ramírez, E. R., Hatheway, W. H., Smith, D. L. & Manglesdorf, P. C. (1957). *Races of Maize in Colombia*, 153 pp. Nat. Acad. Sci.–Nat. Res. Counc. Publ. **510**. Washington, DC.

Timothy, D. H., Hatheway, W. H., Grant, U. J., Torregroza, C. M., Sarria, V. D. & Varela, A. D. (1963). *Races of Maize in Ecuador*, 159 pp. Nat. Acad. Sci.–Nat. Res. Counc. Publ. **975**. Washington, DC.

471

Timothy, D. H., Peña, V. B., Ramírez, E. R., Brown, W. L. & Anderson, E. (1961). *Races of Maize in Chile*, 84 pp. Nat. Acad. Sci.–Nat. Res. Counc. Publ. **847**. Washington, DC.

Wellhausen, E. J., Fuentes, O. A., Hernández, C. A. & Mangelsdorf, P. C. (1957). *Races of Maize in Central America*, 128 pp. Nat. Acad. Sci.–Nat. Res. Counc. Publ. **511**. Washington, DC.

Wellhausen, E. J., Roberts, L. M., Hernández, E. X. & Manglesdorf, P. C. (1952). *Races of Maize in Mexico*, 223 pp. The Bussey Inst., Harvard Univ., Cambridge, Mass.

37. Genetic resources centres – a co-operative global network

O. H. FRANKEL

Plants have been, and will continue to be, collected, evaluated, used, stored, and in many instances discarded or lost by institutions or individuals, irrespective of national or international endeavours or programmes. But in the position of urgency in which the world finds itself, with most of the remaining genetic heritage in actual or imminent danger of extinction, the idea has spread with increasing momentum that an international effort is required to stimulate, assist, complement and link the efforts of governments, institutions and individuals. The goals are to assemble, and in many instances to salvage what is left of the crop genetic resources, to see that it is preserved against loss and deterioration, to make it generally available to those who can evaluate and use it, and to process and publish all available evaluation records for the benefit of all users.

The concept of a genetic resources centre goes back to the USSR Institute of Plant Industry, which in the 1920s under N. I. Vavilov became not only the first centre, but without a doubt the most highly developed and successful ever established, embodying all phases of genetic resources work from world-wide exploration to extensive evaluation and long-term conservation. It gave tremendous inspiration for exploration activities by others and for the establishment of institutions to deal with collected material, probably the first being the Empire Potato Collection in Britain. Plant introduction services in USA, Australia and elsewhere, though mainly concerned with obtaining material needed by their agronomists and plant breeders, increasingly took up exploration, and somewhat later the conservation of their collections.

A global network

In the context of current ideas of a network of genetic resources centres, a centre is defined as an institution primarily concerned with exploration and/or conservation of genetic resources. The proposal for a global network grew out of reports and recommendations by the FAO Technical Conferences in 1961 and 1967 (see Whyte & Julén, 1963;

473

Frankel & Bennett, 1970), the FAO Panel of Experts on Plant Exploration and Conservation (FAO, 1969; 1970; 1973), and of an *ad hoc* expert group which met at Beltsville in March 1972 under the auspices of the Technical Advisory Committee of the Consultative Group on International Agricultural Research. Concepts and definitions presented in this chapter follow in the main those proposed in the Beltsville report (unpublished) and the most recent report of the FAO Panel (FAO, 1973).

In broad terms, an international collaborative network of institutions is proposed to assemble, conserve and evaluate crop genetic resources. Participating organisations will undertake tasks and responsibilities commensurate with their interests and resources, as far as possible complementary with those of other members of the network. The task of liaison and co-ordination will need to be performed by FAO as the appropriate United Nations agency. The aim is the world-wide participation of all institutions and individuals wherever they are located, who can contribute to the objectives of the network. However, special emphasis must be given to participation within the regions of genetic diversity – essentially the Vavilovian centres – where most of the plant material which is to be preserved is situated. Most of these regions are in developing countries, in many of which both human and institutional resources in this field will need to be strengthened, and in some instances established *de novo*.

Definitions

The following definitions are proposed in the 1973 report of the FAO Panel:

> Genetic resources centres comprise either or both of the following components:
> (i) base collections, for long-term conservation;
> (ii) active collections, for:
>> (*a*) medium-term storage;
>> (*b*) regeneration;
>> (*c*) multiplication and distribution;
>> (*d*) evaluation;
>> (*e*) documentation.

These two components are necessary for the continued maintenance and utilisation of germplasm collections. If they are not situated in the same institution collaborative links between them will be necessary.

474

Breeders' working collections are regarded as outside the framework of this system, but they may generate valuable information which should be incorporated in the genetic resources records.

Base collections

The FAO panel recognised the following categories:

(i) substantial collections of a wide range of species;
(ii) substantial collections of a limited range of species;
(iii) significant and original special purpose collections;
(iv) replicates of any of these.

As its fourth and fifth sessions (FAO, 1970; 1973) the Panel specified the following requirements for the maintenance of base collections:

(i) Adequate facilities for long-term conservation, be it seed storage or maintenance of vegetative material; standards to be set and reviewed from time to time.

(ii) Seed stored in base collections not to be used as a source for distribution, which is a function of active collections.

(iii) Arrangements to be made for replicate storage in other base collections to minimise the risk of losses.

(iv) The sample size should be at least adequate to represent the original population and to satisfy foreseeable demands over the expected life of the sample.

(v) Except for replicate collections stored under similar conditions, there should be periodic germination tests.

Base collections for the conservation of seeds require considerable expenditure for facilities and maintenance, direction by experts, and a sizeable staff for receipt and despatch, germination tests, records, and general administration. A relatively small number of such collections would be sufficient to conserve the world's germplasm of all seed-reproduced crops. Yet the number should not be altogether too small, for the following reasons. First, there are limits to the work-load any one institution can shoulder, especially when faced with a multiplicity of species including some with special storage requirements. Second, the distance between basic and active collections should not be so great as to inhibit the essential flow of people, information, and seed material. Third, care must be taken to minimise the interference of quarantine barriers with seed regeneration which should be sited in accordance with ecological requirements. Fourth, convenience as well as safety

475

considerations make the dispersal of base collections highly desirable; distance between base collections re-inforces the safety factor in replication between collections. One might envisage between one and six base collections for seeds in different continents, or a total number of 20–25. These might cover anything from a single crop, such as the rice collection of IRRI (the International Rice Research Institute), to the multi-crop collections of the Seed Storage Laboratory at Fort Collins, Colorado, or the Vavilov Institute of Plant Industry at Leningrad.

Collections of vegetative material present greater technical, organisational, logistic and financial problems. Until tissue or meristem culture becomes a routine operation which can safely be used in long-term conservation, base collections of vegetative material will continue to be few in number, mainly of annual or short-lived crops. Yet, as the Survey of Crop Genetic Resources (Frankel, 1973) shows, there is a great need for the conservation of many tree fruits and plantation crops, only a few of which can be preserved in the form of seeds.

Active collections

Active collections are in the main concerned with operations in the field. Hence they are often, but by no means invariably, part of or attached to an institution concerned with plant introduction, plant breeding or plant protection. This may have the advantage of direct participation of a multi-disciplinary team in the evaluation of a collection, although evaluation can also be conducted co-operatively with the participation of interested institutions.

Active collections, according to the report of the Fourth Panel meeting (FAO, 1970), 'must possess adequate facilities for the propagation and distribution of plant material . . . [They] must include or be associated with a quarantine station responsible for inward and outward clearance . . . [They] are responsible for organizing the regeneration of seed. This should take place in environments as similar as possible to those from which the original material derives, [which] may – and usually will – necessitate collaborative arrangements with a number of institutions.' Such arrangements for regeneration will, of course, be made at the initiative of and in consultation with a base collection.

Finally, an active centre 'is responsible for the maintenance of records of the origin, further treatment and distribution of accessions'.

Existing collections

In the planning and organisation of the network the emphasis is on the preservation of genetic resources in the field which are likely to be lost, and in particular on the primitive cultivars and the wild and weed relatives of crop species. The advanced cultivars in current or recent use are well represented in national collections and in the working collections of plant breeders and agronomists all around the world, hence as a rule are freely available; but some of the older and now obsolete modern cultivars, and especially any which have been used in research or breeding, should be preserved.

It goes without saying that the existing collections of primitive and wild material should be preserved with the greatest care. Although a good deal of material has been lost or has been depleted by deliberate and/or natural selection, by drift and human error (see Chapter 6) or by unsuitable storage conditions, some of these collections contain valuable material which no longer exists in the field. Every effort should be made to prevent further losses. Now that good storage facilities exist in several places, valuable material in institutions lacking such facilities should be lodged for safe storage in a base collection. Attention should also be paid to the conservation of material assembled for evolutionary or cytogenetic studies of crop plants and their relatives, and of genetic stocks used in genetic research or breeding practice. The routine of lodging such material in a base collection – already widely practised in the United States – should be generally adopted.

Organisation and scope

According to their terms of reference and source of financial support, genetic resources centres are referred to as national, regional or international centres. These terms were introduced to describe the functions of the recently established or proposed centres within the regions of genetic diversity. Elsewhere the existing centres are under national control, though some of them are engaged internationally in exploration and exchange and, as base collections, in long-term conservation.

Regional centres are to be designated in areas which are particularly rich in germplasm resources. In addition to activities in their own countries their task will be to stimulate and organise collaborative regional activities in fields such as training, exploration and evaluation. They will have facilities for medium or long-term seed storage, and may

477

provide a data bank for the region as a whole. However, the distinction between regional and national centres is by no means sharp. The regional network of six countries in the Near East, with the Agricultural Research and Introduction Centre at Izmir as regional centre, is at an advanced stage of preparation (see Chapter 34); whereas the proposed centres in Ethiopia (see Chapter 33) and in Indonesia are likely to be fully occupied within their national borders for some years to come before assuming extensive regional responsibilities.

The international crop research institutes, founded under the auspices of the Rockefeller and Ford Foundations and in recent years supported by the Consultative Group on International Agricultural Research, are designated as international centres. Their research responsibilities extend from one to a small number of crops. Several institutes are expected to participate in the genetic resources network, either at regional or international levels.

IRRI, the International Rice Research Institute (Chapter 35) and CIMMYT, the International Maize and Wheat Improvement Center (Chapter 36) have established outstanding world collections for rice and maize respectively. IRRI in particular has assumed a central role on a world scale in all aspects of germplasm work on rice. It should be noted that in addition to the comprehensive activities of the Institute itself, the main emphasis is on stimulating national activities in exploration, conservation and evaluation. IRRI provides training, technical assistance where needed, specialist literature and information, organises technical conferences, and maintains a safe long-term seed storage and a documentation system covering all rice germplasm. In all these aspects IRRI's collaborative network serves as a model for regional or international activities in all phases of genetic resources work.

Irrespective of their designation as national, regional or international centres, the functions of genetic resources centres will include some or all of the following, as defined in the Beltsville report and quoted by the FAO Panel (FAO, 1973).

1. Exploration and collection of material, and collaboration with national centres.

2. Identification and preliminary evaluation of materials.

3. Initial planting of introduced material according to the quarantine laws of the country in which the centre is located.

4. Exchange and distribution of seed and vegetative stocks, including, where appropriate, the introduction of breeding lines and advanced cultivars.

5. Maintenance and storage of seed and vegetative stocks for medium or long-term preservation.

6. Documentation and exchange of information with other centres in the network in an internationally accepted form. Some centres will be able to take advantage of existing local facilities for computerised information storage and retrieval.

7. Organisation of genetic stock rejuvenation by national centres wherever possible, or otherwise by regional centres.

8. Organisation of training programmes for personnel in collaboration with national or international training schemes.

9. Identification of 'genetic reserve areas' in consultation with national centres and the international co-ordinating body.

The Beltsville group proposed institutions which are well placed to serve as regional or crop-specific genetic resources centres. They are quoted from the 1973 Panel report.

1. Mediterranean and Near East: Germplasm Laboratory, Bari, Italy; Agricultural Research and Introduction Centre, Izmir, Turkey.

2. Ethiopia: College of Agriculture, Haile Selassie I University; Institute of Agricultural Research.

3. Tropical America: Interamerican Institute of Agricultural Sciences (IICA), Turrialba, Costa Rica.

4. Rice: International Rice Research Institute (IRRI), Manila, Philippines.

5. Tropical West Africa: International Institute of Tropical Agriculture (IITA), in co-operation with other institutes such as IFCC, IRAT, IRAHO, Moor Plantation, ORSTOM, etc.

6. South-East Asia: Botanic Gardens of Indonesia, Bogor.

7. India: Indian Agricultural Research Institute (IARI), New Delhi.

8. Andean Highlands: Universidad Agraria and Estación Experimental, La Molina, Peru.

9. Subtropical South America: Instituto Agronômico, Campinas, Brazil.

10. Mexican Highlands: Instituto Nacional de Investigaciones Agrícolas, and Graduate School, Chapingo near Mexico City.

International support and liaison

The need for international collaboration in the conservation of genetic resources is now widely appreciated by governments and international organisations. The United Nations Conference on the Human Environment, Stockholm 1972, called for participation from all countries and charged FAO to co-operate with and give assistance to the establishment of an international programme.

This programme – as we have seen in the preceding chapters – envisages the participation of all relevant institutions, in developing and in developed countries, in the regions of genetic diversity and elsewhere, of institutions with world collections or with local or specialised collections or with, as yet, none at all. Many of these are known to be able and willing to co-operate and are waiting to receive the call to participation. Many will require little or no additional facilities or support, but the establishment of the proposed centres in the regions of genetic diversity depends on the availability of substantial financial and in some instances technical assistance. There are encouraging signs that this will be forthcoming through international as well as inter-governmental channels, with moves towards the establishment of the highest priority centres in the Near East and the Mediterranean region, in Ethiopia, and in tropical America.

Important as genetic resources centres within the regions of genetic diversity undoubtedly are, other steps are of equal importance for the establishment of the global network.

(1) *The linking of existing institutions* concerned with germplasm exploration and conservation in the co-opertive network, through entering into agreements on the exchange of material and information with all participants, either directly or through the liaison organisation.

(2) The designation of appropriately equipped institutions as holders of a '*base collection*', under agreements on technical standards, on rights of storage and conditions of access, on germination tests, and quarantine regulations, on procedures for regeneration and for data processing. The recognition of even a small number of base collections at the earliest moment is without a doubt *the most pressing task* if not only existing collections, but those to be assembled, are to be preserved in perpetuity.

(3) The establishment of a *co-operative network of data banks* for genetic resources, covering the accessions of basic and active resources centres and incorporating evaluation data from all sources. This network is second in importance only to conservation itself, since without information on the existence of stocks and their characteristics there can be no rational utilisation. In the short term, information of value to plant breeders is the most practical contribution which the programme can make to developing countries with rapidly evolving agricultural systems.

480

(4) A programme for *emergency collecting* of threatened genetic resources. The targets for exploration are now reasonably clear (see Chapter 6), with priorities suggested in the FAO Panel's reports (FAO, 1969; 1970; 1973). The need now is for action by governments and institutions, based on planning, liaison, and in many instances a modest measure of financial support.

(5) A programme for *training* in genetic resources work, including practical short courses, mainly in the field, at a regional level; university courses; seminars or workshops; and postgraduate research training.

Clearly this programme cannot be brought into being without a strong and broadly supported *liaison organisation* which can only be based in a United Nations organisation. In the words of the 1973 report of the Panel, 'to initiate and to make effective the international network for genetic resources centres will require a high level of co-ordination which is the clear responsibility of FAO'. The decision by FAO to strengthen greatly and to restructure the existing genetic resources unit clearly indicates that FAO is prepared to assume this responsibility.

References

FAO (1969). Report of the third session of the FAO Panel of Experts on Plant Exploration and Introduction. FAO, Rome.

FAO (1970). Report of the fourth session of the FAO Panel of Experts on Plant Exploration and Introduction. FAO, Rome.

FAO (1973). Report of the fifth session of the FAO Panel of Experts on Plant Exploration and Introduction. FAO, Rome.

Frankel, O. H. (ed.) (1973). FAO/IBP Survey of genetic resources in their centres of diversity. FAO, Rome.

Frankel, O. H. & Bennett, E. (eds) (1970). *Genetic Resources in Plants – their Exploration and Conservation, IBP Handbook*, no. **11**. Blackwell, Oxford and Edinburgh.

Whyte, R. O. & Julén, G. (eds) (1963). *Proceedings of a Technical Meeting on Plant Exploration and Introduction, 1961.* Supplement to *Genetica Agraria*, **17**.

Index

486

487